SAUDI ARAMCO AND ITS PEOPLE

A History of Training

SAUDI ARAMCO AND ITS PEOPLE

First Edition, 1998

Edited by
Ali M. Dialdin and Muhammad A. Tahlawi

Researched and written by
Thomas A. Pledge

The Saudi Arabian Oil Company (Saudi Aramco)
Dhahran, Saudi Arabia

A History of Training

ISBN 0-9601164-4-3
Copyright © 1998 Aramco Services Company
Library of Congress Catalog Card Number: 97-62572

Aramco Services Company
Houston, Texas, 1998

Designed in Houston, Texas, by Stratos, LLC
Printed in Salt Lake City, Utah, by Lorraine Press

Printed in the U.S.A. on recycled paper
using vegetable-based inks.

Contents

Acknowledgments

Saudi Aramco and Its People – A History of Training was made possible by the contributions of many people. The task of thanking them all is virtually impossible. The following people contributed in such a significant way that we would like to make special mention of them.

- Editorial assistance in Houston was provided by Michael Pewitt of Stratos, LLC, which also executed the design, and by William Tracy at Aramco Services Company.

- Nimr Atiyeh of Saudi Aramco's Training Department provided background information and supervised the project from beginning to end. Jim Mandaville of Government Affairs was first reviewer, contributing his knowledge of Company and regional history to the manuscript.

- Warren Hodges, Al Lampman and Frank Jarvis supplied information and reviewed chapters on the early years of training.

- William P. O'Grady, Joe Mahon and Dave Steinheimer provided numerous documents and reviewed chapter drafts.

- Wadie Abdelmalek, Ahmad Ajarimah and Dale Saner reviewed and commented on chapters.

- Charlene Howland, daughter of the late Harry R. Snyder, made her father's voluminous records available to the project.

- Sheila Simpson and Alice Sealy assisted in research and organized the files.

- Others who assisted the project included:

Adeeb Yousif Al-Aama	Vince James	William Roof
Abdul Aziz Abdul Latif	Charles Johnson	Shoukry D. Saleh
Mustafa Abu Ahmad	Frank Jungers	Fred Scofield
Jim Adams	Mel LaFrenz	Tim Sedor
Robert Arndt	Robert Lebling	Ibrahim Shihabi
Jun Asistio	Jan Lincoln	Faridah Sowayigh
Sally Aslan	Robert Luttrell	Anne Tandlich
Ali Baluchi	Phil McConnell	John Tarvin
Fahmi Basrawi	Bill Mulligan	George Trial
C. Kenneth Beach	Khalid Najjar	Bill Valbracht
Bob Brautovich	Paul Nance	Ron Visconti
Arthur Clark	Zafer Nashashibi	Dan Walters
Dwight Fullingim	Khalil Nazzal	Sam Whipple
Rady Al-Ghreeb	Bernell Nelson	Kathy Fry Wilson
Tony Harrison	Greg Noakes	'Abd Allah Ali Al-Zayer
Jack Hosmer	Honorio Nor Pangan	
Tareq M. Ibrahim	Zubair Al-Qadi	

For those people who helped in any way in the completion of this book and are overlooked on this page, please know that our appreciation for your help is not diminished by the oversight.

Introduction

audi Aramco and Its People is an attempt to trace the history of training and human resource development at the company that became known as the Saudi Arabian Oil Company, Saudi Aramco, and the influence of this remarkable effort on the people of the Kingdom of Saudi Arabia. It is a story of many parts. In the early years, the company's primary objective was to find oil and bring it to market. Yet, the early pioneers soon discovered that to tap the natural resources of the Kingdom, they must also draw upon its human resources. Even the very first American geologists who surveyed the Eastern Province for the Company had to rely on the navigational ability of their Saudi guides. Individuals from both cultures learned respect for each other's abilities from the beginning.

To create an oil industry where no modern industry had existed before required importing not only the technology, but also the language of technology and knowledge of how to use that technology. The early exploration-drilling crews developed a special language of understood and misunderstood words from English and Arabic to provide a common ground for American engineers and Saudi drilling crews to work together. No two crews used the same language. From these simple beginnings grew the modern era of classrooms with sophisticated models, video programs and computer-assisted instruction.

To a large extent, the history of Saudi Aramco is not just a history of oil, but also a history of knowledge. In helping to transform a country with vast areas of empty desert and barren *jabals* to one of modern cities and thriving industry, Saudi Aramco also played a part in the transformation of a people. The sons of early Saudi guides and laborers are now Saudi Aramco engineers, technicians and managers.

In recounting the history of training and development from the signing of the Concession Agreement until today, this book reflects on the evolution of training, the changes in education, the building of schools, the growth of technology and, above all, on the people of Saudi Arabia. We hope that anyone, whether or not he or she is familiar with Saudi Aramco and the Kingdom of Saudi Arabia, will find this a story worth knowing.

Preface

The concept for a book of this nature originated with Ali M. Dialdin, a prominent figure in Saudi Aramco training for 30 years and the first Saudi to direct the Company's extensive training programs. The idea grew out of conversations over the years between Dialdin and his fellow Saudi employees. "Someone was always mentioning his days in the company school, remembering teachers and classmates," Dialdin said. "I wanted to capture and preserve the memory of those unique times."

Saudi Aramco and Its People touches only the highlights of human resource development in what grew to be the largest oil company in the world. The book follows the rapid development of Saudi employees from largely illiterate common laborers under American management to educated and professionally qualified employees and executives who became, over the years, responsible for the operation and management of the Company. It is a story of how people from different cultures came together to accomplish one of the largest transfers of knowledge and technical know-how in history.

In 1990, as Saudi Aramco approached its 60th anniversary year, Dialdin formed a "History of Training Project Team" under the supervision of Nimr Atiyeh, a long-time teacher and Training Department supervisor. Tony Shehadah, a 35-year veteran of the Training Organization, collected the first research materials. Tom Pledge, an editor of Training documents for Saudi Aramco and former journalist, completed the research and wrote the book.

The story of *Saudi Aramco and Its People* is based on more than six decades of Company records and on information gleaned in a review of every issue of the Company's 50-year-old newspaper, *The Arabian Sun*. The author also conducted some 200 interviews with retirees and current employees living in Saudi Arabia, the United States and other countries, including Australia, Canada, Great Britain, India and the Philippines.

At Ali Dialdin's suggestion, Saudi Arabia's Minister of Petroleum and Minerals, H. E. Ali I. Al-Naimi, and some of the Company's top executives — all themselves beneficiaries of Saudi Aramco training programs — were interviewed and their words incorporated into brief, personal remembrances spaced among the chapters of this book. These individual profiles were written and edited, for the most part, by members of the Company's Public Relations staff.

The author reviewed thousands of photographs from Saudi Aramco's archives to find those photos that were finally selected for inclusion in this volume. Former employees contributed some previously unpublished photographs, including photos of the company's first schools and early Saudi trainees by Jack Hosmer and Fahmi Basrawi.

This book is intended to appeal to a broad spectrum of readers: professional educators, historians, sociologists, trainers and managers, past and present employees of Saudi Aramco — in short, nearly anyone interested in people, education, the oil industry or the Middle East. It is hoped the book will also rekindle the memories of those who participated in the evolution of Saudi Aramco, inspire young people to seek the rewards of learning, and motivate educators to consider how the experiences and achievements at Saudi Aramco might apply to training programs elsewhere in the world.

Space did not permit the inclusion of footnotes in this book. Scholars wishing further information should direct their queries to the Public Relations Department; attention Support Services Division; c/o Saudi Aramco, Dhahran 31311, Saudi Arabia.

SAUDI ARAMCO AND ITS PEOPLE

A History of Training

The Early Days
1933-1944

"Training was our second job, if not our first. ..."
– Early American employee

n May 29, 1933, at Khuzam Palace, then on the outskirts of Jiddah, representatives of the Kingdom of Saudi Arabia and Standard Oil of California (Socal) signed the Concession Agreement that allowed Socal to explore for oil in Saudi Arabia. This book is about some of the events that followed that agreement. Specifically, it is about the growth of an oil company, and the transformation of mostly illiterate and unskilled Saudi laborers into executives and technicians in charge of the largest, and one of the most advanced oil companies in the world.

The Concession Agreement

audi Arabia's negotiators were led by Shaykh 'Abd Allah Al-Sulayman, the Minister of Finance to King 'Abd al-'Aziz Al Sa'ud. He came to the negotiations armed with knowledge of the terms already won by Iraq and Persia for oil concessions to the British. Harry St. John B. Philby, friend and unofficial advisor to King 'Abd al-'Aziz, described Shaykh 'Abd Allah as a polite, shrewd and keenly intelligent man. Born in the Qasim area northwest of Riyadh and trained as a clerk in India, Shaykh 'Abd Allah rose to the position of finance minister on the basis of his performance as a clerk and loyal personal secretary to King 'Abd al-'Aziz. In the lean years of the new country, Philby said, Shaykh 'Abd Allah "frequently found himself in the unenviable position of having to produce loaves and fishes out of non-existent ovens and seas."

The Socal team was led by Lloyd N. Hamilton, a lawyer specializing in contracts, and Karl Twitchell, a mining engineer who had already completed a surveying expedition across the Arabian Peninsula. Hamilton, just turned 40, had traveled to Saudi Arabia from London, where he represented Socal's foreign interests. He was described as a keen judge of character, affable yet discreet, a man whose word could be trusted. After graduating from the University of California, Hamilton served as a U.S. Army infantry officer in France during World War I and attended Oxford University in England after the war. In 1933 these two men and their wives were the only Americans in Saudi Arabia.

Negotiations started on February 19, 1933, with Shaykh 'Abd Allah asking for an initial

Opposite: Dhahran camp with Well No. 1 in March 1935. Below: Shaykh 'Abd Allah Al-Sulayman and Lloyd N. Hamilton sign the Concession Agreement between Saudi Arabia and Socal.

1

loan against royalties of 100,000 English pounds, gold (about $500,000 U.S.), plus rentals and other fees if oil was found in commercial quantities. Hamilton countered with an offer of $50,000 up front and smaller other payments than those sought by the government. His offer was included in a 10-point proposal — a working paper for negotiation — modeled on other Socal concession agreements. Point seven read: "The direction and supervision of the enterprise shall be in the hands of the Americans, who shall employ Saudi Arabian nationals as far as practicable."

Harry St. John B. Philby, 1938.

This passage was the first mention of employing Saudis. Socal stated it would give preference to Saudis in hiring, but indicated it felt no obligation to train them.

On at least one occasion, Shaykh 'Abd Allah pressed for a commitment to train and educate Saudi workers. Hamilton reported to Socal's San Francisco headquarters on May 16, 1933, that there had been: "... an extended discussion, involving the question of whether the company would agree to train a certain number of Saudi Arabians in the oil business and send a certain number to the United States for schooling." He concluded: "I think we managed to avoid these obligations, after much argument."

No further mention of training, human resource development, or the transfer of skills and technology to Saudis is found in Hamilton's correspondence with the company, nor did Philby report any discussion of these topics at the Saudi privy council meeting where terms of the agreement were explained to national leaders and approved by them.

After three and a half months of hard bargaining, an agreement was finally reached. The historic signing is captured in a photograph that shows the chief Socal negotiator, Lloyd N. Hamilton, signing the agreement with Shaykh 'Abd Allah Al-Sulayman. Shaykh 'Abd Allah's private secretary and interpreter, Najib Salihah, watches the signing over Hamilton's left shoulder.

The Kingdom of Saudi Arabia Royal Decree Number 1135 "granting a concession for the exploration of petroleum" was issued July 7 and officially proclaimed on July 14, 1933, in the government gazette.

The concession contained 37 separate articles. It pledged Socal to loans of 50,000 English pounds in gold (equal to about $250,000 at the time), yearly rentals of 5,000 English pounds (about $25,000), and royalties of four shillings (about $1.00) per ton of oil produced. In return, Socal obtained exclusive right to prospect for and produce oil in eastern Saudi Arabia and preferential rights in most of the rest of the Kingdom.

Article 23 of the concession concerned the hiring of Saudis to work for the company. In it Socal had made no commitment to train Saudi workers. The company merely promised to give preference to "suitable" Saudis. The first paragraph of Article 23 read: "The enterprise under this contract shall be directed and supervised by Americans who shall employ Saudi nationals as far as practicable, and in so far as the Company can find suitable Saudi employees it will not employ other nationals."

Casoc's first payment to the Saudi government being counted at Dutch Bank in Jiddah, 1933.

A second paragraph, added at the government's insistence, makes clear the labor laws of Saudi Arabia, not those of any other country, apply to company workers. It read: "In respect of the treatment of workers, the Company shall abide by existing laws of the country applicable generally to workers of any other industrial enterprise."

Article 23 was a small opening — yet wide enough to open the industrial age to thousands of Saudis. It did so partly because of the government's aggressive enforcement of Article 23, no matter how small the issue. For example, the government protested vigorously in September of 1934 when company geologists brought three English-speaking workers (a cook, a waiter and a houseboy) from Bahrain.

The acting minister of finance declared in a letter that the workers were foreigners and could not be employed under Article 23. The company replied this was true, "from a purely technical standpoint," but pointed out the personal nature of their services and the difficulty with servants who do not understand one's language, customs and manner of living.

The three were allowed to stay, but the minister cautioned the company to abide strictly by the provisions of Article 23, since it was a matter in which the King had a particular interest. "Any infraction of that article will create difficulty for the company and the Ministry of Finance," he warned.

California Arabian Standard Oil Company (Casoc)

In May 1933, Socal assigned its concession rights to a wholly owned subsidiary, California Arabian Standard Oil Company (Casoc). The company operated in Saudi Arabia under that name for the next 10 years. With the Concession Agreement signed and the corporate structure settled, the next step was to explore the newly opened territory — some 320,000 square miles of casually mapped, barren desert — an area larger than France or the state of Texas.

Company headquarters in San Francisco.

In the previous decade Socal had spent more than $50 million on ventures in half a dozen countries, from Central and South America to the East Indies, without success of note. Then, in 1928, Socal signed a concession to search for oil in Bahrain, an island nation in the Arabian Gulf just off the coast of Saudi Arabia. Four years later it hit oil on the island, in formations very much like those visible in the lumpy brown *jabals* (hills) of broken limestone visible on the horizon in nearby Saudi Arabia. They struck oil in rocks of the Cretaceous period at the depth of 2,000 feet, to the surprise of British experts around the Arabian Gulf who believed oil in this area could be found only in younger, Tertiary period formations.

Socal operated in Bahrain through a wholly owned Canadian subsidiary, Bahrain Petroleum Company (Bapco), so the first geologists sent to map the new concession on behalf of Casoc were experts from Bahrain Petroleum Company who traveled the few miles from Bahrain to Saudi Arabia on a motor launch.

Early Arrivals

Robert "Bert" Miller and Schuyler "Krug" Henry landed September 23, 1933, at Jubail, 20 miles north of present day Ras Tanura. Karl Twitchell joined them after crossing the Arabian Peninsula from Jiddah, following old camel trails, with two touring cars rented from the government, plus drivers and mechanics. All three men had grown beards and wore Arab dress to avoid appearing too conspicuous to the local people, most of whom had never seen a Westerner. They were greeted by the local Amir, an escort of soldiers, and a throng of curious locals.

The crowd was in a holiday mood, but the geologists were eager to get to work. After a few courtesy calls and the customary cups of coffee, they piled into one of the touring cars and sped off toward a likely looking, nearby hill called al-Jubayl al-Barri (source of the name for the Berri oil field years later). The Arabs

Jubail Harbor, where the first geologists landed.

King 'Abd al-'Aziz reads congratulatory cable handed to him by Floyd Ohliger in Ras Tanura, 1939.

Early Dhahran well with dome of the new mosque in foreground.

Khumayyis ibn Rimthan, Saudi guide for early exploration teams.

Casoc geologists with field cars at 1933 encampment.

Aramco pioneers at Jubail in 1934, left to right: T. W. Koch, J. W. Hoover, R. C. Kerr, R. P. Miller, H. L. Burchfield, S. (Krug) Henry, F. W. Dreyfuss, Charles Rocheville, A. C. White and A. B. Brown.

4

followed them on camels and donkeys and soon caught up. The going was too rough and the sand too deep for the cars, so, amid much laughter, the geologists agreed to leave their touring car and accept a camel ride to the *jabal*.

Mining engineer Karl Twitchell crossed Arabia in 1932.

They spent their first night in Saudi Arabia in a building owned by the Al-Gosaibi merchant family. Five days later they visited the limestone hills known as Jabal Dhahran, a distinctive geological feature they had observed from Bahrain. By the last day of the month they were in Hofuf, where they made a courtesy call on the Amir of the province, and rented office space in another house owned by the Al-Gosaibi family. They kept the Hofuf property as a branch office, but had the Al-Gosaibis engage another house in Jubail as the company's first in-Kingdom headquarters.

Reinforcements arrived quickly. Another geologist, J.W. "Soak" Hoover, landed in October, bringing three Ford V-8 touring cars, a mechanic, a helper and two drivers. Before the end of the year three more Americans joined the party: H.L. Burchfield, Felix Dreyfuss and Dr. J.O. Nomland. By September 1934, there were 13 Americans living at the camp, and they had added a Fairchild monoplane for aerial photography and mapping. One of the new men who arrived during that period, Max Steineke, would play a critical role in the discovery of oil in the Kingdom.

The early Westerners found much to admire in Saudi Arabia and its people. In writings and interviews they recall with amazement the ability of their Bedouin guides to find their way across miles of seemingly trackless desert. One of these guides, Khumayyis ibn Rimthan, later had an oil field named after him. The Americans of that period expressed admiration for the skill of pearl divers and the seamanship of dhow captains. They delighted in Saudi hospitality, and respected the average man's devotion to his religion.

Containers of gasoline being bundled for delivery by camel to remote drilling sites.

At the same time, they found the Saudis they met to be unaware of Western technology. The Saudis still lived much as their ancestors had done for centuries. The Kingdom had only a few privately organized schools, no paved roads and no electric lights.

Many Saudis had never seen an automobile, much less an airplane. They were unfamiliar with such basic hand tools as hammers, saws, screwdrivers or measuring rules. The vast majority of the Saudi people at that time could not read or write. How strange the paraphernalia of an oil company — gauges, meters, routing slips, requisitions, generators and pumps — must have seemed to them. Yet they could not be called uneducated, because they were superbly educated in the ways of living in Saudi Arabia.

The Work Begins

The company used both aerial photography and geological field trips in its exploration and mapping of the concession area. By 1935 the area from the center of the vast southern desert known as Rub' al-Khali, or Empty Quarter, all the way to the northwest boundary of the concession area had been photographed from the air. Between 1934 and 1937, exploration teams roamed from the Yabrin oasis on the south, to Maniya and Lafiya on the

north, and inland as far as Hayil. In the spring of 1937, Max Steineke crossed the Arabian Peninsula in both directions, carefully surveying the geography as he went. The information he and his party gathered became the basis for all future geological profiles of the country.

The coming of these motorized geological exploration parties was an unforgettable experience for many of the Saudis. Nassir Al-Ajmi, who would become one of the top executives in the world's largest oil company and later the president of the government railroad services, was born in a nomadic encampment west of 'Uthmaniyah, on the edge of what would become the Ghawar field, the world's largest oil field. In his autobiographical book, Legacy of a Lifetime, Al-Ajmi recalled his initial contact with the oil men and their machines:

Geological survey field party in 1935.

"The first time I saw a vehicle was a frightening experience. I cannot remember the date, but it was springtime and I was playing with other children next to our encampment when we heard a strange noise. We saw an odd-looking thing rushing toward us with a cloud of dust behind it. We ran as fast as our legs could carry us and hid inside the tents. As we peeked through the holes to observe the noisy, strange-looking creature, we noticed two or three unfamiliar looking people wearing funny clothes and deep plates or funneled pots over their heads! It was an exploration party asking for water and seeking directions."

To keep two geologists in the field required a small army. A well-equipped field team included 15 to 20 cargo camels, two trucks and a touring car, a guide, a cook and a cook's helper, a houseboy, a mechanic and a mechanic's helper, an automobile driver, four camel drivers and an escort of 15 to 30 armed Bedouins.

The government supplied the people and the "military" escort, but Casoc paid their salaries. The company hired its first Saudi drivers and mechanics from those who had served on these field trips.

Training Begins

n the absence of formal industrial training programs, the handful of Americans had to train Saudi recruits as best they could. The first record of on-the-job training appears in the diary of J.W. "Soak" Hoover, one of the geologists mapping the Dammam Dome where Dhahran now stands. In the entry for December 27, 1934, Hoover writes, "I taught 'Abd al-'Aziz how to read the aneroid barometer this a.m." It must have been a memorable experience, for Hoover added, "It's remarkable how easily some can learn, and most remarkable how some cannot get their brains to function at all!"

Training was a necessity for effective operation of the field teams and for the safety of everyone involved. "Training was our second job, if not our first," an American employee of the time said. "But we never thought of them as trainees. We called them hired hands, but you had to teach them everything from the grass roots up."

Dammam No. 1

n June 1934, Hoover and Henry completed detailed mapping of the Dammam Dome, named after the larger of two fishing villages in the area, and made a location for the first test well, Dammam No. 1.

The decision to drill triggered a flurry of construction activity. Casoc headquarters in San Francisco authorized construction of the first offices, the first bunkhouses for Americans, the first quarters for the Saudi camp, the first road to the neighboring coastal village of al-Khobar, and a masonry pier to receive small vessels at al-Khobar. All these were to be built while preparations were under way for drilling Dammam Well No. 1.

An American, Walt Haenggi, who had designed and built housing for Bapco in Bahrain, came over to take charge of the construction. He had permission to bring along several skilled workers, provided they stayed just long enough to train their Saudi Arab replacements. They were the first expatriates brought over expressly to train themselves out of a job.

Haenggi's crew was to build a manager's house, an office building, a mess hall and recreation room, a garage and shop, a storehouse, and two bunkhouses to replace the tents in which the Americans had been living. They were to be constructed of *furush* and *jiss* — local coral slabs and gypsum plaster. In addition the crew was to build the derrick for Dammam No. 1.

Dammam Well No. 1 showed only traces of oil, 1935.

Saudi workmen were quartered in *barastis,* simple structures constructed of wooden pole frames, with woven palm leaf ceilings, walls and floors. Each *barasti* had two or three "partitions" for individual occupants. "The usual habitation of that part of Arabia for the natives" was how a company communiqué described these huts.

Rapid Expansion

he pace of expansion was hectic and could be simply overwhelming. In 1938, 2,400 new recruits were hired, increasing the size of the Saudi work force by some 400 percent. At times harried foremen found themselves trying to run crews of as many as 200 beginners at a time.

By the spring of 1936 the Dhahran camp must have seemed luxurious to those who had seen it grow from a few tents set on barren rocks and sand, but

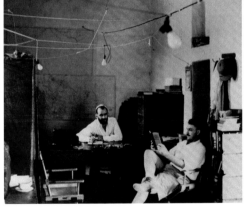

Max Steineke, right, and Felix Dreyfuss at the company's Jubail office, 1935.

to new arrival and ex-U.S. Marine, Ralph Wells, it was still "a wildcat camp."

In his diary, Wells described landing at a rock pier stretching 400 to 500 yards into the Gulf. There he boarded a balloon-tired Ford sedan, which scraped its bottom on the rocks of the pier many times before reaching the shoreline.

"Seven miles of sandy road brought us to the camp," Wells wrote. "It now consists of two bunkhouses with large rooms, a small emergency hospital, an office building with six rooms, a bosses' house of several rooms, a mess house and recreation club combined, a machine and blacksmith shop, an ice plant and a few other minor buildings.

"Two wells are now being drilled not far from camp. There is not a tree, bush or anything but spotted dry bunches of stiff grass anywhere. A few rock points stick up here and there, but the general contour of the land is flat."

On-the-Job Training

n Aramco legend says that regular on-the-job training for Saudis began on the drilling floor of the company's first well, Dammam No. 1, in November 1935. A typical drilling crew of the time consisted of an American driller, an assistant driller and 12 Saudis, compared to an average crew in the States of a foreman and a crew of five.

Saudi workers crush minerals for "mud" to lubricate drill bits.

The assistant driller's job was to get the crew up to speed. If the assistant driller was an American, he trained a Saudi as his replacement. The American assistant would then move to another rig either as a driller or as an assistant driller.

The need for larger than average work crews was based on several factors. Extra hands helped to compensate for the inexperience of the average employee. Plus, some of the Saudis lacked the physical strength to handle certain jobs alone. A third reason was the high rate of absenteeism among Saudis. Every year, in season, many Bedouins returned to their flocks and many pearl divers to the pearling boats. Other men became discouraged and homesick and quit or went on vacation unannounced.

At first it was company policy to automatically drop a man from the payroll after one week's unexplained absence. That was later increased to 10 days, then to two weeks, and then to three weeks.

The American (and usually Texan) crew boss for each rig developed his own way of communicating with the Saudis on his crew. He typically used a mixture of English and pidgin Arabic punctuated by grunts, shouts, gestures and a few indispensable words.

A driller might know '*foq*' as the Arabic word for 'up,' but not know or forget in the heat of the moment the word for 'down.' So, he shouted "*Foq* no." The crew came to understand that as meaning 'down.'

By all accounts, it was a workable system. An unexpected side effect of this system was described by an employee in a written memoir: "Whenever a new Saudi employee was assigned to a rig, he had extreme difficulty communicating, not only with the Americans, but with the other Saudis on the crew. It took the new man several weeks before he fit into the team. An inquiry into this phenomenon was initiated.

"It was discovered that each drilling rig had, in effect, its own language, which a new man would have to learn before he became an effective part of the crew. This 'language' was a mixture of Texas-English and pidgin Arabic, each word meaning what the Texan thought it should mean, regardless of what its actual dictionary meaning was. Once a Saudi discovered that an Arabic word had a meaning on the rig that was far different from the meaning he had always thought it had, or that an Arabic word was pronounced in a new way there, he would accept the situation and conform. Whenever he switched job locations, he had to learn to talk all over again."

Some Americans and Saudis taught words to each other by what became known as "buddy tutoring." It was a simple system — "You teach me Arabic, I'll teach you English" — through which two men learned to communicate well enough to get an idea across, no matter how fragmentary and ungrammatical the sentences. Some examples of the sounds of English commonly spoken at the time by Arabs: "*abwak*" (books), "*akwat*" (coats), "*asyat*" (seats), "*panacil*"

Dammam No. 1, right, and Dammam No. 7, center, in 1937.

(pencils), "*shayyek*" (check or examine) and "*yeshayyekoun*" (they are examining).

Americans, in recalling the early days, often comment on the intelligence of Saudi employees. Phil McConnell, author of a book about that time period titled *The Hundred Men,* might have been speaking for all such Americans when he said: "I was very much impressed by the intelligence of the Saudis. They were untrained people, but there wasn't anything wrong with their intelligence. They could pick things up in a hurry. They were eager, and just naturally very intelligent people."

8

The effectiveness of training depended on the zeal of the trainer as well as the eagerness and intelligence of the people being trained. In late 1939 or early 1940, Cal Ross trained the first Saudi crew capable of dismantling the 136-foot-high steel derrick at a completed well, moving the derrick and reassembling it at a new well site. It was a strenuous and hazardous task that required teamwork and left no room for mistakes.

First Teacher Is Hired

n 1938, at the urging of William J. Lenahan, the company representative in Jiddah, a young college-educated Lebanese, Bisharah Dawud, was hired to teach English and Arabic writing to Saudis in Dhahran and act as an interpreter. He quit after only two months.

Finding a replacement for Dawud, as well as other options for training, were discussed at length between company management in Dhahran and its representative in Jiddah. Because of difficulties related to this matter, the arrival of a trained educator in Dhahran was delayed for another four years, and the company was influenced to hire educators from America rather than from the Middle East.

Although the company in these early years had different rules for handling its American and Saudi employees, the Saudis and Americans seemed surprisingly compatible on a personal basis. As a close observer noted, "Saudis and Americans laugh at exactly the same place in jokes."

Early Accomplishments

hatever the social relationship, the makeshift, haphazard on-the-job-training system got results. A partial list of accomplishments in the period of extravagant growth between 1934 and 1940 is testimony to that.

During those years, American-Saudi work teams built the first stabilization plant at Dhahran, the first refinery at Ras Tanura, marine terminals at Ras Tanura and al-Khobar plus more than 50 miles of pipelines, a score of storage tanks, and permanent roads from Dhahran to al-Khobar and Ras Tanura. They drilled 19 wells on the Dammam structure, five of them in 1939 alone, and seven more were started in 1940. Eleven of the wells produced oil, three supplied gas for industrial use, one supplied water, and two were abandoned.

They also drilled wildcat wells at Abu Hadriya about 100 miles northwest of Dhahran, at Abqaiq, about 45 miles to the south, and near Ma'agala, north-

Dhahran mess hall at the south end of King's Road in 1939.

west of Dhahran. They drilled dozens of water wells at camp sites throughout the concession, all of them available for public as well as industrial use.

In Dhahran they constructed a hospital, two mess halls, storehouses, a commissary, garages, a carpentry shop, an electric shop, machine and welding shops, bunkers for dynamite, a power plant, cooling houses and an ice plant. They installed electric power lines, telephone cables, water purification systems and sewer lines.

Archie Perry is said to have organized the first regular industrial training classes for Saudis. He taught Saudis working in the new stabilizer plant the English names for tools and gave instruction on how to do various plant assignments. His classroom was an empty bay in the engine room.

The first American wives and children arrived in Dhahran in 1937. They settled in air-conditioned cottages between what became 10th and 11th streets. These were the first air-conditioned houses in Dhahran. Some offices and common buildings in Dhahran received air conditioning for the first time that same year. Meanwhile, another new community was taking shape at Ras Tanura. The first permanent housing, with electricity and running water, was built there in 1940, replacing the tents in which 12 staff people had been living.

March 3, 1938 – Oil!

 audi Arabia became an oil producer in 1938. Up to then Casoc had suffered a string of expensive disappointments. Dammam No. 1 failed to meet expectations; No. 2 showed promise, then fizzled; wells No. 3, 4, 5, 6, 7, 8 and 9 produced nothing. Many employees were wondering if the company would pull out of Saudi Arabia. Casoc executives asked Chief Geologist Max Steineke if it was worthwhile to continue operations. His advice: "drill deeper."

Tanker D. G. Scofield prepares for royal visit.

On March 3, 1938, Dammam No. 7, down to 4,727 feet in the Arab formation, came in at a rate of 1,585 barrels of oil per day. By March 22, it was producing 3,810 barrels per day. Gloom quickly changed to euphoria.

On May 1, 1939, King 'Abd al-'Aziz made his first visit to the company to celebrate the arrival at Ras Tanura of the first tanker to be loaded with Saudi oil. That tanker, the *D.G. Scofield*, was named after the founder of Socal. The King traveled from Riyadh in a caravan of two thousand people in five hundred cars. They set up a city of white tents near Dhahran, and after two days of general feasting, the King inspected the facilities and dined aboard the *D.G. Scofield*.

The euphoria lasted only a short time. Two months later, Dammam No. 12 exploded, killing an American and a Saudi worker, and erupting into a spectacular oil well fire. Putting out the fire took 10 exhausting days, and everyone in camp, from office boys to drillers, helped out in some way.

The Saudi camp in Dhahran grew faster than the company could build *barastis*. New men had to live in tents. Laborers walked to work from the Saudi camp or rode in from al-Khobar on flat-bed trucks. They were paid one riyal a day to start, and were given living space in a *barasti* and a food ration of dates, rice and milk, plus some meat every two weeks. In 1939, instead of one day off every two weeks, they were given every Friday off and free transportation to shopping areas at Dammam and Qatif on their day of rest.

Repeated appeals and the pressing need eventually persuaded the government to reluctantly allow the company to import a new class of semiskilled and skilled workers called intermediates. They were mostly Bahrainis and Indians experienced as shop workers, interpreters, clerks, storekeepers, rigmen and masons, people with skills the company said it could not find in Saudi Arabia and urgently needed to continue the expansion.

Saudi drilling crew at Dammam No. 7 with crew chief Les Hilyard.

The First Classrooms

On-the-job training has its limits, as illustrated in the following story told by an American employee of the time: A Swedish-American carpenter was responsible for training a number of Saudis. One day he came to his supervisor. "I'm stumped," he said. "What's wrong?" the supervisor asked. "Well," the carpenter said, "I've taught them how to saw and use a hammer and a chisel, but they don't know where to cut because they can't read a ruler."

The supervisor's response was, "Get a blackboard and teach them numbers." So he did.

In 1940, training began to move from the job site into the classroom. That year the first formal industrial training programs were started and the company opened the first schools to offer English and Arabic instruction to Saudis.

The new industrial training program included classes for 19 Saudis in metal welding. Classes in electric wiring, electric motor winding and accounting were also offered, but for only one Saudi workman at a time. Industrial training at that time meant working on mock-ups under the eye of an experienced man for as long as it took for the trainee to get it right or fail to do so and get sent back to a work crew. In addition, the company sent 26 Saudis to Bahrain on the first out-of-Kingdom (OOK) training assignment — 15 for instruction in refinery operation, five in laboratory testing and six in welding.

Industrial training, however, was not enough. American managers believed that, in order to become part of an effective work force, Saudis needed English language skills. Casoc was becoming a complex organization. It needed the

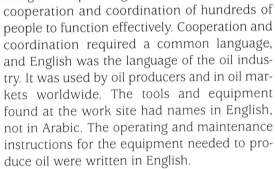

Hijji ibn Jasim's home in al-Khobar.

cooperation and coordination of hundreds of people to function effectively. Cooperation and coordination required a common language, and English was the language of the oil industry. It was used by oil producers and in oil markets worldwide. The tools and equipment found at the work site had names in English, not in Arabic. The operating and maintenance instructions for the equipment needed to produce oil were written in English.

The company executives felt that they could not expect help yet from the fledgling government school system. The first elementary school in the Eastern Province was opened by the government at Hofuf in 1937. By 1940 there were four elementary schools in the city enrolling 639 students. But Hofuf was 85 miles from Dhahran and it would be a number of years before more than a few graduates would be available from these schools and it was uncertain how much English instruction they might receive. So Casoc decided to open schools of its own.

An early government school at Hofuf, 85 miles south of Dhahran.

First Company School

The first company school, named the al-Khobar School, opened on May 11, 1940. The first teacher, Hijji ibn Jasim, was an interpreter and translator from the Government Relations Office. Classes were held in the evenings after work in a rented room of Hijji's home, a relatively large masonry house in central al-Khobar. Hijji was born in Kuwait, joined Casoc in 1933 as the

English and math lessons at Saudi camp school.

first Arab employee to come from outside Saudi Arabia, and stayed with the company until 1945.

The company furnished tables, benches, blackboards, chalk and lamps for the new school. It was open to any man who cared to attend, employee or not. Hijji began with a class of 19 students but very shortly was teaching 50 a night. The class was divided into two sections, and another part-time teacher, Hamzah Salah, also a Casoc employee, was engaged. The original plan was to teach only spoken and written English, but it was soon found necessary to teach Arabic vocabulary to explain and define the English terms and vocabulary.

A second company school opened on July 6, 1940, in a *barasti* built in the Saudi camp, the company compound in Dhahran where many Saudi employees lived. Muhammad 'Aridi, a Syrian-educated employee, was appointed instructor. Unlike the al-Khobar school, the new Saudi Camp School had electric lights powered by its own generator. All equipment was furnished by the company. The school was open to any employee, regardless of classification, as well as to interested nonemployees. On the first night, enrollment totaled 85, but in a few weeks the school had 165 students.

Jack G. Hosmer of the Personnel Department volunteered to supervise the school system and coordinated its activities with Casoc's operating divisions. He was encouraged by the rapid increase in enrollment and interest expressed by Saudi government officials. He was talking about expanding the school system to include a vocational school where Saudi employees would learn not only the mechanics of their work but the theory behind the job as well.

The company was on the verge of becoming one of the major oil producers in the world, with lavish plans for expansion. In 1940 Casoc employed 3,229 Saudis, 363 Americans and 121 other foreign employees, compared to 73 Saudis and a dozen Americans when the first well was spudded five years earlier. The company had invested $28.9 million in Saudi Arabia. Average daily production of crude oil was about 15,000 barrels a day. Company profits, estimated at $680,000 in 1939, were being plowed back into expansion. Abu Hadriya No. 1 struck oil in March, and it looked like a big new producer, Abqaiq No. 1, was about to come in.

Italy Bombs Dhahran

hen, on October 19, 1940, the war in Europe intruded and everything was put on hold. That night Italian planes dropped two or three dozen small 50-pound fragmentation bombs in and around Dhahran and perhaps 80 bombs on Bahrain. The Italian prime minister, Benito Mussolini, later apologized in a radio broadcast, saying his planes bombed Saudi Arabia, a neutral country, by mistake. Their only target was Bahrain, where British naval ships were fueled, he said.

The actual damage in Dhahran was minimal — an oil flow line was punctured and a water main cut — but the psychological impact was considerable. Women and children began to leave almost immediately. Within six

Employees measure two unexploded Italian bombs after 1940 raid.

months more than half the American men were gone and about 40 percent of the Saudi workers were off the payroll.

Instead of headlong expansion, things slowed to a crawl. But the slowdown had its beneficial side. There was more time to look around. Saudis and Americans got to know each other better. More attention was given to training and its scope broadened.

More Training Facilities

In 1941 the industrial training program begun a year earlier was expanded to include instruction in typing and lathe operation. A total of 50 Saudis received special training — 19 in welding, 22 in typing, five in lathe operation, two in accounting and one each in electrical wiring and electrical motor winding.

That same year the company discovered some employees, such as waiters, houseboys and office boys, who wanted to attend school but couldn't because of their work schedules. In January 1941, a *barasti* was built for them and a third school started. A full-time teacher, Ja'far ibn Muhammad, was employed to teach in the new school and assist at the school in the Saudi camp. The new school was soon being attended by more than 100 pupils.

After several months a course in basic arithmetic was introduced. It covered addition, subtraction, division and multiplication. A total of more than 300 people were taking at least one of the courses, English, Arabic or arithmetic, at one of the three schools.

The schools were divided into primary, beginners, intermediate and advanced classes, depending on the English capabilities of the student. A British text, *MacMillan's New English Reader, Primer I,* was furnished to aid students in learning the alphabet. Once the alphabet had been learned, students switched to another British text, *Ogden's The Basic Way to English, Book 1.*

In addition to regular classes, special one-hour classes were conducted each Monday for Saudi telephone operators in order to standardize switchboard procedures and the operator's salutation, "Number, please." There were also special classes taught by Ja'far ibn Muhammad for the two children of the local government representative, 'Abd Allah Al-Fadhl.

On March 29, 1941, the al-Khobar School moved from Hijji's house to a long *barasti* near the government wireless station to take advantage of the generator there to power electric lights for evening classes.

This barasti became the home for the al-Khobar School in 1941.

Monthly report cards were introduced in February of 1941. Students were graded on the A through F system, with an A grade being in the 90 to 100 range, and an F grade anything below 60. The Personnel Office kept a record of each student's grades and progress.

Hosmer was pleased with developments to date. He wrote: "The manner in which the students have adjusted to the educational facilities that have been made available to them has been very enlightening. They have taken their educational opportunity seriously.

"Many of the students take their report cards to their American foremen. They are pleased and proud to show the grades they earned." The Saudis demonstrated an eagerness to learn that impressed and moved those who visited the schools. "We had eight- to 10-year-olds and 30-year-olds sitting side by side trying to learn their ABC's. It was quite touching, actually," Hosmer said.

Strictly speaking, the company could have fulfilled its obligation under the Concession Agreement by continuing to employ Saudis as unskilled or semi-skilled workers. Schools were not necessary. The required English could have been learned on the job. But the Americans heading the company's field operations were practical men, and some of them, at least, farsighted as well. They realized that instead of just using the manpower at hand they could use education to improve it, and thus create something more durable and valuable than shorter-sighted policies could have achieved.

Floyd Oligher, the resident manager, declared in a letter to company headquarters in San Francisco: "The scope possible in the educational field is probably greater than any other single venture that might be contemplated for the country. Aside from the real advantage that the company will gain from better educated employees, there is no other single project that will bring more good will and favorable publicity to the company. It is a project that should have been started much earlier in our program of development."

Early Safety Training

 safety engineer, Adrien Louis Anderson, better known as "Andy," arrived in Dhahran in late 1940. Soon safety posters with stick-figure drawings and slogans in English and Arabic were appearing around camp. He recruited and trained Saudis to give the safety lectures in Arabic to crews at work sites. The effectiveness of this training was apparent on those occasions when Saudis would bring an American foreman up short by calling his attention to a safety violation.

Anderson also made sure workers learned artificial respiration, a must for men working with the "sour" hydrogen-sulfide-laden oil. A few years later, five Saudis using this training applied artificial respiration continually for one hour, 15 minutes to save the life of a Saudi overcome by hydrogen sulfide gas.

Anderson campaigned to get Arabs who worked in industrial areas to wear safety helmets, safety shoes and tighter-fitting Western clothing instead of the long, flowing *thawb* and *ghutrah,* or headcloth, which could easily become entangled in whirling machinery.

The change from *thawb* and *ghutrah* to Western-style clothes was very difficult for Saudis, just as it would have been for Americans if they had been required to wear Arab-style clothing. Nevertheless, Saudi workmen were soon wearing company-imported khaki work pants, although they quickly changed to traditional dress after work.

Safety helmets were another matter. For years, many workmen resisted wearing the hot and heavy safety helmets instead of the usual *ghutrah.* In 1956, 15 years after Anderson started the campaign, the company was still asking the government to help it convince Saudi industrial workers to forsake their traditional headgear in favor of safety helmets.

Saudi camp school, intermediate class, in front of barasti.

Safety shoes were also a lingering problem. It would be years before Arabs and some Americans, accustomed to lightweight sandals, could be convinced to wear the heavy safety shoes. Many pairs of company-issued safety shoes found their way to the shelves of resale stores in Bahrain.

But safety training paid off. An industrial injury rate estimated at 90 injuries per million man-hours in the 1930s was reduced to less than 12 injuries per million hours worked in the late 1940s.

In 1942, employment dropped to its lowest point of the war, about 1,600 Saudis, 82 Americans and 84 other foreign employees at mid-year. No supply shipments were received during the last seven months of 1942. The Ras Tanura Refinery was closed down, drilling suspended and new construction stopped, except for bomb shelters. Still, the company maintained production of 10,000 to 12,000 barrels of oil per day for shipment to the Bahrain Refinery.

Classes continued in the schools at al-Khobar, the Saudi camp and Dhahran, but special training programs were cut back. Only 24 Saudis, less than half the previous year's total, received special training.

Training in commissary operation was initiated, each class enrolling two Saudis. Four employees received special training in metal welding, six in typing, five in accounting, four in lathe operation and one each in electrical wiring and electrical motor winding.

Some new programs having potential benefits far beyond the confines of the oil company were launched about this time. Their actual impact outside of the company is impossible to measure, but interesting to think about.

Private Contractors Introduced

he contracting business, which eventually grew to a more than $1 billion a year industry in the Kingdom, was organized at Casoc during the years of World War II. Private contracting on a piecemeal basis went back to the days when camel caravans hauled fuel to the first geological field parties. In 1937, local merchants hauled palm fronds for construction of *barastis,* and in 1939, local contractors were used to build *barastis.* But now the company wanted to contract out support services on a regular basis in order to free its employees for the main job of producing oil.

Bill Eltiste was pulled off his job as head of the Transportation Department along with Cal Ross to form the Arab Contracting Department. They encouraged some of the more entrepreneurial Arab employees to become independent contractors in the fields for which they had been trained by the company.

At first the department contracted mostly for transportation services, and later for construction, purchasing and food supply. Pinched for spare vehicle parts during the war years, they contracted with Bedouins to haul supplies by camel caravan. Anything that could be divided into small enough parcels was sent by camel, saving wear and tear on cars and trucks.

"We had a big lot out on the west side of the storage yard in Dhahran at that time," Les Snyder recalled, "and it was a great big, noisy place with dozens and dozens of camels." During 1943 and 1944, camel caravans regularly hauled drilling mud nearly 100 miles (160 kilometers) from Manifa Bay to Jauf. The caravans consisted of 10 camels, each carrying a 400-pound load.

Bill Eltiste in 1955 photo.

More than one Saudi contractor started as a laborer at one or two riyals a day and became a multimillionaire. Trained, advised and often financed by the company, they began with a truck and a few men doing pick-and-shovel work, and built organizations of hundreds and sometimes thousands of men.

Western Medical Practices Introduced

he company also began to have an impact by providing access to Western medical practices. From the beginning, the oil men offered what medical help and advice they could to anyone who asked. At the company's Jubail office in May 1934, Soak Hoover noted in his diary: "Every evening

Dr. T. C. Alexander, left, and Dr. Robert Page, right, with Mahdi ibn Ahmed.

just before sundown the courtyard fills with Arabs of all sizes with as many ailments. Bad eyes predominate; bruised feet and worms are also common."

By 1940 the company had a regular medical staff consisting of two doctors — Dr. L.P. Dame and Dr. T.C. "Alex" Alexander, a Texas-born-and-educated physician who joined Dame in September of that year. The remainder of the medical staff consisted of one American first-aid assistant, three American nurses, three doctors trained in India, and three Indian male nurses. During the year they treated more than 700 hospital patients, handled more than 16,600 outpatient visits and gave more than 6,000 vaccinations.

Dr. Dame soon left Saudi Arabia, but Alexander remained with the company in the Kingdom for 21 years. In 1943 Alexander and Dame's replacement, Dr. William R. Flood, were the only American doctors in Saudi Arabia. By the end of 1944, the medical staff had been enlarged to include four American doctors and an American dentist. As the senior American physician in Saudi Arabia, Alexander was called upon to treat the Royal family in addition to hundreds of Casoc employees.

Alexander tackled the pernicious malaria problem. Instead of waiting for people to come to him, he went out into the communities with an antimalaria program. The disease was literally epidemic in parts of al-Hasa, and had forced abandonment of homes in some oases areas. A particularly serious epidemic swept Dammam and al-Khobar in 1943, compounding the hardships caused by war-related supply shortages. Alexander undertook a training program of his own. He visited communities to inform people that malaria was spread by mosquitoes. He advised villagers that mosquitoes hatched from eggs laid in the standing water that they kept in jars around their homes and in their gardens. He urged them to dump out their water jars every week to 10 days to kill the mosquito larvae. But discouragingly few people listened despite government regulations and regular inspections. It was difficult to convince some Saudis that the disease was spread by mosquitoes, or that they should dump out precious water. The campaign eventually succeeded, but largely due to the use of DDT when it became available after the war.

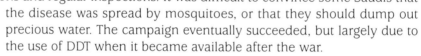

A Casoc employee of that period named Steve Furman had a unique impact on the Saudi economy. He managed a farm established during the war years to help provide company employees with fresh vegetables and some meat. Furman built an incubator to hatch chicken eggs. It was the first such thing seen by Saudi farmers from the area, where today there is a thriving poultry industry. Some thought it was a joke or a trick and stood by amazed as the eggs hatched in 19 to 21 days, just as Furman had said they would.

Steve Furman in 1959 photo.

Another unmeasurable but important benefit of the wartime hiatus was that it gave Saudis and Americans time to learn each other's ways. This enhanced understanding and helped to reduce friction during the great buildup that was about to start.

First Training Supervisor

rospects definitely looked brighter in the oil camps after allied victories in Africa, the Mediterranean and southern Russia removed the threat of attack by the Axis on the Middle East. New men began arriving from the

United States in 1943. One of them was Gilbert McLean Nearpass, known as "Mac." Nearpass arrived in December to supervise training programs and take charge of the company's new Education Division.

A total of 122 employees received special training during 1943, almost 100 more than the previous year. Courses were added in air conditioning, ice plant and power house operation, carpentry, laundry operation, garage repair work, instrument repair, boat operation, gauging tanks and gas-trap operation, stabilizer operation, telephone operation and motor dispatching.

The company began actively recruiting new hands in the United States and Saudi Arabia in anticipation of demand for more Middle East oil during what promised to be a prolonged war effort in the Pacific. In late 1943 the U.S. government officially authorized a massive new construction project at Ras Tanura to be financed by the company. Scarce steel was allocated and men and supplies were moved on a high-priority basis to the building site.

Construction of a blending and transfer pump house in Ras Tanura, 1945.

Company Renamed Aramco

n gearing up for the project, San Francisco decided to give the company a new name. On January 31, 1944, the enterprise became a Delaware-registered corporation called the Arabian American Oil Company. For the following 44 years the company would be generally known by its new acronym: Aramco.

The centerpiece of the Ras Tanura project was a 50,000-barrel-a-day refinery, but it included in addition a tank farm, a marine terminal and a submarine pipeline to Bahrain. The project was to dominate company activities for almost two years. A record 6,100 new employees were added in 1944, and the company was urgently searching for more. At the building site in Ras Tanura there was bedlam. Enlarging the refinery meant enlarging everything else. They needed tents and *barastis* to house thousands of new Saudi employees, a larger pier, more pipeline capacity, more stabilization plants, more storage tanks, more boilers, more welding machines, more trucks, more truck tires, and on and on. Aramco was resuming a metamorphosis, one that had been put on hold by the war. It was changing from its frontier period to the period of great postwar growth.

Lost in the din of the buildup, a new school was opened in Dhahran in the spring. One of the two bunkhouses, built nearly 10 years ago by Haenggi and his crew, was converted to make a school for Saudi boys. They called it the Jabal School, and it became a place where legends were made.

Born Together: Aramco and the Jabal School 1944-1949

*"We were a group of excited Saudis. This was a new adventure—
a new language, a new environment, a new life!"*
– Jabal School student

No generation knows what history will say about it, or what the future will see as symbols of their time. Surely in 1944 no one expected history to remember the humble Jabal School. Yet the little company school endures as a symbol for development — not for the development of an oil company, but for the development of a generation of very special young men.

Many Saudis were introduced to the mystery of letters and numbers at the Jabal School. Among them were future scholars, successful businessmen and powerful executives. One of those young men eventually became the president and chief executive officer of Aramco, and then the Kingdom's Minister of Petroleum and Mineral Resources. The school also stimulated the expansion of government schools throughout the Eastern Province and in the rest of Saudi Arabia. Despite its short life and imperfections, it was, as Wallace Stegner exclaims in his book, *Discovery!*, "the germ of something momentous."

The Jabal School opened on April 8, 1944, to a class of about 70 Saudis between the ages of eight and 18. Within a year, enrollment nearly doubled. Students were mostly office boys, houseboys and waiters classified as Education Trainees. They worked half time, went to school half time and received full-time pay. In the language of professional educators, they were part of the company's first Developmental Training Program.

"*Jabal*" means hill or mountain in Arabic, and early photographs show the school's name transliterated as "*Jebel*." The Jabal School differed in fundamental ways from the voluntary schools. First, it was in a building of its own, not a *barasti*. Second, classes were only for boys under age 18, not men. Third, and perhaps most important, young employees were paid to attend school during working hours. Without those few coins a day, many boys could not have afforded to be in school.

The Jabal School was operated by the new education supervisor, Mac Nearpass, and an Arab scholar,

*Opposite:
Teacher Vince
James leads a
Jabal School
English class,
1947.
Below: Students
and staff of the
Jabal (Jebel)
School, 1946-'47.*

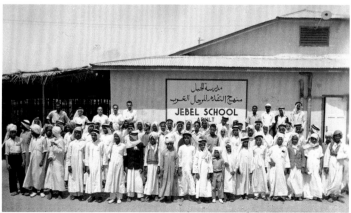

Teacher Fahmi Basrawi supervises recitation by Saudi student at the Jabal School, 1947.

'Abd al-Hafiz Nawwab, who had the distinction of being the first Saudi college graduate hired by the company. Company records show a number of part-time teachers, including George Mandis, 'Isa 'Ashur from Bahrain and Thomas C. Barger, who later became the company's chief executive officer and chairman of the board.

Promising young Saudi employees from the company's voluntary schools at al-Khobar and in the Saudi camp were urged to enroll at the Jabal School. Nonemployees were welcome as well, although they received no pay.

The Jabal School was located on a block of a dozen or so single-story buildings on what is remembered as Main Street in the old office area of Dhahran. The building was constructed in the mid-1930s as a bunk-house in the original Dhahran oil camp. It was made of coral rock *(furush)* and plaster *(jiss)* with an oversized corrugated roof. Since its design was copied from early bunkhouses in Bahrain, its roots might have been an Indian bungalow or the general store of a West Texas town.

The school was across Main Street from the district manager's office and next door to the Accounting Department building, from which there often issued the pleasant tinkle of silver riyals being counted by machine. Main Street was a wide, unshaded dirt road, rutted by tire tracks, and flanked during working hours by dust-covered trucks and sedans parked nose up to pipe guard rails. Usually, the only people moving along the street were office boys carrying papers from one place to another, except on Thursdays, when long lines of laborers waited in the sun outside the Accounting Department to collect their pay in silver riyals from the paymaster's window.

The converted old bunkhouse had two large classrooms on either side of a wide central hallway. Except on special occasions, the two classrooms were subdivided by partitions so that up to four classes could be held at one time. There was an office for the education supervisor, a smaller office across the hall for his secretary, a staff study room and a small noncirculating-library room. A bathroom with running water, but no air conditioning, was located in a separate building at the rear of the school.

The average young Saudi found the Jabal School to be a new and somewhat frightening environment. The new experiences began before a boy was even properly enrolled in school. Prior to being admitted, each boy had to undergo a physical examination. He had to test free of contagious diseases and have at least 6/12 vision. Those with ailments were usually told to report to the hospital for further examination and free treatment.

When the Jabal School first opened, there were two class periods — four hours in the morning and four hours in the afternoon each work day. This soon expanded to three four-hour classes a day at the Jabal School and one four-hour class at the school in Saudi camp.

Subjects taught were English, basic arithmetic and (since there were so few schools outside the company) reading and writing Arabic. Special classes in shorthand, typing and houseboy training were introduced. Hygiene and safety classes were added later. Rather than advancing toward graduation one semester at a time, students worked their way through one textbook at a time until they completed all the books on a subject.

Ibrahim 'Abd Allah Afaleq

"The happiest moment was when I was selected to the university, because all my future depended on that."

brahim 'Abd Allah Afaleq was having no success landing a job with the fledgling Aramco. It was the mid-1940s and the 16-year-old had just arrived in the Eastern Province. Administrators at his local school had told him and the four other boys who had just completed the fifth grade that there would be no sixth grade. Afaleq had heard of Aramco, and he wasted no time in setting out to find work with the company.

Frustrated by failing attempts to get hired by Aramco in Dhahran, Afaleq and some others decided to walk the 10 or so miles to the big compound in Dammam, where they lined up before sunrise. "There were so many, and I was so small," he said. "I was pushed to the floor, when this policeman picked me up and said 'This boy must be recruited first.' I was lucky." With that push began a lifetime association with the emerging oil giant.

Having no benefit of training, Afaleq was put to work first as a kitchen boy where he washed dishes in the dining hall and then as a telephone operator where he earned three Saudi riyals a day to start. His knowledge of English consisted only of knowing the numbers needed for the telephone. It wasn't long before Afaleq saw the need for additional education and he requested a transfer to the training department in Abqaiq. There he studied English and hygiene four hours a day while working the remainder of the time.

Three years later he had reached the equivalent of the first year in high school, and he was selected for an out-of-Kingdom high school summer session. Later, after receiving his high school diploma, Afaleq again worked in the Accounting Department. He was selected for an advanced education program in the United States, where he completed requirements for a business education degree, which included practice teaching in a rural American school in New Hampshire.

Afaleq then returned to Aramco, where he taught math until 1977. After serving as a controlman in the Materials Department, Afaleq was named administrative supervisor of a Materials Supply Organization (MSO) Training unit. In 1983 he was named MSO training superintendent until the department consolidated. Afaleq later became superintendent of the Support Division in the Central Area Storehouse Department and retired in October 1992.

Afaleq said learning English was vital for his success with Aramco. In part, he credited his English fluency to his education in a small American school where there were no other Arabic speakers. "It is the experience that is necessary: eating and talking with them. This is the way to learn a language," Afaleq advised.

He urged young Saudis to get as much education as possible. "Without education, man is nothing. ... When I was a young man, I knew of only one Saudi who had gotten a bachelor's degree, and that was Hassan Masharie," Afaleq said. "He was the first Saudi with a degree to work for Aramco."

Afaleq recalled the lesson of what that degree meant. "He didn't stay long," Afaleq remembered. Masharie later became the Kingdom's Minister of Agriculture.

Team Sports and Textbooks

ac Nearpass introduced physical education classes and team sports such as baseball, soccer and volleyball to teach the fundamentals of teamwork and fair play. Team sports were new to Saudi youngsters. They proved to be indifferent to baseball, but took to soccer as if born to play the game. At first, lacking a soccer ball, they kicked around a rag ball and marked out the field boundaries and goals with rocks.

Students were introduced to English through a textbook, Ogden's *The Basic Way to English*, which the British had used with good results in India. The text contained 850 "essential" English words. "This is my eye" began book one, with an accompanying illustration of a man pointing to his eye. "This is my nose" came next with another illustration, and so on.

When students had advanced to the fourth and final basic English level, they switched to Ogden's *Basic Way to Wider English*, books one and two. They also used Thorndike's *Century Beginning Dictionary* and *Junior Dictionary*.

The arithmetic texts were by Laidlaw Brothers of New York, *Arithmetic Readiness Part I and II*, and *Essential Drill and Practice in Arithmetic*, grades three through five.

Saudi students learn the basics of "dodge ball" in Dhahran, 1946.

These same textbooks were used in classes after work for adult employees at the voluntary schools. Various instructors also wrote and distributed mimeographed workbooks for English-language studies and short texts for the hygiene and safety classes.

To many company executives of that time, the Jabal School may have only been a convenient place to give Saudis a start on English and arithmetic before they reached the age where they could begin industrial training, but to the young faces who crowded the classrooms it seemed much more. As a student recalled: "We were a group of excited Saudis. This was a new adventure — a new language, a new environment, a new life!"

As education supervisor, Nearpass oversaw the Education Division, which included the evening volunteer schools and schools for American children, as

well as the Jabal School. He had arrived at a time of explosive growth throughout the company. During 1944, his first full year, enrollment in Education Division classes rose from 134 to 649 students. The staff grew to include six full-time teachers and as many as 15 part-timers, a ratio of about one teacher for each 30 students.

About the same time, a school for Saudi youths, similar to the Jabal School, was opened in a most unusual "building" in Ras Tanura. A big wooden crate used to ship the community's first fire engine from the United States was outfitted to serve as a one-room school. Clyde

Soccer became Saudi Arabia's favorite team sport.

W. Brassfield, a teacher-coach from Oklahoma, operated the "facility." The school later settled into in a small masonry building with a full-time teacher, Emmet Roberts, who was followed by Larry Emigh.

Other Middle East Training Programs Reviewed

earpass conducted a survey of other Middle East training installations during October of 1944. He visited British-dominated installations at Basrah and Baghdad in Iraq; Abadan, Iran; Damascus, Syria; and Beirut,

Lebanon. He found training at those installations was based on traditional apprenticeship programs. He was not impressed, evidently, for there was no recommendation to initiate similar programs at Aramco.

He did note that the Anglo-Iranian Oil Company at Abadan financed construction of government schools. The company also provided housing and helped pay the salary of some teachers, but it refused to take responsibility for the education of Iranians in government schools.

Similarly, Aramco, when dealing several years later with the Saudi government on this issue, refused to take responsibility for education at government schools, but did agree to pay for the construction and operation of many government-run schools.

Nearpass must have reflected many times during those years on the irony of his being with Aramco. In the middle of World War II, when he was in his late 30s, Nearpass wrote the U.S. Navy asking for a commission. In reply, the Navy gave him an unexpected choice — take the commission, or apply for a job with Aramco in Saudi Arabia.

Someone in the War Department had matched Nearpass' credentials with Aramco's need for an education professional qualified to organize the company's education and trade training programs. Expansion of Aramco's production facilities had a high priority at the time, and the U.S. military was doing what it could to help fill key positions within the developing company.

Silver Saudi Riyals are counted in preparation for payday, 1947.

Nearpass joked that one reason he was picked for the job was that he was already acclimated to desert life. He was the school administrator at Blythe, California, a small Mojave Desert town with a climate nearly as hot and dry as Dhahran's. Although Blythe was best known for its sizzling summer temperatures, often exceeding 110 degrees F, it also boasted the nation's largest school district in area, covering many square miles of sparsely settled southeastern California desert. After talking it over with his family, Nearpass chose the Aramco option "and never regretted it," according to a family member.

'Abd al-Hafiz Nawwab

 deep respect for education is traditional in the Saudi culture, and few teachers have commanded more respect than 'Abd al-Hafiz Nawwab, the first of the full-time teachers at the Jabal School. A dignified man, he insisted on being addressed in correspondence as Mr. Nawwab; otherwise, he returned the message unread. Nearpass and others complied, although it was not customary at the time for Americans to use the title "Mister" among themselves, let alone in correspondence with Saudis.

Mr. Nawwab joined the company in 1936 as a chemist with a bachelor of science degree. He was fluent in English, having learned the language in India, where his father was assigned as a religious official. He was from a scholarly family, and his scientific, religious and linguistic learning earned him a high standing in the community.

Mr. Nawwab was often accompanied by a visitor from the Hijaz, in western Saudi Arabia, 10-year-old Ismail Nawwab. The two were cousins, but were so close they referred to each other as "brother." Ismail was already fluent enough in English to tutor to some Jabal School students.

When the older Nawwab became ill, he was treated in the American hospital — the first Saudi ever to be admitted there. When he died as a young man on a windy day in June 1945, many Saudi workmen laid aside their tools, and the boys in the Jabal School left their schoolbooks to attend the funeral in Dammam in a spontaneous show of respect for the man they called "the teacher."

Young Ismail Nawwab left al-Khobar and Saudi Arabia to study abroad and become a scholar in his own right. Thirty years later he returned to Aramco to begin a distinguished career that saw him eventually rise to the position of general manager of Public Affairs.

Fahmi Basrawi

Aramco recruiters in offices from Hofuf to Jiddah were constantly on the lookout for Saudis with some education or specific skills. Such applicants were seldom turned away. Instead, the imaginative recruiters found them a place, even if they had to place the applicant in a position for which he seemed only vaguely prepared.

This was the case with Fahmi Basrawi. He was a teenage clerk in the police department at Jiddah when he answered an ad in a local newspaper for an unspecified type of job. The recruiter, noting the young man had six years of schooling and the ability to both read and write Arabic, told Basrawi he would be made a teacher of English.

"Teacher of English?" Basrawi said. "I don't know any English! I don't know anything about teaching!"

"We'll teach you," the recruiter replied.

Since the salary offered was twice what he was making as a police department clerk, Basrawi accepted. He crossed the Peninsula to Dhahran riding atop sacks of wheat on the back of a truck for 13½ hot, dusty days. No one taught him English. He learned by memorizing words out of the school textbook.

"I pasted up words on the wall above my bed so they would be the last thing I saw at night and the first thing in the morning. In this way I was able to memorize about 10 words a day and stay two or three lessons ahead of the class." He also taught basic arithmetic and Arabic reading and writing, subjects that he had at least studied earlier in school at Jiddah.

Basrawi was an instant success as a teacher, displaying a natural talent inside and outside the classroom. He threw himself into the organization of baseball and volleyball teams, trips to nearby oases and photography classes. In his spare time he tutored Americans in Arabic, launched the first Aramco taxi service and took photographs for company identification cards. Much later he taught English, Arabic and arithmetic on the Aramco television station and became the celebrated host of a quiz show televised nationally by a Dammam station.

In January of 1945 the first two Saudi Arab instructors assigned by the government arrived at the school. They were Shaykh Hamad Al-Jasir, who was to become a highly regarded author and founder and first editor of the publication *Al-Yamamah* of Riyadh, and Shaykh 'Abd Allah Al-Malhuq, later Saudi ambassador to the Sudan. These two instructors, and the government appointees who came after them, taught classes in religion and ethics and kept a watchful eye on what the American teachers were doing.

During the 1945-46 school year the name of the Jabal School was officially changed to the Arab Preparatory School, although everyone continued to call it Jabal School. It was the first of two name changes for the school, both intended to indicate that this was a company school and not one of the government's public schools.

Many future executives attended Jabal School classes like this one.

24

n 1946 Vince James arrived to take charge of the Education Trainees Program for young Saudis. After 16 years as a teacher, athletic coach and school principal in the New Jersey school system, he had developed a yen for overseas adventure and joined Aramco. He took charge of the Jabal School and ran it for the remainder of its bittersweet life.

The Jabal School staff included James, Basrawi and an able clerk and interpreter, Wadiy' Sabbagh from Iraq, plus a series of part-time instructors. Several advanced students were pressed into service as teachers in lower-level classes.

Students at the Jabal School were supposed to be sons of company employees or young nonemployees approved for enrollment by H.H. the Amir of the Eastern Province, but the rules were difficult to enforce. Some youngsters were cousins or nephews instead of sons. A few, such as young Ali Al-Naimi, just walked in and began taking classes without any sponsorship.

"Usually their guardian or someone would bring them in," Vince James recalled. "The Amir would send some over and the government would send some. We were overwhelmed with applications. It was a popular place."

They were so eager to learn, so well behaved, that American educators couldn't help wishing "American kids would be more like them."

"We really had no disciplinary problems," James said. "The Saudis were there to learn. I would say they were more eager to learn and grasped things as well as or better than the average American kid."

A youngster with some prior education was a rarity who merited the red-carpet treatment when he applied to the Jabal School. When Ali Baluchi applied after graduation from an al-Khobar elementary school, James bundled him into a pickup truck and personally drove him around to finish his processing. "And the reason he did it," Baluchi said, "was that he thought I might change my mind. It was unusual for a Saudi with an elementary school certificate to come to Aramco. The government usually got the 'educated' ones at the time."

James began learning Arabic in much the same way Basrawi first picked up English words, by posting a list of words to be memorized above his bed at night. He later tried "buddy tutoring," trading English for Arabic-language lessons with Basrawi and others.

In 1947 the Jabal School's official name was changed once again, from Arab Preparatory School to Arab Trade Preparatory School, although as before, it continued to be known as the Jabal School. That fall the Jabal School had an enrollment of 220 students and a long list of applicants. The school expanded, establishing classrooms in neighboring Bunkhouse 10, which it shared with the Accounting Department. By the next spring, more than three-fourths of the students were gone, on government orders to dismiss all boys under 15 years of age. Under the nation's labor laws, the government said, boys under 15 could

A Jabal School class, 1947.

not work part time for the company and go to school. Only 55 students had acceptable documentary evidence of their age to continue study. However, by the fall 1948 semester, enrollment was up to 129 and climbing.

On January 31, 1949, in Building Nine of the old office area, the first and only group of youngsters to complete all levels at the Jabal School received certificates. They were listed in the company newspaper as Salih 'Abd Allah Al-Rubayyi' and his brother, Sulayman

'Abd Allah; Fahd 'Abd Al-'Aziz; Khalid Hamad Al-Dossari; Muhammad Ahmad and Hasan 'Ali Ghanim.

The event would have been more festive had not the end of the Jabal School been clearly in sight. In a few months the Jabal School and its smaller twin in Ras Tanura quietly closed and the students were absorbed into the new General Industrial Training Centers.

"It may have been inevitable," James recalled, "but it was sad because we felt something was disappearing that had done some good."

Saudi boys who attended the Jabal School developed differently from any previous generation. They were introduced to Western patterns of thought and the concepts necessary for success in the industrial world, and they were well prepared for further education. Along the way they made special friendships and took their place in a peer group that grew in importance through the years.

Jabal School Student: Ali Al-Naimi

Any roster of illustrious Jabal School students has to be headed by Ali Al-Naimi, the first Saudi president and chief executive officer of Saudi Aramco. Al-Naimi was later appointed by King Fahd ibn 'Abd al-'Aziz to be Minister of Petroleum and Mineral Resources for the Kingdom of Saudi Arabia. In an interview, Ali Al-Naimi talked of his childhood in a Bedouin tribe, and how he happened to enter the Jabal School:

"Sometime in 1944-45, an elder brother, 'Abd Allah, who was working with Aramco, said, 'Why don't you come along with me and go to the Aramco school? There are no requirements and you don't even have to work for the company.' It was very impressive. I saw this teacher; he had a huge red beard. He must have been Irish. I walked in and enrolled. No one asked any questions. The first thing I learned was 'This is a fox.' He was pointing to a picture behind him, and everyone was shouting, 'This is a fox.' So I said, 'This is a fox.' That's how it all started.

"I did that for two years, then my brother died. I was about 11 by then. I took over his job as office boy. In 1947, the government passed a law that said no one below 18 was allowed to work full time, so I got terminated.

"Not long after, I reapplied for work. I told them, 'I may look young, but really I'm 20. I'm just a little Bedouin. I am short, and that's why I look young.' That was partly true. I wasn't 20, but the rest of the description was correct.

Fahmi Basrawi with Jabal School students, including Ali Al-Naimi, second from right, holding a baseball.

"They said, 'We'll give you a chance. Go to the government doctor, and if he says you are 17, we will hire you.' There was only one doctor in Dammam. He examined me and said I was about 12. I said, 'No, there's a mistake. I am really 20, but I will settle for 18.' So we haggled. He said, 'Why do you insist you are older than you are?' I said, 'Because I need a job and need to work to take

care of my family and go to school.' So he said, 'Well, okay, how about 16?' I said, 'How about 17?' and we settled on 17. I'd have been in trouble if he hadn't agreed. Sometimes you have to sell yourself. It's marketing."

Ali Al-Naimi was rehired as a junior clerk in the Personnel Department, went back to school, and was launched on a historic career with Aramco. "We were very useful young boys," he recalled. "We could type, we knew a little bit of math, we knew some English, and we knew Arabic. So we were very handy to do clerical work. The company then was in dire need of this kind of staff."

Other notables from the Jabal School include Ali Baluchi, who became general manager of Community Services; Khalifa Assara, a senior labor relations representative; Mustafa al-Khan Buahmad, former director of Career Development; Ahmad Hazza, former manager of Storehouse Operations; Muhammad A. Salamah, former administrator of the Dhahran Area Personnel Division; Abdul Rahman Dhuwaihi, former superintendent of Television Services; and Saif Yousif Al-Husseini, former supervisor of the Saudi Riyal Personnel Unit.

Aramco Postwar Expansion

 hile the Jabal School had its brief moment, Aramco itself was in the midst of frantic expansion. To nearly everyone's surprise, the anticipated postwar slump in demand for petroleum products failed to materialize, and Aramco continued selling oil as fast as it could produce it. At the same time, despite high priorities for its purchases, the company continued to be hamstrung by shortages of supplies and new equipment.

Caught in a squeeze between demand for oil and the lack of men and materials to produce it, the company adjusted its priorities almost daily. One day men

Aramco Administration Building in Dhahran, 1947.

and equipment were diverted from drilling for oil to building bunkhouses for other men who would soon arrive. The next day they might be diverted from their regular work to unload a freighter. Neither the time nor the materials could be found to complete a pier that extended into the deep water off Ras Tanura. Whenever a freighter arrived, men had to use barges to transfer tons of urgently needed supplies to shore. Often supplies were ruined in the transfer and left to weather on the sand.

During this period, the company hired new men wherever and whenever it could find them. At the start of 1944, Aramco had more than 2,800 employees. By 1949 employment had soared to more than 20,000 workers. In those years, 500 or more Saudi workers might be hired each month, and in one peak month, October 1947, nearly 2,000 were added to the payroll. The years of most spectacular growth were 1944, when more than 6,000 workers were hired, and 1945, when Aramco took on 10,683 new workers, an average of 890 a month.

More Foreign Workers Arrive

 raining, housing and assimilating this mass of mostly inexperienced people proved very difficult. What's more, the manpower situation was extremely unstable. The turnover rate among Saudis rose to almost 70 percent during 1947. Many Saudis, it seems, joined the company out of curiosity about the bustling new oil development. In a month or two, with their

Saudi camp with tents and barastis in Abqaiq, 1946.

curiosity satisfied, they went home. Increasingly, Aramco was forced to rely on the skills of expatriates, not so much Americans, but Indians, Pakistanis, Adenese, Sudanese, Lebanese, Egyptians and, in 1945, nearly 2,000 Italians.

The Italian employees helped fill the desperate need for skilled and semiskilled labor. Most came from Eritrea (then a northern province of Ethiopia), where they or their parents had been lured in the heyday of Mussolini's empire, then stranded when the Italian army withdrew from the country in 1941.

In approving their recruitment, the Saudi government specified that Italians were to enjoy no advantages over Saudis, either in pay or in living conditions. They lived on bachelor status in camps consisting mainly of tents, shacks and aluminum huts at a camp about two miles south of the Ras Tanura Refinery and at al-'Aziziyah on the Gulf shore just southeast of al-Khobar. Signs in the work place were in three different languages: Arabic, English and Italian.

Labor Conditions Improve

The influx of foreign labor coincided with labor unrest, culminating in the summer of 1945 with a confrontation between the company and the Saudi employees, who left their jobs. The employees returned to work after four days, on orders from the King, pending an investigation by the government of their demands.

The workers' demands included a minimum starting salary of two riyals per day and pay raises of up to 30 percent for other employees. They also demanded permanent housing to replace *barastis* and tents in the Saudi camps, improvements to the Saudi hospital at Dhahran, and an end to any discrimination in favor of foreign workers over Saudis. A government committee investigated the demands and, in a written report, supported the employees in almost every instance.

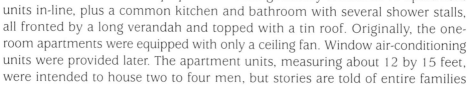

The company approved a 20 percent across-the-board pay increase, established minimum pay of two riyals per day, promised improvements to the Saudi hospital in Dhahran and announced it would try to replace all *barastis* and tents with permanent housing in masonry buildings within three years.

In a short while, the company began constructing one-story masonry apartment buildings to replace the *barastis* and tent cities at Dhahran, Ras Tanura and Abqaiq. These buildings usually contained 10 apartment units in-line, plus a common kitchen and bathroom with several shower stalls, all fronted by a long verandah and topped with a tin roof. Originally, the one-room apartments were equipped with only a ceiling fan. Window air-conditioning units were provided later. The apartment units, measuring about 12 by 15 feet, were intended to house two to four men, but stories are told of entire families

Masonry construction was used for Saudi camp housing in Dhahran, 1952.

living in a single apartment, without the knowledge or approval of the company.

As to the matter of discrimination, the committee in its written report found that "the company's treatment of foreigners is superior in all respects to that of the Saudis." It singled out the treatment of new Italian workers, "a large number of whom have been observed to handle simple and ordinary jobs, such as motor driving, pipe fitting and telephone operating."

The committee didn't accept the company's explanation that these men were doing low-skilled jobs until they could be assigned to the skilled positions for which they had been hired. "If things continue the way they are," the committee declared, "the day will come when foreigners will occupy all jobs under this pretext."

In conclusion, the committee said it trusted the company would correct policies "which have formerly taken place in some instances as a result of increased operations and that everything liable to cause friction and misunderstanding will be eliminated."

A Saudi employee entertains friends in his new Saudi camp room.

King 'Abd al-'Aziz Al Sa'ud Visits Aramco

King 'Abd al-'Aziz Al Sa'ud visited Aramco for five tumultuous days in January of 1947. He arrived in a fleet of six airplanes, accompanied by four of his brothers, eight of his sons, and most of the royal court. Students from the Jabal School sang patriotic songs during the welcoming ceremony. The King responded with gifts to the students and to the school staff.

After spending one night at Dhahran in the newly opened Hamilton House, the King announced he was moving to the temporary tent city where the rest of his entourage was staying. He toured Dhahran and the new Ras Tanura Refinery, and exchanged magnificent banquets with the ruler of Bahrain, who came to visit with a large entourage of his own. On January 25, the King personally received about 150 Americans and their children, occasionally pulling a breathless child onto his lap for private consultation.

King 'Abd al-'Aziz, right, with Floyd Ohliger, center, and P. C. Humphrey, in Dhahran, 1947.

Aramco Continues to Grow

At the time, only about 100 American children were living in Dhahran, with about 43 in Ras Tanura. The company headquarters at Dhahran was in a new building, consisting of what later became the lower floor of the Administration Building South. The first expatriate family housing (portable prefabs) was constructed at Abqaiq in 1947. The Paul Arnot family was the first family to move there. So far, Aramco had invested about $140 million in its Saudi operation, more than $100 million of it in the 1940s.

The years 1944 to 1949 saw a 25-fold increase in oil production and the completion of some enduring construction projects. Aramco celebrated its first one-million-barrels-of-oil month in December of 1944 and its first two-million-barrels-of-oil month in September 1945. Oil was discovered at Qatif in 1945, at 'Ain Dar in 1948, at Fadhili and Haradh in 1949. 'Ain Dar and Haradh later proved to be part of the huge Ghawar oil field.

In 1944 the company had just 24 producing oil wells, one at Abu Hadriya and the rest in the Dammam field. By 1949 there were 80 producing oil wells, 44 in the Abqaiq area, 30 in the Dammam Dome and the rest scattered among the areas of new discovery. By now it was clear Saudi Arabia had some of the largest, if not *the* largest, oil reserves in the world. The company calculated it had already developed more oil reserves in Saudi Arabia than the combined U.S. reserves of the Texas Company (Texaco) and Standard Oil Company of California.

Average daily production went from about 20,000 barrels a day in 1944 to more than 50,000 barrels daily in 1945, and soared to more than 164,00 barrels a day in 1946. Production passed 246,000 barrels a day in 1947, exceeded the 300,000 barrel mark in 1948, and topped 500,000 barrels a day in 1949.

King 'Abd al-'Aziz Al Sa'ud is accompanied by James MacPherson on tour of Ras Tanura, 1947.

Abqaiq, 1948: the start of what became the world's largest crude oil processing center.

Between 1944 and 1949, Aramco's profits increased from $2.8 million to $115 million a year. By 1949 Saudi Arabia had become the fifth largest oil producing nation in the world and was pushing Iran to become the top exporter of oil in the Middle East.

It was an auspicious time to be in the oil business. In 1946, for the first time, petroleum and natural gas topped coal as the main sources of energy consumption worldwide. The demand for petroleum did nothing but increase during the post-World War II boom years, while at the same time the demand for coal declined. In 1948, for the first time, the United States imported more oil than it exported. Although the U.S. retained the title of world leader in oil production, the country had begun a steadily accelerating slide into dependence on foreign oil.

Aramco's dramatic increases in oil production during the mid- to late 1940s could not have been accomplished without equally dramatic increases in the capacity to get the oil to market. The first crude still of the new Ras Tanura Refinery was fired on September 19, 1945 — just 17 days after the end of World War II. The refinery was completed too late to contribute to the war effort, but with the firing of the second crude still in December of 1945, the extraordinary project was concluded, on schedule.

The submarine pipeline to Bahrain went into service during 1945. For a brief time it was the longest underwater oil pipeline in the world — the first of many record-sized projects that Aramco would build over the years. A large-diameter pipeline was laid between Dhahran and Ras Tanura in 1946. Two other pipelines, connecting Abqaiq to Dhahran, 45 miles to the north, were completed in 1949.

By 1949 the Ras Tanura Refinery, originally designed for a daily capacity of 50,000 barrels, had been beefed up, improved and streamlined so that its capacity was increased to 127,000 barrels, two and a half times its original design.

Still another great project was nearing completion. The Trans-Arabian Pipeline, better known as Tapline, would connect Abqaiq with a port near the ancient city of Sidon, Lebanon, on the Mediterranean Sea, a distance of more than 1,068 miles. It was the longest crude oil pipeline in the world.

The Aramco-owned portion of Tapline had been completed by 1949, but political disturbances delayed completion of the western and final portion until 1950. Tapline was the first major project entered into by Aramco and what came to be known as its U.S. partner companies. To fully develop Saudi Arabia's vast oil reserves required greatly expanded market outlets and huge capital investments. In 1948 agreements were made enabling Standard Oil Company of New Jersey and Socony-Vacuum Oil Company (later renamed Exxon and Mobil, respectively) to acquire shares of Aramco. At the time, Aramco already had two shareholders. Casoc, Aramco's predecessor, a subsidiary of Standard Oil of California, and the Texas Company (later Texaco) had formed a partnership in 1937 that helped finance the construction of new facilities. The great expansion of Aramco after the World War II era was, therefore, the result of cooperation between four of the largest oil companies in the world. Their partnership lasted from 1948 until the Saudi government acquired a 100 percent interest in Aramco nearly 40 years later.

Saudi and American workers at the Ras Tanura Refinery, 1946.

A Changing Aramco Neighborhood

n 1946 the King granted permission to build a military air base at Dhahran, adjacent to Aramco. The King insisted that the Americans, who were to build the air base, should train Saudis in airport operations. The $4 million base was completed on March 15, 1946, and training began in June 1947. Up to 100 Saudis at a time took six-week courses in airport administration, aircraft maintenance, operations and weather, air installations, supply, communications, automotive maintenance, hotel and food services. They also took general education classes in English, mathematics, geography, physical fitness, safety and sanitation. In 1962 the Saudi government changed the base to a commercial airport with a Saudi staff.

Malaria control team sprays for mosquitoes in 1948.

The company's antimalaria campaign, begun in the early '40s, became a notable success, thanks largely to spraying with a potent new insecticide known as DDT. Mosquitoes were almost eliminated from the Aramco compounds by 1947, and the number of malaria cases reported in the Eastern Province plunged by nearly 60 percent.

The first locomotive made a run over the new railroad tracks from the port of Dammam to Dhahran on April 25, 1947. A rail connection between Dhahran and Abqaiq was opened in 1949, completing an estimated $10.5 million project linking Aramco's inland districts to the Dammam seaport. The following year, the railroad was extended to Riyadh. Until this time, supplies were still being carried between Hofuf and Riyadh by a weekly scheduled camel caravan.

15 Years

 milestone was reached in December 1948, when the company presented its first 15-year service awards to Saudi employees. The awards went to four Saudis who joined the company in 1933. They were 'Ajab Khan and Ahmad Rashid Al-Muhtasib, translation specialists; Nasir ibn Ibrahim, a driver and mechanic; and Muhammad Amin 'Abd Allah Khaja, a clerk.

Aramco's industrial injury rate, which had climbed from a wartime low to almost 60 disabling injuries per million man-hours worked in 1947, was reduced to an average of about 11 injuries per million man-hours by 1949. The dramatic decline was attributed to the training of all employees in safe, efficient work methods.

In the 1940s, under the direction of Aramco engineers, an 11-mile irrigation canal was dug and large pumps installed for the al-Kharj agricultural project. The project covered nearly 3,000 acres on which vegetables and fruits were grown. An agricultural training school was added later. Aramco's chief engineer, Les Snyder, one of those in charge of surveying the irrigation project, calls it the most memorable accomplishment of his Aramco career.

Saudi flag flies over Dhahran Airport terminal building, 1948.

The American oil men knew the complexities of the oil business could be mastered only by well-trained men. They also knew that because of Article 23 in the Concession Agreement, the labor force must be largely Saudi. Yet everyone had been too busy with the number one priority — getting the oil out of the ground and to market — to develop a strategy for imparting the necessary skills to masses of new recruits. Most training was still done

Saudi inspector walks along the Dhahran-Abqaiq pipeline, 1946. The thermometer reads 140° F.

on the job in a haphazard way — much the same as it had been when the company began — but change was in the air.

Voluntary evening schools for employees and their sons opened in early 1944 at Abqaiq, Ras Tanura and al-'Aziziyah, the new Italian community south of al-Khobar. In addition to classes in English, arithmetic and Arabic, for the first time, they offered instruction in shorthand and typing as well as houseboy training. The company's first true industrial training program was launched during the rush to complete the new refinery at Ras Tanura. An American, Herman Mattson, came out of defense-plant training work in the States to implement Job Instructor Training (JIT) — a train-the-trainer type procedure — for refinery personnel. Mattson left the company after a short time, apparently frustrated by the low priority assigned to training programs. Before he left, he drafted a two-page outline of Aramco's postwar training needs. This paper, although long since lost, is remembered as the first attempt to chart a future course for training in Aramco.

Warren Hodges

The company hired an energetic young industrial-arts trainer named Warren Hodges in March 1944. He was a graduate of San Jose State in California and was fresh from a job as a machinist in the defense industry. Hodges was destined to have a major role in Aramco training for the next 24 years, but first he was assigned (under the "where needed" system of job placement imposed by the times) as supervisor of Senior Staff Bachelor Housing in Dhahran. There he taught some 120 houseboys about changing linen, doing laundry and sweeping floors, all the while fending off complaints from resident bachelors about towel shortages, laundry returned without buttons, etc.

In December of 1945, Hodges finally moved from bachelor housing to training. Later in the same month, he and Frank Tallman of Ras Tanura spent a week studying training at the Anglo-Iranian Oil Company facilities in Abadan, where they didn't see much they liked. Hodges remembers apprentices listlessly filing on iron cubes. Such idle tasks seemed to Hodges "like treading water."

Young Saudi practices carpentry skills, 1950.

Earlier that year, the company had appropriated $100,000 to construct the Trade Training Centers for Saudis at Dhahran and Ras Tanura, a first attempt to move job skills training from the job site into the classroom. The trade schools were to be equipped to train 50 Saudi employees each in basic craft skills. Hodges designed the building at Ras Tanura. It was to be a one-story masonry building about 100 feet long and 40 feet wide, and would cost some $50,000 to build, but it would be nearly 1½ years before ground breaking. Construction of the trade schools had to compete for resources with another historic project, the completion of the Ras Tanura Refinery.

In the meantime, Hodges had his first industrial teaching experience in Saudi Arabia, leading classes for Saudis in a room at the refinery on basics such as how to read gauges, how to use a tape measure and how to measure the amount of liquid in a tank. It was, Hodges said, "the first bona fide off-the-job classroom training in Ras Tanura."

Nassir M. Al-Ajmi

"My father used to ask, 'When is this training going to finish?' He thought you just went for training for maybe one or two months and that was it. I explained to him later that it was a lifetime."

assir M. Al-Ajmi joined Aramco as a teenager in 1950, not looking for a career, but because he "felt the economic need." His father and uncle were already working for the company, and Ajmi had reached an age that local schooling was unavailable.

He traveled by camel caravan from his home near Nariya to join his uncle, who was working at Abqaiq. With his uncle's written recommendation in hand, he was hired by the company recruiting office in Dammam. Al-Ajmi began his training with the Transportation Department in Dhahran, because at the time there was no training program in Abqaiq.

"They took me to the Engine Rebuild Shop," he recalled. "I found blocks and pieces of equipment that were very dirty with mud and oil, and they said, 'This is your job!' I said, 'This is my job?' 'Yes, you hold this steam hose and you clean them up.' And that's how I got started."

Most of Al-Ajmi's initial training was by observation. "You took apart the engines — and got to know the names of the tools, the sizes and which way the thread went." There were no training materials available, and no instructors.

"What could they do with these kids who were not physically capable of doing heavy work or operating equipment? They developed a form of training where they put you with a mechanic, a welder, etc., where you carried tools and learned the names of the equipment." Al-Ajmi found the method "very effective."

After being transferred to the training center in Abqaiq, Al-Ajmi completed within two years all the requirements needed to work as a mechanic. He, along with many other employees, continued going to school on their own time. The most successful of these "volunteer" students were sent overseas for formal completion of high school. In 1959, Al-Ajmi and about a dozen other young Saudis went to the International College in Beirut, where they finished high school.

Al-Ajmi went to the United States in 1963 to study automotive technology and followed that up with two specialized automotive courses in the United Kingdom. He returned to the U.S. and, in 1967, secured a bachelor of science degree in business management at Milton College in Milton, Wisconsin. He began graduate school courses at Marquette University in Milwaukee, but he was soon recalled to Abqaiq as a garage foreman in the Transportation Department.

Over the next two decades, Ajmi's titles included superintendent of Transportation Operations, manager of Land Transportation and general manager of Industrial Services. In 1979 he was made vice president of Community Services and was elected senior vice president of Operations Services later that year. He was named senior vice president of Industrial Relations in 1986 and promoted to executive vice president, the company's second highest position, in May 1988.

His education continued during this period. He completed both the Executive Program in Business Administration at Columbia University and the Advanced Management Program at Harvard University.

Looking back over his career, Al-Ajmi is struck by the differences in training techniques. "Today, we train for different types of skills: we train for high-tech, we train for controls, for computer applications. The [industrial] training we had then was more hands-on experience and by thinking and doing. The training programs today are more effective, more focused and structured. Our training was right outside, not inside. I remember those cold or hot and windy days when we had to check the instruments and valves by hand. That was the type of training we had," he said.

First American Arabists

t the end of World War II, the company hired two Americans who really knew the Arabic language, George Rentz and B.H. "Barney" Smeaton. Rentz had studied classical Arabic at the University of California and had spent the war years in the U.S. Office of War Information in Cairo. He set up the research and translation organization in the company's Government Relations Department.

George Rentz in 1958.

Smeaton was a linguist hired to teach Arabic to Americans. Starting in 1945, he produced a series of booklets for the Education Division called *Arabic Work Vocabulary For Americans in Saudi Arabia.* Using these booklets, Americans learned how to say in Arabic such phrases as "Where is the hammer? Perhaps it is under the automobile."

The company offered a fifty-dollar bonus to American employees who could pass a course based on lessons in Smeaton's booklet. However, few Americans stayed with the program long enough to collect. Two who did were Vince James of the Jabal School and a newcomer, Powell Ownby.

Ownby had been brought over from California to direct the newly consolidated Dhahran Opportunity School for adult Saudis. The original evening voluntary schools at al-Khobar, al-'Aziziyah and Saudi camp were closed during 1946 and combined into the Dhahran Opportunity School, with classes held at night in the Jabal School building.

Harry Ashford and the "Young Lions"

he Education Division became part of the Personnel Department in early 1946. Late that year a veteran trainer from California, Harry T. Ashford, became the first superintendent of education. All company training activities were consolidated under his office. "He has been assigned the task of organizing an extensive education program with special emphasis on the teaching of trades and crafts to Saudis," the company's 1946 annual report said.

In effect, Ashford, a vocational-training expert, replaced Mac Nearpass, the former school district supervisor, as head of training for Aramco. Until this time, training had been caught in a sort of tug-of-war between the San Francisco office, which pushed for general literacy training, and the administration in Dhahran, which wanted to emphasize job training. Now, the pendulum swung to job training.

Ashford was near 60 years of age when he arrived in Saudi Arabia with years of experience in trade schools. The trainers who worked under him were much younger and had different viewpoints. Many of them were industrial trainers straight out of American war plants, men like Don Richards, Francis "Frank" Jarvis and Marcy "Lucky" Luckenbaugh.

Jarvis, who had directed training at General Motors defense plants in

Education Division staff, including Harry Ashford, third from left, and Mac Nearpass, fifth from left.

Tarrytown, New York, reinstated the Job Instructor Training (JIT) program begun by Mattson and developed the first Management Training course in Aramco. He lived in Dhahran and conducted classes in Abqaiq and Ras Tanura, driving to and from these communities over sand roads in rough-riding, dusty vehicles with huge balloon tires. "Air conditioning" in these vehicles consisted of a bottle of water and a damp rag that the driver used to mop his face.

Saudi drillers near Abqaiq, 1949.

In the JIT program, Jarvis explained to selected trainers, who had probably never had any formal instruction in training procedures, how to teach various tasks to workers using four steps — preparation, presentation, try-out and follow-up. Safety factors were covered with each step as well. Once every two weeks he led an open, conference-type Management Training program in each district for the district manager, the head of field operations, the department heads and others in supervisory jobs. For most of them, it was their first management training experience. "All I hoped to do," Jarvis said, "was give them a broad perspective of the organization functions such as planning, directing and organizing, then a review of the basic responsibilities to carry out management functions."

The Senior-Staff School

In 1945 American wives and children began returning to Saudi Arabia. The first group of seven returning wives reached Dhahran aboard a military transport plane in February of that year. They were Kathleen Barger, Dorothy Ohliger, Esta Eltiste, May Beckley, Roberta Scribner, Marie Ross and Gertrude McConnell.

The returning children required schooling. Before the war, schooling had been provided for the sons and daughters of American employees at the home of Mrs. Max Steineke, wife of the famed geologist. Mrs. Edith Chamberlain was the teacher, with up to 10 students. That school had been in operation for only a few months when, in October 1940, Italian planes bombed Dhahran, triggering the evacuation of American women and children from Aramco.

In 1945 the company hired a recently discharged serviceman named Sam Whipple to open a new school, known as the senior-staff school. Whipple had just spent two years with a U.S. Army antiaircraft division, based first in San Diego, California, then at Fort Bliss, near El Paso, Texas. Prior to that he had taught at rural and small-town schools in Washington state.

Whipple, like many men returning to civilian life, was at loose ends. "I didn't know what I wanted to do," he recalled during an interview some 50 years later. "I knew I didn't want to go home. I wanted to do something different."

At age 29, he was living in a cheap hotel in Los Angeles and working at a dead-end job in a paper factory. Whipple later recalled the day his life changed, and the career that followed.

Children arrive for class at the "little red school house" in Dhahran, 1947.

"One day I picked up a copy of the *Los Angeles Times* and saw an ad, 'Teachers Wanted, Foreign Service.' So I pursued it, being an adventurous individual at heart. The guy who interviewed me said, 'Saudi Arabia.' I said, 'Saudi Arabia! Where the hell is that?'" The location of the Kingdom was explained, and in April 1945, Whipple was offered a job with Aramco. He was to establish and be the only teacher at a one-room, kindergarten-to-eighth-grade school in Dhahran.

The recruiter "presented a very, very, very bleak picture to me of Saudi Arabia," Whipple said. "It was considered a hardship post in those days. My salary was to be $250 a month, plus a $50 hardship bonus. That was a pretty good salary. Before I signed the contract, the recruiter asked me, 'Are you sure you want to go?' I said, 'yes.'"

Whipple arrived in Saudi Arabia in June 1945, after traveling for more than a month in the company of a raucous gang of construction workers. He was billeted in a bunkhouse just off what became Gazelle Circle in Dhahran.

"I was very thankful to the recruiter who had painted such a dark picture of conditions at Aramco. When I got there I didn't expect much, and I didn't find much either," he said. "There was no air conditioning. A lot of fellows would take their mattresses and put them up on the roof and sleep there. These guys would lie up there on the roof looking at the stars, thinking of home and family, and probably saying to themselves, 'What the hell am I doing here?' I did that myself many times."

Whipple faced a crisis in September 1945. The new school was supposed to open in a few days, and "I didn't have any textbooks. I had ordered some books, but the ships were late in coming." An appeal by the company newspaper for the return of any schoolbooks that might have been stashed on bookshelves in private homes was not successful.

"I went to Bapco (Bahrain Petroleum Company) in Bahrain just two or three days before the school opened to beg some books. They were very generous. They loaned me some books.

"While I was in Bahrain, I learned that their schools were on the trimester system. So I brought that back and suggested that to my supervisor (Nearpass). It has been in existence ever since. I guess the teachers were somewhat divided on it. We felt it would be good because the kids would be in an air-conditioned room during most of the summer and on break during the cooler spring and autumn months."

Whipple opened the first senior-staff school October 1, 1945. Classes were held in the living room of a duplex house, located where the east wing of the main Dhahran dining hall was later built. Community Services had offices in the same duplex. The classroom was equipped with a blackboard and chalk, some maps supplied by the company, a small desk and chair for the teacher, and two tables and some chairs for the students.

The only student to show up on opening day was Steve Furman, son of the head of Aramco's commissary. Two days later Robin "Pinkie" Alexander, the red-haired, freckled-faced daughter of the company doctor, T.C. Alexander, registered as a fifth grade student. A few days after that, Carol Dunten became the first kindergartner at the senior-staff school.

"By year's end I think I had 13 students," Whipple said. "There were children in all eight grades. I did everything. I taught all subjects in all grades. But I was familiar with the situation. My first school was a one-room school with three students in the Grande Ronde River country of Washington state."

In the spring of 1946, the senior-staff school moved into House 1621, next door to the bunkhouse where Whipple lived. House 1621 became known as "the little red schoolhouse," because it was built of locally produced reddish-yellow bricks. About this time, Whipple's students presented the first school play, a fairy tale titled "The Proud Princess," inviting the entire community to attend a single night's performance in the recreation hall. Steve Furman played a king, and Miles Snyder portrayed an old man. A year later, Miles, son of Les Snyder, the company's chief field engineer, became the first graduate of senior-staff schools.

First senior-staff classroom, Dhahran, 1945. Clockwise from left: Carol Dunten, Miles Snyder, Jimmy MacPherson, Robin Alexander, Stevie Furman and Grace MacPherson.

Whipple stayed in Dhahran for only one year before moving to Ras Tanura, where he opened a second senior-staff school in the fall of 1946. There he presided over 14 students in a makeshift classroom behind a partition in the family section of the Ras Tanura dining hall. Mary Leonardini, described by Whipple as "a very pretty Italian lady," took charge of the Dhahran school. She was joined a short time later by a second teacher, Jane Seeley.

Sam Whipple, left, with students in classroom inside Ras Tanura dining hall, 1947.

In September 1946, workmen in Dhahran completed a new two-room brick school building, the first building designed as a senior-staff school. The new school was to open in mid-September. Construction was completed a few days ahead of time, but the building caught fire the next night. Firemen saved the brick exterior shell, and the interior was hurriedly refurbished so that the opening was delayed only a month. This building also became known as "the little red schoolhouse." It later became the single men's mess hall, the Dhahran Boy Scouts' facility and, eventually, the front section of the Dhahran Recreation Library. Along the way it acquired a cement stucco facade.

In December 1946, a second teacher, Jane Dean, arrived in Ras Tanura to assist Whipple. In June of the following year, Jane Dean, her new fiancé and three other people were killed when the car in which they were riding was crushed by a truck.

Whipple later returned some of Jane Dean's personal effects to her family in the United States. He asked the family whether they would have their daughter's body brought back to the States for burial. "The mother told me that if Jane had to die, it was all right for her to be buried in Saudi Arabia, because she was happier there than she had ever been before." She remains buried in the Dhahran cemetery.

The Ras Tanura school moved into a portable building across from the dining hall in 1947, and "casual teachers" — mothers and wives — assisted Whipple until another certified teacher, Helen Jones, arrived.

Whipple spent 10 years as a teacher and administrator in Aramco schools, eight of those years in Ras Tanura, and has fond memories of those years. "The school was the hub of the community in those days. I was a friend, a teacher and father confessor to my students, and to some of their parents," he said. When Whipple left Aramco in 1954, the enrollment at the Ras Tanura school was about 250.

A third senior-staff school opened in a private home at Abqaiq in October of 1947. It began with three students, Valerie Ridgeway, Charlotte Hubner and Norman Gray. The teacher was Josephine Rose, whose sister, Constance, was married to Dr. Robert C. Page, the medical director of the Texas Company (later Texaco), and later medical director of Aramco. The Abqaiq school moved in 1948 to a portable building near the community swimming pool.

By 1948 the three senior-staff schools had enrolled a total of 94 students and employed seven teachers. Five years later, student enrollment at the three schools topped 800. A high point was reached in September 1983, with 3,677 students enrolled in nine schools employing about 600 teachers and staff.

Another year of instruction was added to the senior-staff school's curriculum in 1949. There were ninth graders for the first time, and several youngsters who graduated from the eighth grade in 1949 stayed another year and graduated again in 1950.

First Saudi Overseas Scholarship Students

n the late 1940s, the Saudi Arab government began sending Saudi scholarship students to colleges and universities in the United States. The first bursary students arrived in San Francisco, then the headquarters city of Aramco, in late 1947. At government request, Aramco employees met the seven young Saudis at the airport and made the necessary arrangements for their time in the States. The first thing the Aramco delegation did for the students was to purchase warm clothing to protect them against the damp and chilly rigors of a northern California winter. These bursary students were the predecessors of thousands of Saudis who would be sent on scholarships to U.S. colleges and universities in years to come. One member of the first group, Salih Al-Fadl, earned a master's degree in economics from the University of California at Berkeley and worked at Aramco for five years before moving on to become chairman of the board of Arabian Drilling Co., and a member of the board of the Saudi Arabian Monetary Agency.

Arab Trade School

n 1947 the company finally approved construction of the Arab Trade School building designed 18 months earlier by Hodges. The building was opposite the government office in Rahimah, the Arab residential area just outside Aramco's gates at Ras Tanura. The concrete-block structure had classrooms on either end, a workshop in the middle, an office and a tool room. It was similar to vocational schools found in nearly every school district in the United States, but it represented a milestone in the history of training at Aramco and in Saudi Arabia. It was the first school dedicated to job skills training, a forerunner of the Industrial Training Shops and the first vocational school in the Kingdom.

The Trade School hadn't been open long when a building inspector noticed the roof was beginning to sag and shift. An examination revealed that the 40-foot trusses supporting the roof were starting to fail because they were put together with nails instead of bolts. The roof had to be removed and the trusses rebuilt. Meanwhile, classes were held in rooms without a roof.

Saudi "helper" and "apprentice" employees learned carpentry, welding, sheet metal work and machine operations at the Trade School. Hodges and his second in command, Elgin Mason, both industrial-arts trainers, trained a staff of Saudi instructors, each of them a specialist in one of the trades. The first half day of training was devoted to what came to be called CORE classes, subjects such as basic measuring that any craftsman would need to know. The second half of the day was occupied in working on actual projects, most of which were items required by the company — kitchen cabinets, for example, to be installed in local housing by Community Services. In one month of 1950, Trade School students turned out 273 articles for company operations.

A second training shop, the Advanced Trade School, opened in Dhahran in 1949. Italian workmen put a fancy stone front on the building. Otherwise, its design was identical to the Ras Tanura school (except the trusses were bolted together). The building later became the studio for Aramco television.

The number of Saudi men enrolled in company training programs went over the 1,000

Vince James found Saudis to be good students.

mark for the first time during 1947. There were about 650 Saudi employees in training programs at Ras Tanura, more than 300 in Dhahran, and about 100 in the new Abqaiq community. The great majority enrolled for evening classes at the volunteer schools. Training was decentralized in late 1948, broken down into three districts — Abqaiq, Dhahran and Ras Tanura — with an education supervisor in each district operating under the Personnel Section of the districts. In effect, the company had abandoned the idea of a training organization under a strong centralized authority, as envisioned when Ashford was named superintendent of education in 1946. "Lucky" Luckenbaugh became education supervisor in Ras Tanura, and Don Richards had the same title in Dhahran. Mac Nearpass was named supervisor in Abqaiq, but he left the company a short time later. He was succeeded by Frank Jarvis.

Instruction in equipment nomenclature in 1951.

The major development of 1949 was a reversal in the upward trend of crude oil production. Production reached a peak of 532,000 barrels a day in February, then tapered off slowly. The trend reversal was attributed to surplus production following new discoveries in Kuwait, Iraq, Canada and South America, as well as to dollar shortages requiring countries to tighten their import and exchange controls.

Aramco Manpower Reduced

Total Aramco manpower was reduced for the first time in the company's history, aside from the war years. It dropped from a peak of 20,254 men in 1948 to 16,084 in 1949. All the cuts came from the expatriate payroll. About one in every five American workers was let go. The oil trade press speculated that Aramco had grown about as large as it ever would.

Two Saudi employees work on lathe in Dhahran, 1950s.

Once again, as it had at the outbreak of World War II, a slowdown had its beneficial side. More time was available to devote to training and to plan for its future. The company hired an educational mission, headed by a former U.S. Air Force officer, Harry R. Snyder, to survey its training and educational procedures and requirements. Snyder later described the challenge he saw facing Aramco: "The problem," he wrote, "was to convert a labor force, virtually devoid of all industrial or technical skills and with practically no formal education, into a team of employees capable of doing large blocks of Aramco's work when given adequate supervision. ..." The conversion had to be made by men who did not share the same culture or speak the same native language. The challenge and scope of the task remains unmatched in the history of human resource development in an industrial setting.

Growing Up Fast
1949-1955

"The avowed purpose of Aramco is to give number one priority to the training of Saudis."
– Harry R. Snyder

he years 1949 to 1954 were remarkable for the speed in which the economic and social character of Aramco continued to change. Rising expectations in the Saudi Arab work force were helping to drive the company into a new age of training. The transition to that new age probably began on February 12, 1949, when Harry R. Snyder arrived in Dhahran. He came with a new corporate title, director of education and Arab training, and an ambitious new agenda for training Saudi workers.

Snyder was the nominal replacement for Harry T. Ashford, who was soon to retire after heading the Education Division for nearly three years. Aside from having the same first name, they were very different men. Ashford had been a teacher and administrator in vocational schools for nearly 40 years, almost as long as Snyder had been alive. Snyder's background was in colleges, publishing and the military. Ashford had no knowledge of the Arab world before he arrived in Dhahran in 1946. Snyder was considered an expert on Arab affairs when he joined Aramco.

They differed in personality as well as experience. Ashford was thought of as solid, conservative and set in his ways. Snyder was characterized as outgoing, enthusiastic and persuasive.

Nearly two years earlier, Snyder and a group of consultants from the Near East College Association of Beirut surveyed Aramco's training programs. The consultants recommended creation of a technical and professional institute in Dhahran where Saudis would learn to handle "large blocks" of Aramco's work under the supervision of American employees. Now, instead of talking about Saudis working under American supervision, Snyder was saying: "Someday this company is going to have a top management of Saudis." The reaction of many men at that time was to ask in disbelief: "How can that ever happen?" They didn't often wait for an answer. When they did, the reply was "through education and training. How else?"

Reasons for skepticism were not hard to find. Despite nearly a

41

decade of formal training programs, about 85 percent of the company's 10,000 Saudi employees in 1949 were unskilled, illiterate laborers in Aramco grade codes one to three. Almost all the others were semiskilled workers in grade codes four and five. Only 80 Saudis had reached the journeyman or skilled craftsman level at grade code six or above. A few Saudis were designated as supervisors in charge of other Arab workers. No Saudis supervised American employees. No Saudis had senior-staff status yet. Every American had senior-staff status, and except for teachers, security guards and a few other positions, most Americans were at grade code 10 or above on the pay scale. Teachers started at grade code eight or nine, at the low end of the dollar payroll compared to other professionals.

The notion that Saudis would one day take charge of Aramco was not something Snyder decided on his own. It was told to him by James Terry Duce, a veteran geologist and company executive, during a job interview at Aramco's New York headquarters. Duce said it would be Snyder's job to prepare Saudis for the day when they would run the company.

Duce spoke from experience. He had been president of the Colombian Petroleum Company, a subsidiary of Texaco, when that company was nationalized in 1939. Between 1941 and 1943 he was director of the Foreign Division of the U.S. Petroleum Administration for War. At the time of his interview with Snyder, Duce was responsible for Aramco's government relations in the U.S., and was on a first-name basis with most of the high and mighty in international oil circles.

James Terry Duce, 1959.

Snyder later wrote of that interview: "... I was told in substantially these words: 'Your task at Dhahran is to train the Saudis as quickly and as soundly as possible to operate the Saudi oil industry. Inevitably, the Saudi government will eventually nationalize the industry. When that occurs, we want young Saudis to have attained the proficiency that will enable them to operate the oil industry efficiently and with goodwill toward Aramco. Thus, they will be serving their country's best interests and will be protecting the interests of our parent companies.'

HRH Crown Prince Sa'ud and other dignitaries visit a Dhahran trade school with Aramco trainers Fahmi Basrawi, Harry Snyder and Don Richards in 1950.

"On my arrival in Dhahran," Snyder continued, "and in all the years that followed, I found that this enlightened business philosophy was the cornerstone of Aramco's relationship with the Saudi government. While our educational policy and procedures might be faulted for not making progress as dramatically as Saudi officials may have wished, my mandate from top management was to prepare Saudis as rapidly and as efficiently as possible to be able to eventually operate the Saudi oil industry in its entirety."

Snyder's Middle East experience dated back to the late 1920s. After graduating from Lawrence College in Appleton, Wisconsin, he joined the American University in Beirut as an instructor of commerce. On vacation in Cairo, he met and married Olive Winifred Somerville, who had been born and raised in Lebanon by Scottish parents. The Snyders lived for 12 years prior to World War II in New York City, where Snyder earned a master's degree and edited college textbooks for McGraw-Hill.

He was deeply immersed in Middle East affairs during World War II as a Middle East intelligence specialist for the United States military and the State Department. After the war, he became assistant director of the Near East College Association, a consortium of eight colleges and universities, including the American University of Beirut. He was called back into the military as an Air Force colonel in 1947 and served in the Middle East. When the King issued permission for construction of a military air base near Dhahran, Snyder established a training mission in airport operations for Saudis at the Dhahran air base.

"He had great enthusiasm, was always optimistic and was a great believer in Saudi culture," a former aide said.

In 1949, as a member of Aramco's Personnel Planning Committee, Snyder helped draft the company's first five-year training plan. The training plan aimed to lift a large portion of the Saudi work force from laborer and low-skill classifications to semiskilled and journeyman classifications by 1954.

Automotive shop class, Dhahran, 1949.

The committee, Snyder later wrote, concluded "It was necessary to adopt, on a temporary basis, a system of training and of work organization that was largely copied from America's war industry. The committee was confident that mass, intensive training in simple, routine, manipulative skills could be as effective in Saudi Arabia as in the United States."

Aramco Production Training Program

The new initiative, announced with great fanfare in July 1949, was named the Aramco Production Training Program. The program called for an immediate "intensification" of on-the-job training for some 8,000 unskilled Saudi workers in grade codes one to three. In addition, the "best qualified" among 1,800 semiskilled Saudi workers in grade codes four and five would receive advanced trade training to perfect their technical skills and develop administrative and supervisory skills. The goal was to raise one-half of the Saudi work force to skilled and semiskilled levels by 1954.

The Aramco Production Training Program was the first step on the long process that later came to be called *Saudization*. The program was the first organized, company-wide effort to qualify large numbers of Saudis for jobs then being performed by expatriates.

That five-year plan was also the first Aramco training program to integrate on-the-job training with formal classroom instruction. The plan called for a balance between work experience, on-the-job training and job-related classroom instruction, an approach that was still in use decades later.

Under the five-year plan, training was organized into separate academic and job-skill activities for the first time. The plan established four principal types of training activities: Pre-Job Training, General Industrial Training, Job Skills Training and Advanced Trade Training.

The heart of the new initiative was Job Skills Training, also known as "one-eighth-time training" because it required one hour out of each eight hours' production time be devoted to training Saudi workers. It placed responsibility for deciding who should be trained, in what job and at what rate, squarely on the line organizations, specifically on the employee's own supervisor, supplemented by the know-how and methods prescribed by the company's Training and Education Division.

The traditional trade school approach to vocational training was abandoned in favor of "one-eighth-time training." The Arab Trade School, opened at Ras Tanura in 1947 under Warren Hodges, abruptly shut down in late 1950.

Working with a scale model helps trainees learn construction techniques, 1950.

The school, with a maximum enrollment of only about 50 job skill trainees at a time, seemed totally inadequate to cope with the numbers of workers who needed training. Job Skills Training was to be done in the shop bays, where trainees could handle the tools of their trade. Plant operation trainees would learn while they worked as helpers in the plants; drilling workers would train at the drilling rigs.

The line organization supervisors for whom these Saudi trainees worked were busy men, often operating under pressure to meet escalating production quotas, and usually completely inexperienced in the training field. Many of them did not appreciate being asked to accept the added responsibility for a major part of the company's training program.

"The line guys weren't happy with training," a Training Division employee remembers. "We told them, 'You're going to have everyone give one-eighth-time training, period.' Top management gave their full support, but there was an intermediate group whose resistance you had to overcome. They were the ones who had to work out the mechanics of getting their jobs done, plus getting the training done without any extra manpower."

Snyder went on the road in an effort to sell the program to the expatriate work force, and he was a masterful salesman. As an associate put it, "On the darkest days he could make you think this was the greatest sunshiny day that ever came." He touted the new training agenda in meetings in every district. He brought division heads and division training coordinators together with the foremen, supervisors and unit trainers who supervised Saudis on the job. In what listeners described as a "pep rally" atmosphere, he exhorted them to concentrate on upgrading the Saudi work force. "The avowed purpose of Aramco," he declared, "is to give number one priority to the training of Saudis."

On June 1, 1950, Snyder was named director of Training and given responsibility for the company's Training and Education Division. He had been the head training man in the field prior to that, but the final responsibility for training had rested with Roy Lebkicher, a geologist and Aramco executive, stationed in New York.

New Training Programs Introduced

he five-year plan was the most extensive of several new training programs launched between 1949 and 1954. Not all of the new programs were innovations of the Training and Education Division, and not all of them were successful. A summary of these new programs follows, beginning with the four training activities encompassed by the five-year plan.

Job Skills Training (also called One-Eighth-Time Training) • This program formed the backbone of the five-year program. Between 1949 and 1954, enrollment in Job Skills Training averaged more than 3,000 Saudis a month. The key man in this program was the foreman or experienced craftsman who did the actual training, or who selected people to do the training and wrote a Unit Training Program. He was assisted by a division training coordinator and, in larger units, by a unit trainer, backed by professional help from the District Training Division.

The foreman was allowed great flexibility in the use of training time. He could schedule training at whatever hours he liked as long as the equivalent of one hour out of each eight was given over to training. He could schedule training at the work site or in one of the district training centers.

Each unit employing Saudis was required to submit a written Unit Training Program to the Training Division for approval. The program was to include: 1) a specific plan and schedule under which Saudis would advance to higher grade codes; 2) a plan for training new Saudi recruits; 3) training targets for each six-month interval; 4) estimated cost of the program. But, by the end of the five-year plan, only 38 out of 90 units had an approved Job Skills Training Program. None of the approved programs was in full operation.

Pre-Job Training • This program was basically a 90-day orientation program for new recruits, launched in 1950 under the name Vestibule Training, but soon renamed Pre-Job Training. The first two weeks were devoted to full-time instruction in such things as hygiene, safety, employee benefits, the nature of Aramco's work, and company rules and procedures. A period of gradual introduction to a job followed. Classroom training might be prescribed. Saudi counselors were available to advise the new-hire. Almost 3,000 Saudis completed this program in 1952.

Advanced Industrial Training • This program opened in 1950 at Dhahran as an experimental program with 31 Saudi employees. It was a two-year, full-time, full-pay technical training center of a secondary-school level for selected Saudi employees from all three districts. The center offered training in crafts, clerical skills and cookery. The training of nurses at the Dhahran Medical Center was an adjunct of Advanced Industrial Training. The training center was located in the old office area, in a building that doubled as the cashiers' building on paydays. Enrollment, exclusive of nurses' training, peaked at 162 in October 1952.

Helen Stanwood, the first woman teacher hired by Aramco, headed Clerical Training, the unit with the largest enrollment. A former teacher in Redding, Massachusetts, she first came to Saudi Arabia as a member of a New Hampshire congressman's staff. Soon afterward, she applied for a job at Aramco and started to work at Dhahran in 1950.

In the spring of 1953, management decided advanced training duplicated much of what district craft training centers were doing, or should be doing. New rules required students to complete all district-level courses in order to qualify for advanced training. Advanced training was no longer full time, but would alternate between periods of full-time training and full-time work in the districts. Using the new criteria, only 23 students qualified for classes in 1954.

Teacher Helen Stanwood in a clerical training class, Dhahran, 1952.

General Industrial Training (GIT) • This forerunner of Industrial Training Centers began operations in 1950. Employees were assigned to GIT centers part time during work hours for training in English, arithmetic, science, typing, drafting, office practices or hygiene. The majority of assignments were two hours per day, usually for one hour of instruction in English and one hour of arithmetic lessons.

In the evenings, employees could attend GIT centers on their own time. These so-called Voluntary Schools provided instruction in English, Arabic and typing, with English being by far the most popular course. The courses were divided into six levels corresponding approximately to the levels in Middle East elementary schools. Classes met between 5:15 p.m. and 10:00 p.m.

GIT enrollment varied widely according to season and workload. In Dhahran, as many as 2,101 employees, or 43 percent of the Saudi work force, were assigned to GITs during a peak enrollment period in 1951. Ras Tanura, with better and more conveniently located facilities, recorded the highest attendance of any district for the evening voluntary classes. About 880 students, or 24 percent of the Saudis employed at Ras Tanura, signed up for Voluntary School in 1952. This compared to Voluntary School enrollments of 14 percent at Abqaiq and 13 percent at Dhahran.

A number of ambitious employees attended both assigned and voluntary classes, often staying up until midnight to complete their homework, and coming to work at 7:00 the next morning. This group included people like Ali Al-Naimi and Nassir Al-Ajmi.

The former trade school building at Ras Tanura was subdivided into academic classrooms and enlarged until it served as a GIT center for several hundred Saudis. Larry Emigh was the first principal. The assistant principal was Bill O'Grady, who later became head of Aramco training. For several years Ras Tanura was the only district with a building actually designed as a training center. The GIT center at Abqaiq was in a portable building next to the Transportation Department. Jerry Ripberger was the first full-time principal. The Dhahran center was in a one-story building in the old Jabal School area. Former Jabal School students were more or less absorbed into this facility.

Aramco scholarship recipient 'Abd Allah Ahmed Sa'id studies at Aleppo College library in Syria, 1951.

Saudi Scholarship Program
• Aramco inaugurated this program in October 1951, in cooperation with the Saudi government. Ten full scholarships were awarded each year for the next five years to Aramco employees for study at the American University of Beirut, Aleppo College in Syria, or other accredited Middle East institutions.

Scholarships were awarded on the basis of past work performance, intelligence, adaptability, educational background, and ability in company training programs. Scholarship students worked for Aramco in the summer, but were under no obligation to remain with the company once they completed their studies. The program cost Aramco about $1,800 per student a year, including tuition, room and board, and pocket money. Married men on scholarship were accompanied by their families. The first 10 scholarships were awarded in 1951. (The 1951 scholarship winners were 'Abd Allah Ahmad Sa'id, Hasan 'Ali Ghanim, Mustafa Husam Al-Din, Rashid ibn Ibrahim Al-Rashid, Salih 'Abd Allah Al-Rubayyi', 'Abd al-Qadir Bubshate, Salih Sa'd Al-Zaid, Muhammad 'Abd al-Wahhab Salamah, Sulayman ibn 'Abd Allah Al-Rubayyi' and Ihsan Tawfiq.) By 1954, 30 scholarships had been awarded but only 13 of the recipients were in college. The others had not yet passed the high-school equivalency test required for admission to college.

Aramco employees on the campus of the American University of Beirut, 1953.

Saudi Supervisory Training
• In early 1949, the company recognized the first Saudi supervisors to qualify as trainers of other Arab supervisors. Fourteen Saudis received certificates of graduation from a 36-hour Supervisory Training Conference leadership course. (The 14 were 'Abd Allah Uthman, Khalifah Mijdal, 'Ubayd Allah Nafail, Muhammad ibn Ali, Muhammad Salih, 'Abd al-Jalil ibn Ishag, 'Abd al-Rahman Muabid, Ahmad Muhammad Jum'ah, 'Abd al-'Aziz ibn Salih, Moaula Owaidh, Said Al-Radhi, Muhammad ibn Hasan, 'Abd Allah ibn Ali and Ahmed Rahman.) They, in turn, conducted a series of 15 conferences on supervisory techniques for other Arab supervisors. More than 100 Arab supervisors enrolled for these conferences. The "*muqaddam,*" or first-line supervisor, training program began with a class of 18 Saudi employees at Abqaiq on December 19, 1954. The program consisted of a two-part course in supervisory techniques and company procedures. A monthly average of 114 Saudis enrolled in the program during its first full year.

Ibrahim S. Al-Kabour

"My advice to young Saudis is that they have to work very hard. You should not expect to be given a position; you have to earn it."

Ibrahim S. Al-Kabour began his 42-year career with Saudi Aramco in 1951 as an office boy with the Accounting Department. "When I started I was 14 years old," Al-Kabour recalled. "It was quite scary. In Jubail, where I lived, most people worked in the fish market or worked for Aramco, so you had lots of relatives and friends working for the company. You were proud to join Aramco."

Al-Kabour worked mornings in Accounting and then studied English and math at the ITC in the afternoons. However, his schoolwork continued past the end of the workday, when other subjects were taught at the Company's volunteer evening schools.

"I went to the ITC on my own time from five o'clock to nine o'clock, where I took typing, Arabic and other subjects," he said. "At the time, Aramco taught you English and math on assigned time and then other subjects on your own time."

In the early 1960s, Al-Kabour studied English at Bucknell University and then accounting at Philadelphia's Pierce Junior College. From 1965 to 1968, he attended Armstrong College in Berkeley, California.

"I was very active in Berkeley," Al-Kabour noted. "I was president of the International Club and of the Accounting Club. I was also vice president of the student body. Because I was so active in school, I was also named student of the year. I remember an incident when I brought my wife and son to the college. The president of the college held my son and said, 'Listen, your father's name is Ibrahim Al-Kabour and he's the student of the year for 1967.' It was very emotional to see this man carry my son and say, 'We are proud of your father.'"

After returning to Aramco from college, Al-Kabour moved from Accounting to Office Services. For two decades, until retiring in 1993, Al-Kabour served as general supervisor for Office Services Divisions in Ras Tanura and Dhahran. In 1980, he spent six months on assignment in Houston with Aramco Services Company's Saudi Arab College Relations Program.

"I went around to many different universities telling young Saudis of my experiences to encourage them to join Aramco," he said.

Al-Kabour's recipe for success is hard work. "The young Saudis have to learn to be patient and to work harder and harder," he said. "They should work on their own time. Any assignment they are offered, they should take it, because it will help them."

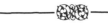

Summer Courses • In July 1949, the company sent 40 Saudi clerks, teachers and engineering aides to the American University of Beirut (AUB) for 10-week training courses. A committee composed of representatives of the Personnel Department, Industrial Relations Department and Training selected the students from a larger group nominated by supervisors in the three districts. An additional 50 Saudis, selected by a similar process, attended the special AUB summer course in 1950. This group included 20 Saudi clerks, 20 teachers and 10 engineering aides. In 1954 the company suspended the program until the student selection process could be put "on a sounder basis." A training report to management said the summer sessions "had a stimulating and broadening effect upon the Saudis" and helped to close the gap in high-school-level education.

Saudi employees about to depart for summer courses at the American University of Beirut, 1950.

Ali Al-Naimi attended AUB in 1953 under the summer courses program and discovered he had an aptitude for science. He often referred to the experience as "the turning point in my career."

Government Schools • In negotiations concluded during a visit to Dhahran by the then Crown Prince Sa'ud, the company agreed to build and pay the operating expenses of 10 public schools in the Eastern Province. The agreement, signed February 7, 1953, settled an issue first raised by the closing of the Jabal School two years earlier. It concerned the company's obligation to educate the sons of its employees. The Saudi government had notified the company in July 1951, a few months after the Jabal School closed, that under Saudi labor law the company was obliged "to open schools for the education of the workmen's children. ..." The notice led to negotiations that culminated in the 1953 agreement. At the time there were 21 elementary schools and one secondary school in the Eastern Province, with a total enrollment of about 3,500 boys, less than half the estimated number of eligible students. The great majority of students were in lower grades. Only 65 boys graduated from Eastern Province elementary schools in 1953. Most existing schools were overcrowded, understaffed and lacking in basic equipment.

The 10 Aramco-built schools were supposed to accommodate a number of students equal to the number of school-age sons of Muslim employees — a total at the time of about 2,400 boys between the ages of six and 14. The government would select the teachers and the curriculum. Sites chosen for the first six schools were at Dammam, al-Khobar and Hofuf for schools with space for 300 boys each, and at al-Mubarraz, Rahima and Saihat for schools with space for 200 boys each. The first Aramco-built school opened in Dammam on December 7, 1954. The second school opened three days later in al-Khobar.

Campaign Against Illiteracy • Aramco joined this government-sponsored program in June 1953. The government supplied the textbooks and specified the curriculum; Aramco provided the facilities and teachers for evening classes in all three districts. It was a two-year program, comparable to the lower grades in government primary schools, with instruction in Arabic, arithmetic, geography, history and religion. Classes for Aramco employees met in converted Saudi dormitories, separate from facilities devoted to other company training programs. Cash awards were given students when they completed portions of the course and passed tests. Initial enrollment was 1,070 in Abqaiq, 1,230 in Dhahran and 497 in Ras Tanura. Only about 20 percent of the students passed the course in 1953. Enrollment company-wide dropped to 1,127 in 1954.

Instructor Mohammed Omar conducts English class for Saudi employees in Dhahran, 1951.

Results of the Five-Year Plan

he five-year plan was a great success, at least on paper. By 1954 the number of Saudis in higher grade codes was 14 times larger than in 1949. However, the announced goal to raise half the Saudi work force to the semiskilled level or above had not been achieved, mainly because the work force was significantly bigger than had been forecast in 1949. For the same reason, the number of Saudis in lower classifications did not decline as much as hoped. The results of the five-year plan looked like this:

	1949 Actual	1954 Goal	1954 Actual
Total Saudi work force	10,944	10,000	14,182
Grade 6 and above	80	975	1,131
Grades 4 and 5	1,564	3,765	5,401
Totals above grade 4	1,644	4,740	6,532
Laborers and low-skilled	9,300	5,260	7,650

The years 1949 to 1954 saw new highs in the number of employees in training, and in the number of people doing the training. The peak year was 1953, when 8,200 Saudis enrolled in training programs and Training employed 413 full-time and 224 part-time people. About 60 percent of all Saudi workers were enrolled in training programs in 1953, compared to 24 percent in 1949. Training costs in 1953 totaled $4.75 million, an increase of more than $500,000 from the previous year. All these numbers declined sharply in 1954, a year of general cutbacks at Aramco.

The total number of Saudi and expatriate Aramco employees increased by more than 6,500 during the life of the five-year plan. Nearly two-thirds of the new employees were Saudis. The largest increase had been in the number of Saudis above grade code four, the semiskilled level, an indication Saudis were taking some job openings that would in the past, by necessity, have been filled with expatriates.

Job Skills Training Critiqued

ot everyone gave the Job Skills Training high marks, despite the impressive results claimed for the program. A consultant from Columbia University, Dr. Robert King Hall, surveyed the Job Skills Training program in 1951, and wrote a stinging critique. He found "amateur and incompetent instructors," "a serious lack of teaching aids" and "an expensive and confusing duplication of effort."

Hall was especially hard on the policy of using production men as training supervisors and unit instructors. He acknowledged that they had a high degree of technical knowledge, "but a considerable percentage of these men are floundering in the unknown seas of training procedures. They are not trained teachers. When faced with the problem of running a training program for the first time, they tend to take refuge in paperwork. They try to write down [for themselves, not for Saudi trainees] what seems to them the essentials of each job. They are amateurs in textbook writing and their manuals exhibit nearly every mistake in methodology that can be imagined. Few of them ever are really used in actual training of Saudi Arab personnel.

"As for teaching aids, there are practically no mock-ups of the complicated and costly machinery used by Aramco, virtually no visual aids with the exception of some 'rudimentary' charts and diagrams, and no evidence that the few cutaway models available are being used. Some of the charts that are in use are amateur affairs, almost incomprehensible to a university-educated scientist, and

Saudi assistant drillers study safety and first aid, Dhahran, 1951.

obviously unsuited to instruction of Saudi personnel. Instruction by means of simulated operation is nowhere apparent.

"What's more," he continued, "the lack of an 'official company method' for doing many operations resulted in a duplication of effort. There are different training manuals, completed or in preparation, in the three districts that presume to teach exactly the same skills. Separate achievement and aptitude tests used in the three districts attempt to test the same qualities.

"The construction of a training manual, or the construction and standardization of a psychological examination, are extremely technical matters that should be entrusted to a highly trained team of experts."

Roy Lebkicher referred to Hall's comments in a report submitted sometime later to management. He said the main thrust of the five-year plan was toward program planning and development, not methodology, where the brunt of Hall's criticisms were aimed. Yet he acknowledged, without apology, the validity of some of Hall's comments. He wrote: "Many people were placed in training jobs without being trained in training methods. Moreover, they had to develop their own training 'tools.' ... It is no criticism of these people to recognize that much of the methodology and the training materials have been below proper standards for most effective results. ... In view of the magnitude of the task there was no other way to do it."

GIT Academic Program

he General Industrial Training academic program had problems of its own. Inevitably, the regular daytime hours of classroom instruction, which trainees needed in order to advance, did not always dovetail with a far-flung oil company's 24-hour, seven-day-a-week production schedule. Many workers followed a 28-day shift schedule, during which they worked days for one week, evenings for a second week, and the overnight shift during the third week. Their hours changed every nine days. It was an awkward situation for trainer and trainee.

Shift workers were assigned to academic training on overtime, also known as "R" time. This meant they were paid to attend class, like regular day workers, but unlike day workers, "R"-time trainees went to school during their off-work time. The shift worker usually had to change his classroom schedule every time his work hours changed, meaning a new teacher in a new class every nine days.

Instructor Ahmed ibn Fahad conducts class in GOSP operations in Abqaiq, 1951.

What's more, his off days did not often coincide with the normal weekend holiday, so he missed classes if he went home to his village on his days off.

"R"-time trainees didn't mind the overtime pay, but they were hard pressed to keep up their grades. Trainers universally disliked "R" time. A former operations manager explained: "They [instructors] were faced with students who were sleepy and tired because they had just come off shift after working all night. There was not much student-teacher rapport because the students were popping in and out of different classes. The academic performance of these

students suffered. There were lots of flunks and dropouts. Even very smart, talented trainees got mediocre grades following this regimen."

Organizations with operations in remote areas such as Drilling, Exploration, Transportation and Marine had an even tougher time with academic training. They had to take a trainee off the job and assign him for weeks or months at a time to an area with an academic training facility. It was difficult to do, impossible in many cases.

Academic training activity was consolidated in GIT centers. Because of past problems, Aramco was cautious about naming its new training facilities. The word "industrial" was in the name of most new facilities, and they were called "centers" instead of schools. Officially, the only "schools" in Aramco were those attended by the children of expatriate professionals. Saudis attended a "center" and received "training" rather than schooling. Outside Aramco, "schools" for Saudis were run by the government.

The Importance of People in Training

Success or failure in training often depended as much or more on the zeal and imagination of the trainer as on the training program. For example, John Hoss made his name synonymous with training in gas-oil-separator (GOSP) operations in the Abqaiq area. Hoss, a power plant operator, was assigned in January 1950 to train a group of Saudis as technicians and supervisors for the new GOSP No. 3. By May, the 100,000-barrel-a-day plant was operating around-the-clock with a staff of 30 Saudis, the first such facility to be operated completely by Saudis. Within a year, Saudis were running three more Abqaiq-area GOSPs, putting to rest any doubts as to their ability to perform intricate technical operations.

GOSP training progressed rapidly in part because Hoss compiled a detailed manual with step-by-step instruction for GOSP operation. It included drawings of scores of instruments with their names, as well as detailed drawings of instrument parts. He also managed to sidestep the teaching of complicated theories and equations. GOSP operators normally worked algebraic equations to determine when it was time to adjust the GOSP machinery. Instead of taking months to teach Saudis these equations, Hoss had reference cards made that listed the answers based on various instrument readings.

First all-Saudi GOSP operations crew with supervisor Khalil ibn Assa, front, at GOSP 3, 'Ain Dar, 1952.

Nurses' training was another success. The first Saudis to qualify as practical nurses received their diplomas at ceremonies in Dhahran on September 22, 1951. Nine students graduated after training eight hours a day — four hours in class, four hours in wards and clinics — for three years in a program run by John Miller at the Dhahran Health Center. (The nine were Ibrahim ibn 'Abd al-'Aziz, Nasir ibn Sultan, Ibrahim ibn Ali, Muhammad ibn Ibrahim, Abd al-Rahman ibn Muhanna, Ahmad ibn Makki, Tahir ibn 'Abd Allah, Mubarak ibn 'Abd Allah and Sa'id ibn 'Abd Al-Karim.) As practical nurses they were qualified to perform fundamental medical procedures such as taking a patient's temperature; reading thermometers; taking X-ray photographs; and applying bandages, dressings and tourniquets.

Nurse trainees in anatomy class at Dhahran Health Center, 1952.

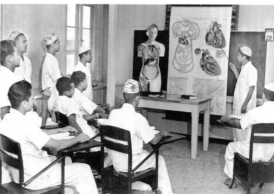

A mosque was among the first masonry structures in Dhahran's Saudi camp, late 1930s.

By 1948 Dhahran was a thriving family community.

Building a rock pier at al-Khobar was a priority project for the company, mid-1930s.

Trucks line Ras Tanura's main street, late 1940s.

Employees learned to rebuild engines through on-the-job training.

Dhahran was a modest oil camp in 1938.

In 1952, after more than a year's training in the Marine Division, the first Saudis qualified as firemen-watertenders on a steam tug. They were Ahmad ibn 'Abd Allah Al-Amiri, Hashil ibn Mubara Al-Dosari and Ali ibn Buseain Trafi. Here was additional proof that the skeptics were wrong; properly trained Saudis could be relied on to perform complicated jobs safely and in a timely manner.

Trainees learn the working parts of a drilling rig, 1951.

That same year a team of advanced Saudi carpenters, plumbers and sheet-metal worker trainees built a prefabricated, Swedish-made duplex at Dhahran in 5,300 man-hours, nearly besting the fastest construction time turned in by teams of experienced expatriate workmen.

Formal driller training started in Abqaiq with "Red" Kindell as coordinator and Al Flick in charge of training at the job site. By 1952 the Oil Operations Division in Dhahran had the most elaborate driller-training equipment. The division assigned a full-scale drilling rig exclusively to Training. A company publication called it "probably the only drilling rig in the world calculated *not* to strike oil." The National 100-1 rig was set up a few yards from historic Dammam No. 7, the first well to produce oil in commercial quantities in Saudi Arabia. The rig was used to train assistant drillers, rigmen, diesel mechanics and other Saudi workmen. Up to 30 trainees spent seven to 14 weeks in the course, splitting their time about evenly between the classroom and the drilling rig. They pulled and coupled pipe, drilled a 200-foot-deep hole, and went through all the other steps involved in creating an oil well. Robert Cooney, the division-training supervisor, devised the course, and Art Osborn, a unit trainer with 33 years' experience as a driller, did most of the actual instruction.

The Problem of Know-How

ne of the most difficult tasks in training had to do with the evaluation of know-how, or *al-khibrah al-fanniyyah*. The problem went to the heart of training. It raised questions such as: What's the difference between a skilled and semiskilled carpenter? What does a grade-four pipe fitter know that a grade-three pipe fitter doesn't? What skills does a journeyman welder have that welders in lower grades don't have?

In the fall of 1951, Richard Holmes transferred from Ras Tanura to Dhahran as coordinator of Wages and Salaries, and decided to tackle this problem. He asked all the supervisors to list the jobs required for their organizations. Then he asked them to describe all the knowledge and skills needed to perform each job, listed in the approximate order a trainee would acquire the skills and knowledge while on the job.

The answers to these questions produced an ordered list of the skills and knowledge a fully trained worker needed for a job. It was a basic job ladder containing all the steps that made up what came to be called the whole job. "It sounds simplistic now, but it had not occurred in Aramco up to that time," Holmes said.

Measurements are matched to drawings in machine shop training class, Dhahran, 1950.

Holmes suggested employees be rewarded for each step taken up the job ladder. The steps were initially called "training progress steps," later "wage progress steps," and finally "job progress steps." The "step" approach to job evaluation was adopted in 1952. The authority to define jobs, to oversee the training and to reward the trainee's attainment of skills with pay increases was vested in unit supervisors. It was a fundamental step in establishing a proper basis for employee training.

Palestinians Recruited by Aramco

In May of 1949, at the urging of the Saudi government, Aramco began hiring Palestinians through recruiting offices in Lebanon and Jordan. Thousands of Palestinians had fled their country after the forced incorporation of a Jewish state in Palestine and the ensuing Arab-Israeli war. In five months, the company had more than 5,600 applications from Palestinians for work with the company in Saudi Arabia.

Aramco was able to choose from among the best educated and the brightest of that cosmopolitan Mediterranean nation, including skilled craftsmen, experienced teachers, and other professional people with English-Arabic language skills. Palestinians helped fill the communications gap that separated Americans from Saudis, and that, for practical purposes, had restricted the training of Saudis by Americans to relatively simple, step-by-step learning. Further, the arrival of bilingual teachers provided a morale boost for the hard-pressed Americans in charge of the Aramco training system.

The first Palestinians arrived in January 1950. By September, they were the largest group of expatriates working for Aramco aside from Americans. At year's end there were 826 Palestinians on the company's payroll, about five percent of the total work force.

A display of tools and precision parts made by students at the Advanced Training Center, Dhahran, 1952.

Palestinians, along with smaller numbers of bilingual Lebanese, Jordanians, Sudanese and Somalis, quickly came to the forefront in training. In Abqaiq, for example, the GIT centers employed six to eight Palestinian teachers. Two of them were immediately made head teachers. In Ras Tanura, Palestinians headed three of the nine on-site shop training units, and taught in six of the nine units. The remaining three units employed a Pakistani and two Sudanese instructors.

At the Ras Tanura Refinery, an expatriate remembers, "Unit trainers were mostly Palestinian, Lebanese or Jordanian. They were taskmasters to their charges, but also, in many cases, their counselors and advocates within the organization."

Before long, Palestinians and other Middle Easterners involved in training outnumbered Americans by a considerable margin, although Americans continued to occupy leadership positions in the Training Organization. By 1953 there were 118 Americans employed full time or part time in Training, compared to 307 non-Americans. Those classified as non-American were almost all natives of Middle Eastern countries.

One of the early Palestinian hires, Zafer Nashashibi, was taken aback by what he saw of the Saudi trainees and the Westernized training they received at Aramco. "Not one percent of the Saudis could read or write," he recalled. "It was shocking to see these Saudis being taught to read and write English before they could read and write in their own language."

Nashashibi had observed one of the unforeseen consequences of the training process. Many Saudis who joined Aramco with little or no prior schooling learned to read and write basic English at Aramco training centers, but remained illiterate in Arabic. More than one Saudi trained by Aramco knew how to write his name in English, but could not do the same in Arabic. The company offered Arabic-language training to Saudis during volunteer school hours, but these courses never really caught on because learning to read and write Arabic did not seem to be related to advancement on the job.

As new-hires arrived at Aramco from the Middle East, Italians were released. Italians held many of the skilled craft jobs to which newly trained Saudis aspired. The number of Italian workmen declined from more than 1,200 in January of 1950, to 693 in December of 1950, and to 110 at the end of 1957.

Italians left their mark on Aramco. The Ras Tanura Refinery is a monument to their collective skills. Their artistry is shown in attractive stonemasonry buildings and walls in Aramco communities. Examples of their stone work include the dining hall in Dhahran and buildings on the American Consulate General's compound in Dhahran.

Aramco Headquarters Moves to Dhahran

On March 5, 1952, in what the company called "a significant policy change," Aramco's headquarters moved from a 20-story office building at 505 Park Avenue in New York City to Dhahran, Saudi Arabia. Fred A. Davies, chairman of the board, moved to Dhahran on that date. Aramco's new president, Robert L. Keyes, took up residence in Dhahran on April 23, 1952.

Four months later, on July 1, Roy Lebkicher was named director of Training. Snyder, who had held the title of director of Training for two years, was appointed coordinator of General and Industrial Training. In his new post, Snyder had charge of the GIT centers in all three districts, and in the daily routine, acted as a stand-in for Lebkicher. Much of Lebkicher's time was taken up with his duties as a member of the Executive Management Committee, the top policy-making committee in Saudi Aramco.

Concurrent with Lebkicher's appointment, a new department with three divisions, the General Office Training Department, was formed within the Industrial Relations organization. Snyder headed the General Industrial Education Division. The Job Skills Training Division was led by Lou Skidmore. A leader for the third division, Supervisory Development, had yet to be chosen. In conferring department status, the company rec-

Roy Lebkicher, 1952.

ognized the increasing size and importance of training to Aramco. At the time, about 45 percent of Saudi workers, nearly 7,000 men, were enrolled in training programs.

The General Office Training Department, headquartered in Dhahran, was formed to coordinate training policy and develop standard operating procedures in the three districts, which had been operating more or less independently of each other. The General Office had no administrative control in the districts; the district superintendents of Training reported to the manager of Industrial Relations in their districts. The General Office was essentially an advisory group, providing functional guidance to the district organizations. The districts could accept or reject this advice. In almost all cases the advice was followed. When it was not, the case could be referred to a higher authority. Training superintendents in the three districts at the time were Frank Jarvis, Abqaiq; Don Richards, Dhahran; and Marcy "Lucky" Luckenbaugh, Ras Tanura.

Aramco's new headquarters in Dhahran, 1952.

Official Training Policy Established

 he company's first official policy statement on the education and training of Aramco employees was issued with President Keyes' signature on January 25, 1953. The statement read:

"The company is to conduct employee education and training in which all personnel of all levels are expected to cooperate and participate under approved programs which are adequate to accomplish the following purposes:

- To offer Saudi employees encouragement and assistance, through improvement of both skill and knowledge, to develop their ability to perform company work of all kinds and levels; and to do so in sufficient numbers to provide for the maximum practicable utilization of Saudis in every organizational unit and every type of work.
- To promote ease of communication, cooperation, and mutual understanding within and between employee groups and between the company and the government and communities of Saudi Arabia.
- To assist all employees of all levels who supervise the work of others to understand the company philosophy, policies, methods, and procedures applying to dealings with training and utilization of employees and to any other methods and procedures applicable to their own jobs; and to improve their own skills of leadership and administration.

- To assist individual employees, on a selective basis, to develop their special interests and talents for the greatest mutual benefit of themselves and the company and to provide for the future filling of responsible company positions by the planned development and promotion of employees as far as possible."

R. L. Keyes, 1957.

Keyes' statement spelled out policies that were to remain central to Training's mission for years to come. He made it clear that training was everyone's job and that Saudis were being trained to occupy a maximum number of jobs at *all* levels of work. Lebkicher called the publication of this policy "one of the important accomplishments of 1953 ... a meaningful guide for administration and for program formulation."

"Saudization"

 he word "Saudization" was already in the vocabulary and on the minds of trainers. Al Lampman, who was on the Training Department's General Office staff, recalled when the word first came into common use among trainers.

"'Saudization' was a general term in the early 1950s. It became more specific about the time they moved management to the field (Dhahran). Somewhere in there it moved from a kind of vague object out in the far distant horizon to a very deliberate, close-up 'Let's go to work and get these people trained' kind of thing. But the word never had a specific meaning, as far as I know, as to how many Saudis should be working and in what positions, and so forth."

Training Operations

 n 1954 the General Office of the Training Department operated out of offices on the second floor of the U-shaped administration building in Dhahran (later called the Administration Building South). The top figures in Training, beside Lebkicher, Snyder and others already mentioned, included: Powell Ownby, administrative assistant to the director; Emmet Roberts, GIT advisor; and George Trial, supervisor of Advanced Industrial Training. Trial, who had succeeded Snyder as director of the training mission at the Dhahran

Powell Ownby, 1956.

Air Base, followed in Snyder's footsteps and joined Aramco in March 1950.

The former head of the Jabal School, Vince James, became superintendent of the senior-staff schools in 1954. At the time, there were 421 American school students in Dhahran, 221 in Abqaiq and 185 in Ras Tanura, all record highs. Sam Whipple, the first teacher at the American schools in Dhahran and Ras Tanura, announced his retirement that same year. He returned briefly to Dhahran in 1956 as a fourth-grade teacher, and later took a teaching assignment in Iran.

Training's first library — a facility called the Training Materials Workshop — opened in Dhahran in April 1954, next to the Advanced Industrial Training building. It housed a gradually growing supply of books, magazines, training manuals, maps and charts, most of them acquired from colleges and universities in the States. The man responsible for collecting training materials and sending them to the Dhahran workshop was C. Kenneth Beach, a former Cornell University professor who joined Aramco's staff in New York and later moved to Dhahran as coordinator of Management Training.

Vic Greshman ran the Training library with the help of Musa Hawa, a job trainer intern. The workshop was created to help line-organization trainers design and develop their own training programs, using the materials collected in the States as a model. Trainers in groups outside Dhahran had materials sent to them on loan. Students from the Advanced Training Center used the workshop's reading room and library to study for training projects. The workshop included a Training Aids Construction Unit, which supplied illustrations and other training aids, and a Publications Unit, which produced finished printed copies of courses that were designed and written by Aramco trainers.

George Trial, right, meets with American University of Beirut students, 1951.

1950s: Busy and Tumultuous Times

 he list of modern utilities, conveniences and recreation projects completed with Aramco's assistance in the Eastern Province between 1949 and 1954 included the huge Dammam port project, the 360-mile-long railroad line connecting Dammam to the capital in Riyadh, the first electric lights for Dammam and al-Khobar, and a network of blacktop roads that began at Dhahran and reached out toward Abqaiq and Ras Tanura. During those years al-Khobar grew from a fishing and pearling village of half a dozen *barastis* to a community of largely masonry buildings stretching about three-quarters of a mile along the waterfront and extending eight to 10 blocks inland. The first department store had opened in al-Khobar, a yacht club had been formed at Ras Tanura, and golf courses with sand fairways and "greens" were developed at Dhahran and Ras Tanura.

The late 1940s and early '50s were very tumultuous years in Middle East history. The United Nations vote in 1947 to partition Palestine into Jewish and Arab states ignited a struggle that was to engage the diplomats of the world's superpowers, the delegates to the United Nations, and the armies of Israel and the Arab nations for decades to come. Many in the Arab world were angry, and Aramco was an easy target for that anger, even though the company had publicly warned against any hasty United Nations action.

The dissolution of the British Empire (Commonwealth status for Ceylon in 1947 and for India in 1949, independence for Burma in 1947 and Egypt in

Aerial view
of Ras Tanura
Refinery, 1955.

1952) did not go unnoticed at Aramco. In May 1951, the government of Iran nationalized the Anglo-Iranian Oil Company and all non-Iranian personnel were evacuated from Iran. This virtual shutdown of the Anglo-Iranian Oil Company left Aramco as the largest oil producer in the Middle East. These events also reinforced the feelings of farsighted executives such as James Terry Duce, that Aramco should move ahead quickly with plans to develop Saudis.

Other political events increased restive feelings among Aramco's work force. The Cold War between the U.S. and the Soviet Union deepened after the first successful test of an atomic weapon by the Soviet Union. The Korean War began in 1950, and along with it the anti-Communist hysteria of the McCarthy era. Radio Moscow broadcasts in Arabic alleged "organized discrimination in the oil fields" of the Middle East and charged that Arab workers were being "exploited by the imperialistic companies." A U.S. State Department policy paper cited Aramco as a bulwark against the "threat of Communist aggression in the Middle East" and a force for "economic and political stability."

Aramco Vice President Floyd Ohliger felt obliged in October 1950 to reassure expatriate employees that the company was "taking all possible steps to ensure the continuing safety and well-being of you and your families.

"Although at this time the likelihood of a serious emergency seems remote," Ohliger said in a statement printed on the front page of the company newspaper, "plans have been made to meet all anticipated situations."

That same month, Aramco drillers discovered the first oil field beneath the waters of the Arabian Gulf. The field, named Safaniya, was later recognized as the world's largest offshore oil field.

The Korean War triggered an increase in demand for petroleum products. Aramco's annual production doubled, from 174 million barrels in 1949 to 348 million in 1954. Production averaged more than one million barrels a day for the entire month of May 1954, a new record. Total production from the Abqaiq Field passed one billion barrels in the third quarter of 1954.

In 1954 Aramco briefly held the title as the world's largest oil producing company, according to the Associated Press. The newspaper wire service quoted "industry sources" as saying that, although official figures were not published, Aramco's average production of 953,000 barrels per day most likely pushed it slightly ahead of the Kuwait Oil Company, which had taken the "world's largest" title from Aramco in 1953. The Kuwait Oil Company took the title back in 1955 and held it for many years.

Trainee plumbers
in the One-Eighth-
Time Training
Program, 1952.

On October 17, 1953, Saudi workers at Aramco had another confrontation with the company. A committee composed of a dozen young Saudis, all enrolled in training programs at one time or another, presented a long list of demands ranging from pay raises to ice water. They sought equal pay for equal work and "all the privileges that the Americans enjoy."

King 'Abd al-'Aziz died on November 9, 1953, after serving 51 years as leader and king of the country he had unified. He was succeeded by his eldest son, Sa'ud ibn 'Abd al-'Aziz Al Sa'ud.

Lathe work at Dhahran's Advanced Training Center.

King Sa'ud visited Dhahran, and on January 14, 1954, issued a Royal Proclamation that granted Saudi workers substantial benefits and brought that confrontation officially to an end, although the great majority of Saudis had returned to work of their own accord several weeks earlier.

Aramco's relationship to its Saudi employees and to the Saudi government evolved along with the stormy events of the early 1950s. In 1950 the company agreed to a new financial arrangement that gave the Saudi government a 50-50 share in the company's net operating income. For the first time, Saudi government income became tied through Aramco to the world oil market. Until then, Aramco had paid a fixed royalty and changes in oil prices had no direct impact on the government's income. Two years later, the 1950 agreement was modified to give the government a 50-50 share in the company's gross income. Under this arrangement, royalty payments to the Saudi government leaped from an estimated $57 million in 1949 to $170 million in 1952.

Starting in 1950, the company eliminated the old categories that had divided Aramco's living and recreational areas along ethnic lines between Saudis, Americans and other foreigners. The term "Saudi camp" was dropped and "senior staff" no longer meant Americans only. Instead, three job categories were established: senior staff, intermediate and general, with the job, not nationality, determining an employee's eligibility for housing and recreational facilities.

The first three Saudis reached senior-staff status in 1950 and became eligible for housing in the areas previously reserved for Americans. The company started a thrift program in 1951 under which the company contributed to the savings of participating Saudi employees. An interest-free home loan program for Saudi employees was introduced in 1951. The company's first Arab-language periodical, a monthly magazine called the *Oil Caravan (Qafilat al-Zait)*, began publishing October 29, 1952, under editor Hafiz Al-Baroudi.

New intermediate camps costing more than $10 million opened in all three districts in 1952, with housing in permanent, air-conditioned buildings, and separate recreational and dining facilities. Pay raises were given in four of the five years between 1949 and 1954. Minimum pay went from 1.75 riyals a day to six riyals a day. The work week was reduced to 5½ days or 44½ hours, and vacation time was increased to one month.

The number of Aramco employees in Saudi Arabia surged from 15,314 in 1949 to 24,120 in 1952, a mark that would not be reached again for 25 years. The percentage of Saudis in the total work force dropped from 84 percent in 1944 to a low of 59 percent in 1952. The trend was reversed in 1953, and the percentage of Saudis in the work force began to slowly climb.

In the five years from 1949 to 1954, Aramco accepted significantly more responsibility for the well-being and education of its Saudi employees. The company seemed to recognize that its continued success would depend on the goodwill of the Saudi people, who served both as a readily available work force and as a host with growing expectations.

Aramco recruiters in Dammam gave an eye test to each job applicant, 1951.

The Search for Community 1955-1960

"The happiness of people is not improved, nor their good will obtained, by welfare or philanthropy. But education and training, which enable a man to better himself through his own efforts, give him an entirely different slant on the nature of material progress."
– Roy Lebkicher

An American visitor in the late 1950s found Dhahran much like home. "It is an American community set down in the Arabian desert, complete with weekend gardeners, women's clubs, Parent-Teacher Associations and television," he wrote. The visitor, a *New York Times* correspondent, saw Dhahran as "distinguished by bright, air-conditioned homes, each with its own yard and hedge. The inhabitants are recognizable as genuine suburbanites. Their children are equipped, according to age, with the usual American toys and means of locomotion from scooters to hotrods."

Much the same was true of Abqaiq and Ras Tanura. These two proud communities, one the center of production and the other the center of refining, liked to boast that without them Dhahran could not exist. But media attention focused on Dhahran because it was Aramco's largest community, the headquarters for operations and services, and the home of Aramco's administrative complex.

Abqaiq, Dhahran and Ras Tanura in the latter half of the 1950s were far removed from the wildcat camps of the 1930s, even from the camps only a decade earlier. Streets and sidewalks had been paved. There were grass lawns, trees and flowers, libraries, golf courses, tennis courts, bowling alleys, baseball diamonds, yacht clubs, horseback riding stables and swimming pools. The company built Saudi Arabia's first swimming pool, golf course, bowling alley and baseball diamond.

Aramco communities were like no other communities in Saudi Arabia. Where else could one go to the movies, dance on a moonlit patio to live music by popular U.S. dance bands, or listen to a celebrated pianist like Jose Iturbi play the works of Mozart, Chopin and Liszt? Where else could one attend lectures by scholars such as historian Dr. Arnold Toynbee, Arabist H. St. John Philby and, under the sponsorship of the Training Department, anthropologist Dr. Margaret Mead?

Opposite: The growing city of al-Khobar as seen from the air in July, 1959. Below: Student patrolman halts traffic for students returning home from school in Dhahran, 1952.

What role would Saudi employees play in this very American-like community? By 1950 only a handful of Saudis had achieved senior-staff status and moved into one of the nearly all-American communities. They were not enough to change the character of the towns, but their presence gave rise to an important question for the future: Could people of widely different nationalities, religions, and cultures learn to work and live together in partnership and harmony?

Working in Two Cultures

Saudis employed by Aramco split their time between two cultures. At work they spoke English, frequently wore Western-style clothing and interacted with Americans. After work they returned to the Saudi community, used the language and followed the customs of their native country. Saudi employees needed flexibility and tolerance to function in such a setting, as the following two stories suggest:

Ali Baluchi faced a dilemma while working as an office boy in the Training Department's General Office. Marsha Miller, a secretary returning from a vacation in Spain, presented Ali with his first pair of long, Western-style trousers. Until then he had always worn the traditional ankle-length Arab *thawb*. Ali, who lived with his parents in al-Khobar, recalled: "I didn't know how to explain this to my parents, changing my Arab clothes for Western clothes. But I talked to my father about it and we worked out a solution. When I go to work, I wear my *thawb*, but I tuck it inside the pants, like a long shirt. When I go home, I just pull the *thawb* out and it will cover everything. I didn't need to take off my clothes or go to the men's room to change. I just pull it out when I leave work."

Hamad Al-Juraifani never forgot the embarrassment he felt when he put on Western clothes for the first time. When Al-Juraifani joined Aramco in 1951 at age 15, he was told he would be expected to wear Western-style clothing to classes at the Ras Tanura ITC. Furthermore, he had to buy the clothes himself; they were not provided by the company. He purchased some cheap pants with flared legs and a too-large shirt. Years later he recalled that when he put on his new clothes and looked in the mirror, "I couldn't go to school because I was so embarrassed at the way I was dressed. It was embarrassing not to wear a *thawb*. To wear a *thawb* was the accepted way, and you would be surprised how traumatic it was to change the way you looked." Al-Juraifani eventually overcame his embarrassment and began attending classes. He taught a night-school class in English while wearing Western clothes, but he never forgot his original embarrassment and never lost his preference for traditional Arab-style clothing.

Two cultures, but one world, for two daughters of Aramco employees at their Dhahran schoolroom in 1952.

Ali Baluchi, seated, with Dhahran training superintendent Don Richards, 1956.

The company had made its first move toward developing a partnership between the expatriate and Saudi work force in 1949, when it adopted the five-year Aramco Production Training Program. The goal of the program, as announced by the company, was to develop "an efficient, coordinated and closely knit team" of expatriate and Saudi workers. This was to be achieved by training the largest number of Saudis possible in all types of work, at all levels, over a five-year period.

When the five-year period ended in 1954, the company could claim with some justification that the program had been a success. The gap had been narrowed. The number of Saudis classified as journeymen had increased from just 84 men in 1949 to more than 1,100 in 1954. The number of Saudi employees at the semiskilled level or above had quadrupled, from about 1,600 to more than 6,500.

Saudi Accomplishments

Saudis had already taken over a large percentage of the company's day-to-day work. For example, when a Consolidated Shops building opened in Dhahran in early 1956, it combined under one roof 15 smaller shops employing sheet metal specialists, pipe fitters, electricians, welders, boilermakers, blacksmiths and similarly skilled or semiskilled workers. Of the 162 workers who occupied the building, 70 percent, or 113, were Saudis.

Aramco spent $6.85 million on training programs plus school construction and operation in 1954 and $6.6 million in 1955, about one-third more than 1952 and 1953 expenditures. Job Skills Training was far and away the most costly program at that time. It accounted for about 45 percent of total training expenses. The second most costly program, General Industrial Training, accounted for only about 15 percent of the total.

Buried within these figures are stories of individual and collective advancement through training and experience. The following items are a few Saudi "firsts" from the period.

On February 23, 1954, Shedgum Well No. 12 became the first oil well in Saudi Arabia brought in by an all-Saudi crew. Four Saudis in training for the job of rotary driller "A" handled or directed all the reporting, recording and operating of controls during the drilling. They were Ali Al-Dhamin, 'Abd al-Latif Al-Makahaytah, Hamad 'Abd Allah Al-Makahaytah and 'Abd Allah Jassim Al-Kishi. The well was drilled to 8,942 feet in 121½ days without a lost-time accident. Two years later one member of the crew, 'Abd Allah Latif Al-Makahaytah, became the first Saudi rotary driller instructor.

On June 28, 1954, Ibrahim Al-Muhtasib became the first Aramco scholarship student to earn a four-year college degree. On that date Ibrahim received a bachelor's degree in commerce from the American University of Beirut. Three other Aramco scholarship students, Mustafa Husman, Rashid Al-Rashid and 'Abd Allah Sa'id, took two-year degrees at Aleppo College in Syria that same month.

Ali ibn Husayn was the first Saudi supervisor *(muqaddam)* to take charge of a gas-oil separator plant. Before joining Aramco, he had been an illiterate Bedouin tribesman. With training and work experience, he was able to take charge of 30 men at the $2 million plant located 20 miles from the nearest company office in Abqaiq.

The front page of the company newspaper in January 1954 announced that Jabr ibn Muhammad of the Dhahran carpentry shop had become the first Saudi to hold the title of assistant foreman in Aramco.

That same year Muhammad 'Abd Allah Abu Sharifah became the first Saudi operator in the Ras Tanura Refinery. In 1958, Abu Sharifah was promoted to supervisor *(muqaddam)*, making him the first Saudi supervisor in the refinery.

In 1955, Sa'id ibn 'Abd Al-Karim was selected to be the first Saudi supervisor of nurses at the Dhahran hospital. Sa'id, one of the original nine Saudis to qualify as a practical nurse, had taken advanced training on an Aramco scholarship at Kennedy Memorial Hospital in Tripoli, Lebanon, and at the American University of Beirut.

In Abqaiq that same year, 17 Saudis completed the *muqaddam* course and became the first Saudi supervisors in the crafts field. They were responsible not only for overseeing the work of craftsmen and their helpers, but also for

By 1956 a majority of the craftsmen at Dhahran's Consolidated Shops were Saudis.

handling discipline and personnel problems, and for giving guidance and work assignments. The Saudi supervisor's title of *muqaddam* was later changed to *mushrif* (first-line supervisor), and a second title, *muraqib* (foreman), was added.

By late 1954, the company had 18 senior-staff Saudis, an increase of 15 since the first three Saudis achieved that status in 1950. Ahmad Rashid Al-Muhtasib, Aramco's employment relations representative in Jiddah, made senior staff early on. (His son, Ibrahim Al-Muhtasib, was the first Aramco scholarship student to earn a four-year college degree.) Salih Al-Sowayigh, Aramco's relations representative in Dammam, was also among the first Saudis promoted to senior-staff level. His family was the first Saudi family to move into senior-staff housing in Dhahran. The Al-Sowayighs took a house at 609 Sixth Street, then on the perimeter of the residential area.

Instructor reviews piping fundamentals with Saudi students, 1955.

Salih had been educated in India. He spoke English, Urdu and Turkish in addition to Arabic. He had been an interpreter and private radio operator for King 'Abd al-'Aziz. He received the first wireless message transmitted to the Najd area of central Saudi Arabia. For many years, Salih ran the company employment office out of his home in downtown Dammam. That building later became the passport office. On the morning of January 24, 1947, Salih stood beside King 'Abd al-'Aziz and introduced some 150 starry-eyed Americans and their children, one at a time, to the King during an unprecedented royal reception in Dhahran.

Production Training Program Reconsidered

y 1955 it was clear that the Aramco Production Training Program and one-eighth-time training were not long-range solutions to the company's personnel needs. The initial success could not be sustained. The programs belonged to a different time and place.

The programs had been modeled on the emergency industrial training system adopted by the United States at the outbreak of World War II. This system proved so successful for the United States that Hitler credited it with turning the tide of war. However, few, if any, of the conditions that made the system work so well in wartime America were duplicated in Saudi Arabia. Most Saudis in those days did not have the education or the experience of American workers. Nearly all 13 million Americans who trained for work in war plants during World War II were high school graduates, whereas, in 1955 less than five percent of Saudis in Aramco's work force had as much as one year of schooling before they joined the company. Americans had the further advantage of being familiar with the products of an industrial society before they began war plant training. Nearly all Americans drove a car and knew how to operate household appliances. Many could make simple appliance repairs. Such experiences were almost completely absent in Saudi Arabia.

The key to the success of the American wartime program was extreme specialization. Applied to Aramco, this meant training workers for only one target job. Beginners might be taught to hang doors, for example, or pack boxes. The system quickly converted a completely un-skilled man into a useful worker. American war plant workers could patriotically tolerate such routine work for the duration of the war, but it would not satisfy Saudis who expected to be with Aramco all their working lives. Dissatisfaction with such a high degree of specialization surfaced, particularly among lower-level employees during Aramco's 1953 labor problems.

On-the-job clerical training.

Furthermore, Saudis could not be expected to progress at a pace similar to that experienced in American war plants. Starting from scratch, both in technical knowledge and academic learning, was a huge handicap for Saudi trainees. Not only did they have to learn new skills, they had to pick up the basics of the English language at the same time. The idea that Saudis could acquire the basics of a new language, learn new skills, absorb new experiences and become trained in the same time as an American war plant worker proved unrealistic. These difficulties, and a larger than expected influx of new Saudi employees, resulted in about 30 percent more Saudis remaining in low-skill positions at the end of the five-year program than anticipated.

The Aramco Production Training Program had left it up to each line organization to decide who should be trained, at what rate, and in what job. In time, line organizations split into two different camps, with divergent ideas about the goal of training. This split proved to be one of the most persistent and vexing problems encountered in Training.

One school of thought held that Saudis should be trained to their maximum potential. This often meant keeping trainees on assigned-time training during regular work hours until they graduated from a training center or dropped out. Most organizations supporting this approach were on a regular, five-day-a-week work schedule. They produced a large proportion of the Saudis who finished Aramco training and qualified for college-level scholarships.

Other organizations — primarily those in the field with shiftwork schedules — argued that training should continue only until a Saudi was able to fill a productive job. It was not the company's responsibility to run an ongoing, open-ended training system, they argued. As one Aramco executive remarked, "I don't see any need for our carpenters and rig mechanics to know how to read Shakespeare."

What's more, many unit trainers were not especially eager to train Saudis, even if given the time and the means. The goal of training was to prepare Saudis to replace expatriate workers. This goal had been company policy all along, reiterated when the Aramco Production Training Program was announced. Expecting an expatriate trainer to enthusiastically support such a policy, however, was asking a lot. A unit trainer might be training his own replacement or the replacement of a friend.

"The hardest job I ever had in Saudi Arabia was to force good career American craftspeople to train themselves out of a job," a former Aramco executive said.

It was especially difficult to find time to train employees in organizations with jobs that needed to be manned 24 hours a day, seven days a week. Once a Saudi trainee had qualified for a job and replaced a foreign contract employee in such an organization, he was always needed on the job. If further training was needed, it was done informally between shifts.

Problems with the Aramco Production Training Program were aggravated by the discovery that some organizations apparently hadn't held training sessions as scheduled, or had cut down on the time they were supposed to give to training, then used the money allocated for training for other purposes. A document

65

Paul Arnot, 1968.

about training during the period declared: "Manpower and money, sincerely allocated by management to training, too often were diverted to cover inefficiencies in operations or to protect vested interests. The costs continued, the hours of 'training' were reported, but real training too frequently did not take place."

On the other hand, some Saudi employees, especially older ones, elected to stay in the jobs they knew, rather than train for advancement. One case involving a teenage Saudi recruit was particularly disappointing for Paul Arnot, then administrator of the Abqaiq area. Arnot had decided to have a mental-abilities test administered to some 90 new recruits. He planned to put the top scorer in school full time.

Arnot recalled: "One young man stood out above the rest. He looked like he hadn't had a square meal in his life, but he was sharp mentally. He was about 16. He came in with an interpreter and three elderly men, who I assumed were relatives.

"I talked about sending this boy to school full time. I told them of the great opportunities before a young man of his intelligence. The old men talked among themselves, and it finally developed that they didn't know anything in the world about the concept of training. It was completely foreign to them. The boy said through the interpreter that he wanted to be a tractor driver. He came here from the Najran area (in southwestern Saudi Arabia). Some people from Najran were already working here as drivers, so it was natural that he should come here and become a tractor driver.

"I couldn't get across to them the concept of training. I finally gave up. He was assigned as a tractor driver. I thought if I went to see him six months later he'd probably jump at the chance to go to school full time. I'm sorry to say I didn't follow up six months later to see if he was still around. It would have been a great opportunity for him. He might have been a vice president now."

By 1955 most Aramco trainers were ready to declare one-eighth-time training a failed experiment. They echoed Dr. Robert King Hall's earlier assessment that the line organization men responsible for doing the training were "floundering in the unknown seas of training procedures." One former Aramco trainer had an even more caustic appraisal: "One-eighth-time training was the biggest disaster. Classes were held behind a partition at the work site. There were no programs. The trainer would ask, 'What shall we talk about today?' The trainees just came to class because the supervisor said they had to. Many slept once they got there."

Although the problems with one-eighth-time training were apparent, it seemed that no one in the field could stop long enough to fix what was wrong. Nothing could be allowed to interfere with rapid increases in production. They reasoned that any loss in meeting production targets was a loss far greater than could be justified by any improved quality in training effort.

The training burden had been placed on already hard-pressed line organizations because the main job of teaching job skills to up to 13,000 Saudis was too big for the Training Department to handle. It may be, as critics said, that job skills training would have progressed faster and more efficiently if it had been done by professional trainers. However, given the situation at the time, the results weren't all that bad.

American supervisor reviews operation of Ras Tanura Refinery distilling apparatus control panel with Saudi employee.

66

Records show that between 1949 and 1954, in spite of many problems, the company did meet its production targets. A Saudi work force capable of doing much of the total work in the company was developed, Saudis were trained for advancement into jobs to which they could immediately aspire, and a setting was created in which the whole economic and social character of the Eastern Province was changed with dramatic speed.

By 1952, the end of the company's massive post-war construction program was in sight. Production facilities had nearly caught up with demand. Most of the critical infrastructure—pipelines, refineries, roads, housing—had been built. Production continued to climb, but fewer employees were needed to keep the company running. More emphasis was put on replacing American and other expatriate workers with Saudis. (After 1955, the number of Aramco employees declined for each of the next 16 years. It would be another 22 years before employment climbed above the 1952 level of 24,120 employees.)

The training problem also shifted somewhat, from a concentration on training unskilled and uneducated new recruits to be semiskilled workers, to training Saudis in the more complicated and higher-level technical skills involved in the production, refining and delivery of Aramco products. At the same time, more Saudis needed to be trained in sophisticated levels of supervision and management if the long-range goal of creating a mostly Saudi work force and progressively eliminating foreign contract employees was to be achieved. This goal could be met only by identifying Saudis with leadership potential and providing them with a broad educational base in both theoretical knowledge and technical skills. Such a process would be lengthy and relatively costly.

Although the training needs were apparent, the means to achieve them were not. In the absence of any other program, the company readopted the Aramco Production Training Program in 1955, although it was by then fast running out of steam. "Everyone knew it wasn't working, but no one had a better suggestion," a former member of the Training Department general office staff said.

Al Lampman, right, and Howard Scott with a stack of manuals produced by Ras Tanura District Training, 1961.

Training Facilities Enhanced

In an absence of ideas for improving training programs, the company set about improving training facilities. Between 1955 and 1957, the company built spacious, air-conditioned buildings for Saudi training in all three districts at a total cost of more than $1.1 million. These buildings replaced older facilities that housed the General Industrial Training Centers (GITs). The GIT classes at Abqaiq and Dhahran had met in crowded portable buildings. The Ras Tanura GIT had become a maze of jerry-built rooms added on to the older, not air-conditioned, structure originally built as a shops training facility in 1947.

The new facilities were called "Opportunity Schools" rather than General Industrial Training Centers — the title in use since the first such centers opened in 1950. But the old concern about government intervention in anything called "schools" soon resurfaced, and in 1956 the name was changed to Industrial Training Centers (ITCs), with emphasis on the word "industrial." Selected Saudis attended assigned classes in job-related subjects at these centers during the day. In the evening

Saudi employees leave the Abqaiq Industrial Training Center after attending classes during working hours, 1956.

the buildings became voluntary training centers, where any Saudi employee who wished to could attend classes in his spare time.

The new ITCs at Abqaiq and Ras Tanura were identical two-story, masonry block buildings, with 20 general classrooms, a science laboratory, a drafting room, a business-machines classroom, a projection room, a visual-aids workshop, a library and staff offices. The Dhahran center was just like the other two, except it had 24 instead of 20 general classrooms. The three centers had a combined capacity of more than 2,000 students per day. The Dhahran and Abqaiq buildings were completed in 1955, and the Ras Tanura facility opened in 1957. The first full-time principals in the new buildings were Larry Emigh at Abqaiq, George Grebin in Dhahran and Bill O'Grady at Ras Tanura.

The company also built a new school for children of the senior-staff community. The Dhahran Junior High, containing 20 classrooms, opened April 30, 1955, at the corner of Third and "M" (now Ibis) streets. Completion of an adjacent building to house the first full-sized gymnasium/auditorium in Saudi Arabia was delayed until July because steel girders to be used in its construction failed to arrive on time. The latter project was a tribute to the perseverance of Vince James, who tried three times before he won management approval for the construction of the combination gymnasium and auditorium.

Aramco-Built Government Schools

or many years, the Saudi Arabian government had insisted that Aramco should be responsible for the education of the children of its Muslim employees. Precedents for this arrangement had already been established by other oil companies with concession agreements in the Middle East, but Aramco argued that the general education of young people was not the business of an oil company. Under a compromise struck in 1953, Aramco agreed to build the schools and pay for their operation, while the government supplied the curriculum and teachers.

Harry Snyder was selected to represent Aramco in the Aramco-Built Government Schools project. He was named acting coordinator, educational liaison-Saudi Arab government, and later coordinator, SAG/Aramco Educational Services.

It was a hybrid organization. Snyder and the directors who followed him were Aramco employees, but the remainder of the staff were contract employees working out of offices in al-Khobar for the Saudi director of education for the Eastern Province.

Al Lampman remembers the day, December 7, 1954, when King Sa'ud ibn 'Abd al-'Aziz Al Sa'ud came to Dammam to cut the ribbon to open the first Aramco-Built Government School. Lampman, then an assistant to the director of training, and Snyder had just moved the 300 student desks and other furniture into the building. As they nervously awaited the King's arrival, Lampman and Snyder went from room to room measuring the space between each desk to make sure all 300 of them were placed exactly the same distance apart.

Saudi children watch King Sa'ud ibn 'Abd al-'Aziz cut the ribbon opening the first Aramco-Built Government School in Dammam, 1954.

Hamad Al-Juraifani

"There is no limit to what people can do and learn. It is just limitless."

 new world opened for Hamad Al-Juraifani in 1949 when a convoy of trucks roared into his native village of 'Unayzah north of Riyadh. To his 11-year-old eyes, the huge trucks and the pale-skinned foreigners aboard them "looked like aliens" coming out of the desert. The trucks came in response to a government request for help in eradicating swarms of locusts. Al-Juraifani never forgot the convoy's dramatic arrival nor the blue-eyed foreigners and the strange language they spoke. "I never really realized there were other people who spoke other languages," he said.

Two years later, at age 13, he left home to seek work at Aramco's offices in Ras Tanura. He walked and hitchhiked for three days to reach Riyadh, and three more to the Aramco office in Ras Tanura. "I came to Aramco so I could learn to speak English," he said.

Al-Juraifani was one of the few job applicants of that time who could read and write Arabic. His father, a camel caravan driver with no formal education, insisted that Al-Juraifani attend school. He graduated from the Qasim district's first elementary school in 1951 and placed 11th out of 540 students that year in the Kingdom-wide elementary-certificate examination.

At Ras Tanura, an Aramco recruiter recommended that Al-Juraifani be hired as a trainee/teacher. He studied English and English-math at the Ras Tanura Industrial Training Center during the day and taught Arabic reading, writing and Arabic-math to Saudis in the evening at the company's volunteer school.

In 1957, Al-Juraifani was selected by Aramco for out-of-Kingdom training. When he returned to Aramco in 1965 he had a master's degree in chemical engineering from Purdue University, one of the most prestigious engineering schools in the United States.

As an engineer at Ras Tanura, Al-Juraifani spent hours in the refinery. "I really enjoyed working outside in the plant area," he said. "In the evening I would come home tired and dirty to the point where my wife would clean my clothes outside! I had these shoes and they smelled, and she wouldn't let me in the house. So, I just loved my work, and promotions came automatically."

By 1982 Juraifani was vice president of Northern Area Manufacturing. In 1988 he went back to the United States as president of Saudi Refining Inc., a shareholding company of Star Enterprise, Aramco's first downstream joint venture. In 1990 Al-Juraifani became president of Aramco Services Company, the parent company of Saudi Aramco's U.S. subsidiaries. He returned to Saudi Arabia in 1993 as vice president of Corporate Planning.

"Aramco is very active and recognizes accomplishment," he said. "My advice to a student is to just work hard and be involved, and the rest will come automatically. Aramco has provided me with the background and information that I wouldn't find anyplace else. The opportunities here are just so great and seem to renew themselves all the time."

The King used a special pair of solid-gold scissors provided by Aramco in the ribbon-cutting ceremony. The ceremony and royal inspection of the school went off without a hitch. But afterward, Snyder and Lampman could not find the golden scissors. They made a frantic but fruitless search. So, three days later, when the King opened the second Aramco-Built Government School in al-Khobar, the best Snyder and Lampman could provide for the ceremonial ribbon cutting was a pair of ordinary store-bought scissors.   •

Gradual Changes in Training Curriculum

Saudi enrollment in training programs dropped during the mid- to late 1950s. Between 1952 and 1957, Aramco's work force declined by about 5,800 employees, a drop of about 24 percent. In the same period, the number of Saudis in training programs fell by more than 5,000. In 1952, 51 percent of all employees were enrolled in training programs. By 1957 the employees in training had dropped to 38 percent.

The Job Skills Training Program and its one-eighth-time training schedule had nearly vanished. Enrollment in the program plunged from a high of 4,800 trainees in 1952 to fewer than 200 by the end of the decade. The emphasis in training moved away from teaching basic skills to large groups of employees and toward giving specialized instruction to selected employees who had mastered the fundamentals of their craft. Aramco's 1956 annual report explained: "There is less need now than formerly to teach basic job skills, since enough individuals to fill the present needs in many types of jobs have already received the necessary training."

More and more trainees were reaching the place where they required high-school-level courses to prepare them for out-of-Kingdom training at technical schools or for enrollment in colleges and universities. In recognition of such needs, ITCs in Abqaiq, Dhahran and Ras Tanura added specialized high-school-level courses in subjects once offered only at the Advanced Industrial Training Center in Dhahran. They included physics, chemistry, biology, history, geography, mathematics through trigonometry, and commercial office subjects like bookkeeping, typing and shorthand. The full line of courses became available in all three districts on both assigned and voluntary time. About a dozen or more senior staff instructors, mostly Americans with high school teaching credentials, were added to the ITC staffs.

Close supervision is provided at a typing class in the Abqaiq ITC, 1957.

With these changes, the Advanced Industrial Training Center, which had opened at Dhahran in 1950, passed into history. It was absorbed into the Dhahran ITC, just as the Pre-Job Training Program, also launched in 1950, had faded into the ITC curriculum several years earlier.

Warren Hodges was promoted to superintendent of Training at Ras Tanura on July 1, 1955. Frank Jarvis continued to hold the superintendent's post at Abqaiq, and Don Richards remained as superintendent of the Dhahran district.

Saudi Development Policy Committee

An idea that was to change the fortunes of hundreds of young Saudis grew out of an informal afternoon meeting in the Training Department offices sometime during 1955 or early 1956. As Frank Jarvis remembers it, he, along with George Trial and Ed Thompson, spent several hours kicking around ideas on how to help young Saudis reach their full potential. They decided young men with high potential should be identified as early as possible, a long-range development plan created for them, and their progress, both at work and in training programs, closely monitored. Those who continued to live up to their potential would be targeted for advanced training, possibly at a university.

Someone, it may have been George Trial, later sold the concept to Roy Lebkicher, the director of Training. Vince James and Paul Case wrote general instructions defining the new policy.

As a result, in 1956 the Saudi Development Policy Committee was created to administer the Saudi Development program. It was the first program designed to identify and develop a vital human resource, the people who had the potential to one day take over professional and management positions in the company. A Saudi Development Committee was created in each district and given the responsibility of locating high-potential people, setting job targets for them and developing the individual training programs necessary for them to reach their target. The committee was comprised of the district manager, several district superintendents, and representatives of the district Training Division. In its first year, the program placed more than 60 Saudi employees in supervisory training, industrial training, special individual training and Job Skills Training programs, including 22 out-of-Kingdom assignments.

Ken Beach, second from right, was among the participants at a Management Development Committee meeting in 1954.

About the same time Ken Beach, the coordinator of Management Training since 1953, moved out of Training to take over the newly created Management Development function, a job in which he reported directly to Robert Keyes, then company president. In this job, Beach targeted Americans for key management positions.

Up to that time, Aramco had no system for identifying prospective managers and preparing them as replacements when key management positions came open. "There was a big tendency," Beach said, "to replace the man who left with the man who reported to him. That's no way to run a show. What you've got to do is find the best guy to fill the job."

Beach had been a professor in the School of Industrial and Labor Relations at Cornell University. At Aramco he saw an opportunity to put his teaching to work, but first he had to convince upper management to give him the opportunity. He used the Engineering Department as a "guinea pig" and worked up a replacement chart for key positions in that organization. Then he made a presentation to top management, "including the suggestion that I be moved out of

A class at the Aramco Foreign Service Training Center on Long Island, 1950.

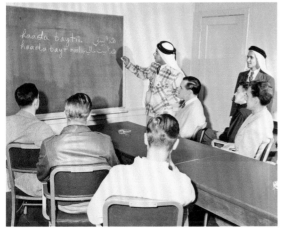

Training and report to Bob Keyes. That was a hell of a lot of nerve, to suggest that I report to the president, but that's what happened."

He proceeded to identify and rank personnel to fill key positions throughout Aramco, all the way from the supervisor level to the top job in the company. He had three candidates for each position. The performance of each man was periodically appraised and development plans drawn up detailing what experience and training it would take to prepare him for his target job. They called it the "green folder" treatment. Beach also established Management Development Committees in each district. He acted as chairman of individual appraisal committees and arranged rotating assignments with the parent companies in the United States for high-potential people. In those days, the confidential rankings of personnel for key jobs contained only American names. It would be some time before the first Saudi names appeared in the rankings, but at least the system had been established through which Saudis with managerial potential were eventually identified and groomed to move into executive positions.

At the other end of the Training spectrum, during 1956, Sultan ibn Salim, a 16-year-old English and mathematics teacher at the Ras Tanura ITC, was promoted to grade code six, the skilled-worker level. He was probably the youngest Aramco employee to ever reach that grade code. Sultan, a graduate of Darin Junior School on Tarut Island, had been teaching at the Ras Tanura ITC since he was 12 years old. He was one of about 15 young Saudis, teenagers or younger, who had been designated as trainee-teachers to fill in during a shortage of bilingual teachers at the Ras Tanura ITC. They taught one or two beginner-level classes a day and attended classes as students for the rest of the day. The youngsters were all related to Aramco employees and had been students at the ITC before being designated as trainee-teachers. Sultan was the only one of the group to become a full-time ITC teacher.

Orientation of New Employees

Training had operated special orientation centers for American employees since 1948, first on an abandoned military base at Riverhead, Long Island, New York, and later at Sidon, Lebanon. At these centers, newly hired Americans got acquainted with one another and with the Arab world in two- to four-week programs. They spent part of the day in classes on the culture and customs of the Arab world and part in Arabic-language programs led by Saudi teachers.

The first crop of 12 carefully chosen Saudi teachers traveled to Riverhead in 1948 to teach Arabic to new American employees for one year. In return, the Saudis received instruction in English, mathematics, geography and history at the local Westhampton High School. The first group included Sami Husayn Ashabib, Khalifah ibn Yousuf, Saif Uddin 'Ashur and Sa'ad 'Azman ibn 'Azman. During their year in the States, the Saudis experienced their first snow storm and visited New York City, Washington, D.C., and industrial facilities in cities such as Buffalo, Rochester and Niagara Falls, New York. They concluded their trip by touring the Ford Motor Company plant in Detroit, Pittsburgh's steel mills, and the Museum of Fine Arts in Cleveland.

In July 1951 the Aramco Orientation Center was moved from Riverhead to a new Aramco Training Center, near the ancient Mediterranean port city of Sidon. For more than a year, new employees arriving in Sidon for orientation endured lectures held in a large metal building without air conditioning. The new-hires were housed in metal huts, four men to a hut, also without benefit of air conditioning. The uncomfortable metal buildings were replaced in 1952 by a modern air-conditioned, two-story building with space for up to 100 employees. It was located a few hundred yards from the sea, where tankers waited offshore for cargoes of oil pumped through the Trans-Arabian Pipeline from Aramco fields in eastern Saudi Arabia more than 1,000 miles away. Orientation lasted between two weeks and a

Saudi instructor, Ahmed Juma'a Awadh, teaches basic Arabic to American new-hires at Aramco's Sidon Training Center in Lebanon, 1952.

month, depending on the nature of the new employee's job and how well he needed to be able to communicate with Saudis. In a month, a student would have about 96 hours of language instruction, learn about 600 Arabic words and master several dozen Arabic phrases.

The Riverhead facility had been established for Aramco by Jay Milton Cowan, head of the Division of Modern Languages at Cornell University. He continued as an adviser and recruiter for Aramco until the Sidon facility was shut down in 1957. For years, Ray Sullivan, a colorful retired New York City policeman, headed the cultural and customs section of the Sidon program. The Arabic-language program was headed from its inception by Merril Van Wagoner, a Ph.D. from the Department of Linguistics of Brigham Young University at Provo, Utah, and a former instructor in the U.S. Army's Intensive Language Program.

Van Wagoner and another linguist, Arnold Satterthwait, drew up lesson plans for the new-hires and developed a usable dictionary of Eastern Province Arabic. The two linguists devised special phonetic spellings of Arabic words, using standard typewriter keyboard symbols to represent the unique sounds used in Arabic. But constant repetition by the student of words and sentences spoken by Arabic instructors remained the backbone of the Sidon system.

Former students recall being amused by Arabic phrases that had a quaint ring to them when converted into English, phrases such as, "Bring me something good to eat," taught in both present and past tenses, and a famous greeting, "Good morning, my fellow workers!"

The Sidon facility was not only for new-hires. It was also a setting for special, intensive courses in training programs and techniques. The first of these involved seven Training Division employees, led by Emmet Roberts, in a two-week course on the new Pre-Job Training Program in September 1951.

The special Training Center for new American employees at Sidon shut down in 1957, victim of an almost total standstill in the hiring of American employees. The Suez Canal crisis and second Arab-Israeli war, coupled with an economic downturn in some major consumer nations, discouraged any increases in production and employment.

Aramco Television Used for Training

ramco linguists also developed a language program used to teach Arabic reading and writing to Saudis in the Eastern Province by way of Aramco television station, HZ-22 TV, Channel 2. When Aramco TV came on the air on September 17, 1957, a small English-language station was already operating out of the Dhahran air base.

The Aramco station was the first Arabic-language television station in Saudi Arabia and the second in the Middle East, predated only by a station in Baghdad. The station occupied the former Advanced Industrial Training Center building. Fahmi Basrawi, the one-time Jabal School teacher, was the first performer on opening night of Aramco TV. He gave a talk about the educational and recreational purposes of the Aramco station. This was followed by a showing of a feature-length movie, *Jazirat al-'Arab (The Arabian Peninsula).*

One-third of the station's air time was devoted to educational subjects. It aired instructional programs on arithmetic, English and a "Learn Arabic Through TV" series adopted from the Sidon program.

Ali Hassan Shihab cues up an Arabic-language instructional video tape for broadcast, 1963.

Basrawi hosted educational programs on Aramco television for 17 years. He declared: "About half the ladies in the Eastern Province learned to read and write Arabic from my classes. I have certification for that from people who have written me in their own hand saying how much they appreciated learning Arabic. Many of the letters were from ladies. They had never had the opportunity to learn, because there were no girls' schools at that time."

Soon television antennas sprouted from rooftops all over the Eastern Province. Saudis apparently loved cowboy programs. The Aramco station used Arabic sound tracks dubbed in by the station's staff. For the first year or so, this meant using men's voices for women's parts. When the Makkah radio station started using women's voices, Aramco TV followed suit. Americans and other non-Arabic-speaking people could listen to an English-language sound track broadcast over local radio simultaneously with the television program they were watching.

Every Sunday evening Basrawi hosted the "Tri-District Quiz," a popular program in which Saudi employees of Aramco from Abqaiq, Dhahran and Ras Tanura competed against one another. Individual contestants could win up to SR300. The districts took pride in a representative who did well in the quiz. H.H. the Governor of the Eastern Province wrote Basrawi praising the program's education value, and saying he watched it every week.

One evening Basrawi asked a contestant to name the capital of Spain. The man answered in Arabic "Ma adri," meaning "I don't know." Basrawi thought the contestant said "Madrid," which is the capital city of Spain. "Correct! You win!" Basrawi enthused. The contestant looked bewildered, and everyone in the studio laughed. Basrawi, acknowledging his own mistake, credited the man with a correct answer.

Effects of Aramco Training Spread

As with television, other Aramco training programs had an impact, directly or indirectly, far beyond the company's fences. The Local Industrial Development Department (LIDD) is one example of the process. LIDD was created to help develop home-grown, Saudi-owned and Saudi-operated, private industry. The people in LIDD encouraged businessmen to form stock companies, to assemble capital and to bid on contracts. They offered advice, and they helped stimulate the local imagination—in brief, they gave the local economy support and encouragement.

Aramco helped introduce modern farming techniques at al-Kharj, resulting in greatly increased crop yields.

Many firms trace their ancestry to a single, Aramco-trained workman with the entrepreneurial gusto to go into business for himself, usually starting as a contractor for Aramco. Early on, the company recognized the benefit of using contractors—men who could be hired to do a specific, one-time job. In 1946, a forerunner of LIDD, the Industrial Development Division, had been formed to organize and channel contractor activity. Bill Eltiste, Cal Ross and others in this division introduced Arab laborers to tools and techniques from wheelbarrows to cement blocks to workmen's compensation funds. The growth rate was extraordinary. The company paid about $2.5 million to local contractors in 1947. In 1958 Aramco paid Saudi contractors more than $11.5 million for services to the company. During the rapid expansion era in the 1970s and early 1980s, contractor payments were measured in billions of dollars a year.

**Mubarak
Al-Hajri**

*"To me, Aramco
is the best
company in the
world. I'm proud
that I'm a Saudi.
I feel that it is my
own company."*

Mubarak Al-Hajri came to Aramco in 1952, fresh out of sixth grade and eager to continue his schooling. "I was young, and I liked to study," he said. The benefits of the job were obvious: "Good salary, training, and you are going to learn English and math and Arabic all together. We didn't want to miss it!" he said.

The Industrial Training Center at Aramco was something new, Al-Hajri remembered. "The way the teachers taught was different from the Arab schools, and the guy was talking in English." Language and teaching style were not the only differences.

"I looked at the teachers," Al-Hajri said, "and the way they were dressed, in trousers. In our culture, we wear a *thawb*: the white *thawb* in the summertime and a different one in winter. I looked at those guys and I saw blue, white and other different colors, which was something unusual. I got used to it, though."

In the 1960s, Al-Hajri completed his high school education by attending evening classes after spending part of his workday in Aramco classes and driving for the Transportation Department. "I went to work and to school at Aramco during working hours," he said. "Afterward, I had to go direct to the government school. I studied for four hours every night, and I reached home at 9:00 p.m. I was young and looking to my future."

In 1976, Al-Hajri went to the United States for a developmental assignment with a trucking company. He also completed a six-month course at Arizona's American Graduate School of International Management and went on to Lee College in Texas for a one-year course, during which he learned the art of public speaking.

"Our teacher told us to stand up," Al-Hajri said. "He gave you a little subject and said, 'Continue.' You had to continue for about five minutes on that subject. You couldn't look down. I thought, 'Oh, my gosh, what's this?' The first time was a little bit difficult.

The second time, I asked the teacher, 'How long do you want me to speak, five minutes? I can give you one hour now!' I really enjoyed it. It gave me more confidence to know that I could speak to anybody, wherever. Later, when our company president came, I had to stand up and talk. No problem."

Al-Hajri, who retired as superintendent of the Southern Area Transportation Division in 1993, after 40 years of service, believed supervisors must ensure that younger employees also benefit from education. "I felt training was part of my responsibility," he said. "I felt that I had to send people who were qualified to school."

The need to learn was also a constant in his career. "When I first joined, my goal was to learn to study, to increase my knowledge, to find out about the oil business. It was interesting, and we learned a lot," Al-Hajri said, noting that Saudi Aramco's tradition of training continues. "All over the world we have people working, being trained. I don't think there is a company anywhere like it. To me, it is number one in the world."

The growth of al-Khobar indicates the results of such assistance. By 1958 it was a bustling boom town of about 33,000 people, with a small department store; an auto dealership; a soft-drink bottling plant; restaurants; banks; grocery stores; medical offices; and stores selling the latest in televisions, washing machines, refrigerators, and so on. Twenty years earlier, al-Khobar was less than

a dozen palm-thatched *barastis* housing the descendents of two families. The men who achieved this transformation were not oil men. They were hardworking Saudis who, not many years before, had been untrained, barefoot boys.

The Transportation Department's driver-training program was another example of a program that helped to change the Eastern Province. Saudis who were taught by Aramco to drive a vehicle taught their friends and relatives to drive. As the motoring public grew, so did the demand for paved city streets, more and better highways, and service stations. A market developed for automobile sales, rentals, repairs, parts

Aramco trainee takes a reaction-time test at Dhahran's driving school in 1951.

and accessories. The first Saudi-owned car rental services, featuring new Ford sedans for rent by the hour, day, week or month, opened in Abqaiq, Dhahran and Ras Tanura during 1956.

Aramco's first formal driver-training school opened in 1944 at Ras Tanura, during construction of the huge new refinery. The school was moved to Dhahran in 1947, amid pleas from upper management for a reduction in the "appalling number" of vehicle accidents. There were 140 accidents in the company's small fleet of motor vehicles during 1946, and five more automobile accidents in the first six days of 1947.

Driver candidates were given physical examinations and tested for depth perception and reaction time. Those who passed were trained in one of three categories: light equipment, meaning passenger cars and small pickup trucks; medium equipment such as bobtail tractors and buses; and heavy equipment such as the giant Kenworth trucks, the largest off-road vehicles made. Robert Modreau wrote that when he arrived in Dhahran in 1953, on special assignment as a driving instructor, there were about 100 students and only five three-quarter-ton trucks for driver training. This meant each student got about 15 minutes of actual driving-time practice a week.

First Volkswagens for Dhahran Transportation Division receive Aramco identification markings, 1959.

With more equipment and better screening of candidates, it became possible to develop in as little as seven weeks a driver with skills both for normal highway travel and for desert driving, where one might go for hundreds of miles without a hint of a road. But it required about three years of specialized training to master the complex cranes and huge trucks in Aramco's service.

Aramco reported 5.3 vehicle accidents per million kilometers (620,000 miles) of driving during 1959, a better safety record than that of drivers on U.S. highways. By this time, Aramco had more than 3,000 vehicles of all types scattered over the Eastern Province. In an economy move that same year, the company replaced the Mercedes-Benz sedans and other large cars and vans in its motor pool with Volkswagen "beetle" sedans, and Volkswagen buses and vans.

To help relieve the chronic shortage of semiskilled workers available to contractors, Aramco in late 1953

An obstacle course at Dhahran's driving school tested the ability of trainee drivers to maneuver their vehicles, 1951.

opened a training school for contractors. The facility, located alongside Dammam Road about three miles from Dhahran, began with 10 welder trainees and expanded until, by 1957, more than 300 Saudis a year were being taught the basics of plumbing, carpentry, metal working, electrical work and welding. The six-month-long courses were given free of charge. The only requirements were that trainees be Saudis between the ages of 18 and 25, and in good health. Most graduates went to work for contractors or other local companies, although a few were hired by Aramco.

In 1957 the Aramco scholarship program was amended and expanded under an agreement reached with the Saudi government. This program was in addition to the company's scholarship program for employees established in 1951. Aramco agreed to fund up to 20 scholarships in each year from 1957 through 1959 until a total of 60 scholarships had been awarded. From then on, funding at the 60-scholarship level was to be maintained. The company provided the money, but the Saudi government chose the scholarship winners. Any Saudi with a secondary-school certificate, employee or not, was eligible to apply to the government for one of the Aramco scholarships. A 1961 recipient of one of these yearly scholarships was 'Abd Allah S. Jum'ah of al-Khobar, a Dammam High School graduate, who would later become president and chief executive officer of Saudi Aramco.

Roy Lebkicher Succeeded by Robert Hall

n June 1957, Roy Lebkicher, director of General Office Training, retired. He was a geologist by education who became, in the words of one who worked under him, a "visionary" in the field of training. He was unique among training directors in that he served at the same time on the Executive Management Committee, the company's top policy-making group. He was able, therefore, to champion the causes of training at the very highest level.

Lebkicher encouraged Harry Snyder to launch the Aramco Production Training Program, the first coordinated Saudization effort in company history. He influenced the company to create a separate Training Department. He was widely believed to have been the guiding force behind a top-level training-policy statement issued in 1953 by Aramco's president, Robert Keyes. The statement declared that it was the responsibility of all personnel, at all levels, to participate in the training of a Saudi work force.

Lebkicher also became a writer of some note on training subjects. One often-quoted passage suggests the fervor of his belief in his work: "The happiness of people is not improved, nor their good will obtained, by welfare or philanthropy. But education and training, which enable a man to better himself through his own efforts, give him an entirely different slant on the nature of material progress."

Lebkicher was succeeded by Robert King Hall, certainly the man with the most advanced degrees and the most international experience in education yet to occupy the post of Training director. As an Aramco consultant in 1951, Hall investigated the then-new Job Skills Training Program and wrote a scathing critique calling the line organization trainers "amateur and incompetent" and the program itself "expensive and confusing." He had joined Aramco in 1955 as assistant director of Training, leaving a post as professor of education at Columbia Teachers College in New York City.

Harry Snyder, left, and Robert King Hall, right, watch Shaykh 'Abd al-'Aziz Turki sign documents opening an Aramco-Built Government School in Safwa, 1958.

Hall was born in Kewanee, Illinois, the son of a minister. By the time he reached Aramco, he had accumulated numerous degrees, including a Ph.D. from the University of Michigan and master's degrees from the University of Chicago, Columbia University and Harvard. He had conducted educational research in Asia, Africa, Europe and Latin America, and his writings had been published in many educational journals. From 1952 until 1957, he was a co-editor of the *Yearbook of Education*, published in London.

Soon after taking office as director of Training, Hall got off on the wrong foot by declaring that no one without a doctorate degree should be working in Training. Almost no one in Training had a doctorate degree. Hall riled other executives by having his name and title, Dr. Robert King Hall, painted on the door to his office. He was told to have the title removed. He irritated callers by answering the phone, "Doctor Hall." Many Aramcons in other departments had doctoral degrees and did not call themselves "doctor." The use of such titles was not in harmony with the feeling of easy companionship in the expatriate community. Hall tended to be brusque in manner, and pedantic. Once he sat a group of trainers in Abqaiq down in front of a blackboard and proceeded to diagram and lecture about the workings of a 45-caliber revolver. The impression was that he did this for no reason other than to show that he could.

At the same time, those who worked with him say he was a brilliant educator. He accomplished much that was good for Training. Until Hall arrived, there had been no uniform set of standards. Course materials depended on the interests and abilities of the individual instructor. The length of time an employee needed to reach a certain skill level varied from organization to organization. Training personnel experienced a high rate of turnover, with the result that training materials were constantly being altered. Instructional materials that had been standardized for years in the United States or Europe were written, then rewritten, adopted, then discarded, seldom getting fully utilized. Under these conditions, it was difficult to keep records and nearly impossible to measure the actual performance of a program. Hall saw to it that training programs were codified, written and arranged in the same manner for all three districts.

New Job Skills Training Program

Hall's investigation of the Job Skills Training Program ultimately contributed to a historic change of direction in Aramco training. Between 1955 and 1957 the company made five detailed surveys of Job Skills Training by line organizations in a total of 123 different activities. These surveys determined the staff time needed to prepare manuals, design lessons, write teaching instructions, coach line organization trainers and make appraisals of

Saudi pipe fitters discuss metal work with Henry Schrader in the Dhahran Job Skills Training Shop, 1958.

each employee's training progress. The overhead and unit costs were evaluated, the efficiency of the program appraised. Finally, on March 10, 1958, management announced a new Job Skills Training Program. In essence, it was a return to vocational shops training, the direction in which the company had been going before the Aramco Production Training Program was adopted in 1949. The new program contained six main features:

- Industrial Training Shops (ITSs) would be opened in each of the three districts.
- All employees enrolled in the ITSs would be trained in basic skills common

to all trades, such as use and identification of tools, measuring, and blueprint reading. This was similar to the Core Training concept used at the vocational training facility opened in Ras Tanura in the late 1940s. Trainees would learn the manual and mechanical skills needed for such jobs as machinists, electricians, instrument repairmen, plumbers, pipe fitters, welders, air-conditioning and refrigeration mechanics, and machinery maintenance mechanics.

- Training would be done by fully qualified instructors. Professional vocational training teachers would lead classes in a number of different crafts. There would also be some specialized instructors. Qualified craftsmen or operators from line organizations would be assigned full or part time for specialized training in areas such as electronic controlled instruments.

- The ITSs would operate under the full control and administration of the Training Divisions in each district. Previous experience showed that the majority of line organizations were too busy meeting the daily demands of their operations to give adequate time to training. Greater efficiency and accountability could be expected by putting responsibility for the program on the training professionals in each district.

- Standards of training and achievement were to be relatively uniform in all parts of the company.

- The ITSs would prepare employees to meet the requirements of a full craft, not an isolated job skill. Training would aim to develop those abilities needed to perform company work. Course curriculum would be modified from time to time as company needs changed.

The new Job Skills Training Program paralleled the program that had been tested and proved effective in academic training at the ITCs. Employees would be assigned to an ITS on company time until they either achieved a targeted skill level or failed to reach that level in a reasonable amount of time. Training would be on a part-time basis, alternating with work either daily or, occasionally, on a quarterly basis.

The new ITSs were scheduled to be in full operation by year-end 1959. They were located in rather makeshift facilities, since Training no longer had buildings specifically designed for shop training. In Ras Tanura the common Shops Building was nearly doubled in size to make space for Job Skills Training. Reginald Strange was supervisor of the unit. At Abqaiq,

Saudi trainees in the metal working section of the Advanced Training Center, Dhahran, 1952.

Job Skills Training operated at first out of the former Saudi cafeteria, but soon moved to the Maintenance and Shop Building. Nicholas Bruey ran the Abqaiq ITS. An ITS was established in the Consolidated Shops building at Dhahran under the supervision of Paul Lightle, a veteran of 20 years in the industrial training field. Lightle wrote and brought with him to Saudi Arabia a primer on blueprint reading.

Aramco Approach to Training Evolves

Aramco entered the decade of the 1960s with a two-pronged training approach. Each of the two approaches, academic training in the ITCs and vocational training in the ITSs, had its own special focus. ITC programs featured courses similar to those offered in American public elementary and high schools. ITS programs were narrowly focused on specific job skills, much like traditional vocational schools. On average, more than 2,500 employees a month attended classes on company time at those facilities in 1960.

79

Of these trainees, nearly 20 percent were enrolled in courses comparable to those offered in secondary schools. About 2,400 more Saudi employees were enrolled in voluntary classes held at the training centers after working hours.

The company continued to build and fund schools for the government. Under the original 1953 agreement, Aramco agreed to build 10 schools, but by the end of 1959 it had already built 11 schools, and a 12th school, the first intermediate school in Dammam, was on the drawing board.

The new Saudi Development

Paul Case, center, confers with four Aramco U.S. scholarship students: (left to right) Ali I. Al-Naimi, Mustafa al-Khan Buahmad, 'Abd Allah S. Busbayte and Hamad Al-Juraifani.

Program sought out gifted young Saudi recruits and guided them through training programs leading toward top-level professional careers in the oil industry. During 1959 Aramco sent seven Saudi employees to colleges and universities in the States. In addition, the company funded 60 scholarships for Saudis who were chosen by the government's Ministry of Education for college and university training.

On July 14, 1959, four young employees left Dhahran to attend colleges and universities in the United States under the Saudi Development Policy Committee's scholarship program. It was a noteworthy group comprised of Ali I. Al-Naimi, Hamad A. Al-Juraifani, Mustafa al-Khan Buahmad and 'Abd Allah S. Busbayte. All had distinguished careers before them, Al-Naimi, Al-Juraifani and Buahmad with Aramco, Busbayte as a businessman. But at the time they were just four "very, very excited" young men. They stopped for a few days in New York, visited Aramco's office and toured the city, by far the largest city any of them had ever seen. Among the sites they visited with an Aramco host were the United Nations, the Empire State Building and the Statue of Liberty. Next they visited Washington, D.C., as guests of Aramco's Government Relations Office. The group went to Bucknell University at Lewisburg, Pennsylvania, for a six-week orientation session and to brush up on their English. Then they dispersed to their individual schools. Al-Naimi went to Lehigh University in Bethlehem, Pennsylvania, to study geology. Al-Juraifani stayed at Bucknell University and studied mechanical engineering. Buahmad studied business administration at Antioch College in Yellow Springs, Ohio, and Busbayte went to Allegheny College at Meadville, Pennsylvania, as an engineering student. The company paid all their expenses, and, in addition, each student received his full pay as an Aramco employee.

By the end of 1959, more than 1,000 Saudis had completed Basics of Supervision training, a prerequisite for Saudis seeking supervisory positions. There were 327 Saudi first-level supervisors and 200 more in training. Five years earlier, only 17 Saudis had completed the first-level supervisor course. Nearly 72 percent of the work force was Saudi, and 76 percent of them held semiskilled, skilled and supervisory or professional jobs.

The company devoted more than one million man-hours to training for the first time in 1959. In 1960 the company allocated nearly $5.5 million for educational programs. The total spent on training was about $1.2 million a year higher than it had been in 1952, even though the company had 10,000 fewer employees in 1960 than it had in 1952. Upon the results of this investment depended the future of a Saudi-American partnership.

First Saudis Named to Board of Directors

s Aramco grew during the 1950s, so did the Saudi government. Along with the gradual development of a modern state bureaucracy, there grew the notion that the government should not only receive money from the oil industry, but also have a hand in running it.

The first small step in this direction was taken in San Francisco on May 20, 1959, when two Saudis were elected to the company's board of directors. One was 'Abd Allah H. Al-Turayqi, a geologist with a master's degree from the

Aramco Board of Directors with first Saudi members, Dhahran, 1965.

University of Texas, formerly the government's Director of Oil Supervision in the Eastern Province and then Director General of Petroleum and Mineral Affairs in Jiddah. Only a month earlier he had headed the Saudi delegation to the First Arab League Petroleum Congress at Cairo. The second Saudi elected to the board was Hafiz Wahbah, an elder statesman, Saudi ambassador-at-large to the Arab world, former representative of Saudi Arabia in Britain, and a delegate to the 1945 United Nations Conference in San Francisco.

A Time of Transition 1960-1970

"Perhaps the biggest challenge we face in the next few years is in training and developing Saudi Arab employees. We have done a commendable job in training Saudis for performance-level jobs and for first-level supervision. Continuing aggressive action is required to achieve similar accomplishments in moving Saudis into professional and management-level positions."
– Tom Barger (1963)

en Beach was only two years short of the company's mandatory retirement age of 60 when he reluctantly accepted the job as director of Training for Aramco. He was completely satisfied being director of Aramco's Management Development. It took some arm twisting by Norman "Cy" Hardy, chairman of the board, and Tom Barger, president of Aramco, to convince him to give that up and move back into Training.

"I wasn't a bit anxious to make that move," Beach recalled some years later. He first came to Saudi Arabia in 1952 as coordinator of Management Training, but he left Training in 1955 to head the newly created Management Development office, reporting directly to Robert Keyes, then company president.

Beach became director of the General Office Training Department on August 15, 1960, replacing Robert King Hall, who had already returned to the United States. G. Edward McSweeney, the acting director of Training, was named assistant director of Training.

There were no mandates or statements of goals from top management to guide Beach when he took over the Training Department. The fact that Beach, a management trainer, was chosen to lead the department indicated the direction Training was expected to take.

"I pretty well knew what ought to be done, and what should be done," Beach said. "They weren't doing enough to bring along Saudis. We had to identify high-potential Saudis, educate them and build them up so they could fill key management jobs." So he devised training programs designed to do just that.

Beach wanted his trainers to conduct supervisory training conferences with Saudis, but "I couldn't find anybody who had time to do it, so I started doing it myself." He formed discussion groups composed of 15 or so high-potential Saudis. They talked about common supervisory problems. In one session they made a list of all the duties

Opposite: Aramco employees at the Dhahran ITC study English using the Learning Laboratory System. Below: Senior-staff employees in a Basics of Management course in Dhahran, 1962.

83

An electronic instruments class at Ras Tanura, 1969.

and responsibilities of a supervisor. In another they analyzed tasks such as handling discipline problems, breaking in new employees and informing new people about Aramco's organizational structure. After each meeting, the main points were typed up, put into a manual and distributed to conference participants.

In the autumn of 1960, Beach brought Frank Jarvis into the General Office at Dhahran from Abqaiq, where he had been superintendent of Training for more than 10 years, and put him in charge of Management Training. Jarvis had developed the company's first Management Training program in the late 1940s. His classes had been attended by most of Aramco's top executives, including the new president, Tom Barger. Jarvis appointed Jim Allen to work with the various districts on advanced training for Saudis, and made George Larsen (no relation to the vice president of the same name) responsible for preparing Saudis for out-of-Kingdom training assignments.

Larsen's counterpart in the United States was Anthony "Bob" Brautovich, an Aramco recruiter based in New York City. He was assigned to locate suitable schools for Saudi students in the States and to watch over the students during their out-of-Kingdom training. At the time, there were only about 30 Saudis on out-of-Kingdom training assignments. The job was "an IAOD assignment," Brautovich said. "That means 'In Addition to Other Duties,'" but it grew to become his main job for 17 years.

Transition to a Saudi Work Force

Aramco was under increasing pressure from the Saudi government and the owner companies to replace foreign industrial workers with Saudis, both as a cost-saving measure and to improve external relations. By 1960, Saudis made up 75 percent of the work force, the highest percentage since the World War II years when operations were virtually shut down. A sluggish oil market caused the company to lay off thousands of expatriate workers. Aramco had 15,310 employees worldwide in 1960, 10,000 less than in 1952, the year of peak employment. In the previous eight years there had been a 22 percent decline in the number of Saudi employees, a 33 percent drop in U.S. personnel and a plunge of 63 percent in third-country workers. Foreign workers had been laid off, or "surplused," but the reduction in Saudi employees was entirely due to normal attrition, the company said.

Saudis had taken over many jobs vacated by expatriates. Jobs now being handled mostly by Saudis included work in building trades and metal crafts, and at drilling rigs, fire stations, pump stations, gas-oil separator plants, gas injection plants and most refinery units. Warren Hodges, superintendent of Training at Ras Tanura, called it Saudization by the "vacuum theory." The theory stated that the vacuum created by the departure of fully trained American and other foreign-country workers would by necessity be filled with trained Saudi workmen.

In 1960 the average Saudi had been with Aramco for 12 years and received an annual salary of $1,740 (SR7,830 at 4.5 riyals per dollar). This was the last generation in which the majority of Saudi employees who came to work for Aramco were illiterate. An estimated 40 percent of Saudi employees were illiterate in 1960, compared to an illiteracy rate of nearly 85 percent little more than a decade earlier.

The 11,100 Saudi employees at Aramco included an elite corps of 50 senior-staff Saudis, equal in status, but not pay, to Americans. The number of

84

senior-staff Saudis had increased by 47 since the first three Saudi employees reached that level a decade earlier. In 1960 this group included three division heads, the top-ranking Saudis of that time. They were Saif Aldeen Ashu, head of the Arabic Press and Publications Division in the Public Relations Department; 'Abd al-Fattah Kabli, head of the Customs Division of the Material Supply and Traffic Department; and Muhammad Younis, head of the Claims and Compensation Division, Industrial Relations Department. Most senior-staff Saudis were assigned to Government Relations, Industrial Relations or Public Relations. Three Saudi senior staffers worked in engineering, but none were in exploration, at the refinery or in other organizations vital to the production and distribution of oil.

Industrial Training Shops class in Dhahran.

The company's new top man, Tom Barger, was an agent for improvement in Saudi prospects. Barger was named president of Aramco on May 20, 1959, and chief executive officer on December 1, 1961. He came to Saudi Arabia in 1937 as a geologist, but from 1941 on he worked mainly in Government Relations. Barger was a man of intellectual vitality and diversity. He was fluent in Arabic, a scholar of Middle East history, an amateur archeologist, a photographer and a business executive who could tear down a motor and repair it. He was uniquely popular with Arabs and Americans. As the U.S. Ambassador to Saudi Arabia, Parker T. Hart wrote of Barger some years later, "That an American could have a social evening with half a dozen Arabs, at which Arabic was the only language used, is remarkable. His popularity was due in no small measure to his personal humility, his eagerness to listen and a complete absence of personal arrogance."

Tom Barger, 1957.

In the early 1940s Barger compiled an 83-page, 20,000-word document explaining why the company's very existence in the future would depend on the quality of its relations with the Arabs. He never lost that conviction or his sensitivity to the aspirations of Saudis. As Barger moved up through the ranks of Aramco management, his beliefs and feelings began to permeate the company with increasing force.

Men like Barger and Frank Jungers saw the common sense of moving Saudis up in the ranks. Jungers was one of a new breed of "technocrats" who joined the company after World War II. With Barger's support, they were more willing than older hands had been to push Americans and other expatriates aside when necessary to make room for qualified Saudis.

As early as 1952, Jungers, then a 26-year-old superintendent in Maintenance and Shops at Ras Tanura, ordered a one-quarter reduction in foreign contract labor by year's end and a corresponding increase in Saudi workmen in the shops. The warnings of foremen that the quality of shop work would decline was proved wrong. "The Saudis stepped in with a will and surprising speed and flexibility," Jungers said. "They had, in fact, been well trained by the supervisors who were reluctant to use them."

By 1960 Jungers was in Dhahran as general superintendent of Mechanical Services and Utilities, in charge of more than 1,000 workers. That's when Bill O'Grady, then principal of the Dhahran ITC, got a firsthand look at Jungers' particular style of Saudi utilization.

One day O'Grady was summoned by George Trial, then superintendent of Dhahran District Training, and told to attend a meeting Frank Jungers was having in the conference room at the Transportation Yard. "I was told by Trial to go there and find out what Jungers wanted and give it to him," O'Grady said.

"I went to the meeting and found that Jungers had assembled not just transportation people but all of the supervisors, almost exclusively Americans, who were involved in the operation of the main shops area. Jungers told them that we were going to make the best use of the Saudi talent we had."

A list was developed with names of Saudis who had missed out on training for one reason or another, but still had a good many working years in front of them. They were people who might be candidates for promotion if given some more English or mathematics. Jungers told O'Grady to prepare appropriate courses especially for these people.

"I set up classrooms and picked who I thought were the best teachers to handle these people, in spite of the fact that we might have just one, two or three people in the classroom," O'Grady

A visiting U.S. diesel-engine maintenance specialist speaks to industrial trainiees.

said. "These students were not tested at the regular time; they were tested whenever the teacher who was tutoring them felt they were ready to take a test for a particular level."

O'Grady attended Sunday staff meetings during which supervisors reviewed the progress of all Saudis in the division. "Jungers made them keep their noses to the grindstone, kept seeing to it that these people were properly enrolled, saw that the test results were made known, saw that there was progress on the job and learned who was being promoted and who could be transferred and promoted. I was able to see that Maintenance Shops had the best Saudi development program in existence, and that it was producing results. And I saw supervisors who had been fairly disinterested in the first instance who were now knowledgeable and motivated to go ahead and do a proper job in the development of Saudis."

From the Maintenance and Shops Division, Jungers singled out a welder, 'Abd al-Monim 'Abd Allah, for special training and made him a foreman *(muraqib)* — the first Saudi craftsman to reach that level in the industrial work force. In all, he promoted seven of the first 12 Saudi craftsmen to make foreman. All seven were advanced to senior-staff status effective December 1, 1961. In addition to 'Abd Allah, they were Hilal ibn Ali, Salih Salem, Farhan Husayn, Salim A. Bawazir, Khalil Younis Ali and Sa'ad Al-Subayt.

Jungers "did an outstanding job and deserves every commendation," O'Grady said. "But when Jungers moved on to other assignments, the program ran out of steam."

Jungers went on to become board chairman and chief executive officer of Aramco. He headed the company during some of the most difficult times in its history, including the period of the so-called oil embargo in 1973. His post was described in *Fortune* magazine as "one of the most delicate positions in all industry." But Jungers continued to believe his leadership in Saudi development outweighed all his other accomplishments.

A Training Status Report (1962)

n the summer of 1962, Beach brought members of Training's General Office staff to a meeting of the Executive Management Committee and department heads in Dhahran for a detailed report on training activities.

Ali Ahmad Saleh

"I liked the challenge and the competition with the other students. That made it fun."

 s a young boy growing up in Jiddah, Ali Ahmad Saleh heard exciting stories about Aramco from two relatives who worked in the Eastern Province, one for the railroad and the other for Aramco. In 1955, equipped with a third-grade education and accompanied by an uncle, Saleh left Jiddah for Dhahran and Aramco. He went with the encouragement of his mother and despite the protests of his father.

Over the next seven months, Aramco recruiters consistently rejected his job applications, telling Saleh he was too young to work for the company. Weary of rejection, he traveled to Abqaiq, where his Aramco job application was accepted.

On May 14, 1956, Saleh began work as an office boy in the main office of the General Construction Department. Two weeks later, he enrolled at the Abqaiq Industrial Training Center (ITC), where he studied English, math and science.

When Saleh completed the basic ITC courses he asked his supervisor for permission to take advanced ITC classes. The request was denied so, driven to acquire all the education he could, Saleh enrolled in the Company's volunteer school, where he attended classes for three hours each evening after a full eight-hour workday.

In 1965 Saleh qualified for the Company's out-of-Kingdom scholarship program. He was sent to Temple University High School in Philadelphia to complete his high school education and improve his English-language skills. He was in the first group of Saudis to go directly from ITC classes at Aramco to the United States, without taking intermediate stage courses at other schools in the Middle East. Saleh completed high school at Temple and entered Arizona State University at Tempe, Arizona, where he earned a degree in electronic engineering.

He returned to Aramco, completed Training's five-year-long Professional Development Program (PDP), and advanced through a series of management positions.

In 1992 Saleh became senior vice president of Manufacturing, Supply and Transportation, and later senior vice president of the Refining and Distribution Department.

Saleh credits his ITC teachers with having a dramatic influence on his career. While Saleh remembers Saudi schools being more rigid than classes at ITC, that was not to say that ITC was easy — far from it, in fact. He thought the Abqaiq ITC staff, including the principal, Larry Emigh, was tough but fair.

Saleh knew what he needed to learn in order to qualify for the next-higher-level job. It was a powerful motivation. "Since I was in the Professional Development Program, as soon as I made the next ITC level, the company linked it to a job move. I knew that, and I worked hard for that." In doing so he realized his goal of moving into higher levels of management.

Looking back over his successes in training, Saleh advises younger Saudis to study hard. "Make sure you have a good education and work hard in everything you do, and things will fall into place," he said.

It was a rare opportunity to present Training's message firsthand to top management. The various reports amounted to a status report on training in 1962. The staff members and the subject of their reports to management were as follows: James Mileham, Industrial Training Centers; Don Richards, Industrial Training Shops; Ed Thompson, Management Training; Bill O'Grady, Saudi Development Program; Vince James, senior-staff schools; and Harry Snyder, Saudi Arabian Government/Aramco Education Services. A summary of those reports follows.

Industrial Training Centers: The three district ITCs had 2,600 students in assigned training, 1,400 in voluntary classes, and a staff of about 130. ITC enrollment had leveled off and was expected to decline as the work force continued to shrink. There were a total of 79 classrooms in the three district ITCs, and 20 other classrooms in outlying areas. The ITCs offered a job-related curriculum including eight years each of English, Arabic and mathematics; five years of science; two years each of history, geography, drafting and blueprint reading; and one-and-two-thirds years of typing. The upper grades were considered to be at the same level as the average U.S. high school. A major concern: "Reduction of American teaching manpower ... is undoubtedly reducing the quality of our service," Mileham said.

Job Skills Training and Industrial Training Shops: The three district ITSs had a total of 328 students in 42 shop classes. The majority of students were experienced Saudi workmen taking specialized short courses in craft work such as bottom-hole testing, instrument applications, sketching and blueprint reading, and electrical theory. For high-potential employees with at least six years of public-school education, the ITS offered an accelerated two-year course in general mechanical subjects. A person completing this course along with ITC level four programs in English, math and science might be classified as a beginning helper. One more successful year of ITC and specialized craft training at the ITS, along with on-the-job experience, qualified the trainee as a full job craftsman. Major concerns reported by Richards were: getting trainees prepared academically for ITS classes, training craftsmen who were used as instructors how to teach, and following up at the job site to make sure ITS teaching was practical and usable.

Crane operator training.

Management and Special Training: There were now five basic management training courses covering a broad range of Saudi employees, from helpers with supervisory potential, to supervisors *(mushrifs)*, up to full-fledged foremen *(muraqibs)*. The education level in this group ranged from illiterate men with no schooling to high school graduates. The beginning management training course, Essentials of Leadership, was for men with supervisory potential. The most advanced course, Management Responsibilities at the Unit Level, described the duties and responsibilities of foremen as outlined in the company's organization manual. Part-time supervisory training courses for foremen, previously confined to Dhahran, were offered at Ras Tanura and Abqaiq for the first time during 1962. As yet there were no training programs for Saudis with a college education. Most of their training was being done on the job by a supervisor. But it was time, Thompson said, to develop a management training course for college-educated Saudis. Although the number of college-educated Saudis in Aramco was small, about 50, the number was expected to grow rapidly in the near future. In addition, Thompson said, each college graduate should be included in the Management Development Program, where they could be appraised and where development plans could be drawn up

88

for them. Thompson expressed concern about the lack of well-qualified American management trainers. The only experienced American management trainers in the districts were Frank Jarvis in Dhahran and Al Lampman at Ras Tanura. The districts were trying to conduct management training using Saudi instructors.

Saudi Development Program: The aim of this program was to develop high-potential Saudi employees. There were 56 Saudis on out-of-Kingdom training assignments in 1961, an increase of 27 over 1960. The list included 29 Saudis in college courses, 21 in one-year preparatory training for college work, and two in short-term work training programs in the Middle East. In 1961, for the first time, Saudis went out of the Kingdom (other than to Bahrain) on job assignments — three for on-the-job training in U.S. oil fields, and one to the Beirut area to gain experience in public relations. Major concerns were the development of programs to assimilate returning college graduates into the work force, and, what O'Grady called "the much more difficult" problem of handling young men who failed to successfully complete out-of-Kingdom training.

Senior-Staff Schools: Enrollment at Aramco's senior-staff schools totaled about 1,150. The number of American students in senior-staff programs, including students attending colleges and high schools outside Saudi Arabia, nearly equaled the number of American employees on the payroll. Major concerns included the cost of education and the rising number of Saudi senior-staff employees with children in Aramco senior-staff schools. There were 30 Saudis enrolled in senior-staff schools and 90 expected by 1964. There was already pressure, James said, to modify the curriculum to more closely fit Saudi Arab cultural needs.

Saudi Arabian Government/ Aramco Education Services: Since the program began in 1953, the company built 15 government schools with a capacity for 3,900 students at a total cost of $5.25 million. To meet its obligation to provide schools for the sons of Muslim and Arab employees, the company would have to build 12 more elementary schools in the 1960s. In addition, the company had reached an agreement with the government to construct intermediate schools for boys, as well as both elementary and intermediate schools for girls. Snyder estimated that the company would have to build five boys' intermediate schools, 22 girls' elementary schools and nine girls' intermediate schools. The total of 48 additional schools would cost about $23 million.

Beach was ebullient following these presentations. The training reports had been well received by management. Beach felt the relationship between the Training Department and upper management, which had been strained during the Hall era, was improved and the morale of department employees had been given a boost at the same time. Beach invited trainers to his house for dinner. During the evening, O'Grady remembers, "Beach was very upbeat. He was happy and he was effusive in his comments and compliments to the group. It was my clear impression he was delighted at the way things had gone."

Later that same year, 1962, Beach resigned to take a job with the Compensation and Executive Development Department of Standard Oil Company of New Jersey. "I knew I was going to be out of Aramco when I was 60. I took the Standard Oil job so I could work until 65," he said. Although his tenure as director of Training was only two years, it had been an effective transition period. He had inherited a dispirited organization. When he left, a measure of enthusiasm and respect had been restored.

John Pendleton Becomes Director of Training

s a replacement for Beach, the company selected a distinguished Rhodes Scholar, John Pendleton. He had been a professor of English at the University of Rochester, an Army Air Force officer in the European theater during World War II and an academic assistant to the president of the University of Alabama. Friends described him as an erudite and cultured man, with a dry, friendly wit expressed in the measured accent of his native Virginia.

Pendleton joined Aramco in 1956 and came to Dhahran as coordinator of Arabian Research in the Government Relations Organization. He was administrator of the Government Relations Policy and Planning staff prior to taking the training job. Pendleton had impressive credentials, but he had moved in Government Relations circles and so was not known to the people in Training. When the announcement of his appointment was made on October 3, 1962, some trainers asked, "Who's Pendleton?"

Although he had little industrial-training experience, Pendleton was a seasoned administrator and a man of stature within the company. There was some tension, even hostility at times, between Training and the line organizations. Training needed a capable defender. O'Grady likened it to finding a man to fill a key position on a football team. "Management was faced with the need to come up with someone to run the Training Organization who would be of sufficient stature to hold the organization together and back the line people off the line of scrimmage."

John Pendleton, 1962.

Saudi Roles Continue to Grow

By 1962, Saudis held one-third of the company's supervisory positions — 467 out of 1,416 — an increase of 150 Saudi supervisors during Beach's two-year term. There were the three Saudi division heads, men recognized more for their technical knowledge than administrative skills. All other Saudi supervisors, however, were either foremen *(muraqibs)* or segment-level supervisors *(mushrifs)*, the two lowest supervisory rankings in the company. No Saudis held positions as section leaders, department heads or executive-level employees. Saudis held only 15 percent of the foreman positions in the company. Saudis dominated at the supervisor level, the lowest-ranking management position. Fifty-one percent, 406 out of 799, of the segment supervisors were Saudis. It was Saudis supervising other Saudis in almost all cases.

Many Americans, especially executives of the owner companies, still pictured the Saudi-American partnership at Aramco as a pyramid with Americans at the top and Saudis in the lower-level positions. In their view, Saudis might advance to foreman or occupy middle-management jobs, but key management positions could be handled only by Americans. This attitude began to change as events outside Aramco made it increasingly certain that Saudis would one day move into top levels of the company.

These events took place on an international stage. In 1959, and again in 1960, Aramco and other major oil companies cut the posted price of oil — the price on which royalties and taxes were figured — without consulting the governments of the oil-producing nations. The price of oil fell from $2.12 a barrel in early 1959 to $1.84 a barrel by mid-1960. Oil industry experts estimated the cuts would cost Saudi Arabia and the three other major oil-producing countries of the Middle East about $235 million a year in lost revenue.

A Saudi student at a company library.

The Arab oil-producing countries were united in their alarm over the loss of funds and at the unilateral nature of the cuts. The oil companies' explanation, that the price was reduced because of a surplus of oil in the world market, did not satisfy the Arab community, nor did it explain why the cuts were made without consulting the oil-producing states. Arab producers were determined to have more say in the policies of the foreign-owned oil companies operating within their borders. In 1959, members of the Arab League organized the Arab Petroleum Congress, the first forum where delegates of Middle East countries could meet to discuss oil-related policies. This was followed by the creation of the Organization of Petroleum Exporting Countries (OPEC) in August 1960, a coalition open to all oil-producing countries in the world.

Specially designed simulators were used to help train refinery operators in Ras Tanura, 1965.

The emergence of these organizations was a cause for concern in Aramco management. Some of the oratory during the Third Arab Petroleum Congress, at Alexandria in 1961, left Aramco executives convinced the company was in a battle for its life. Mel LaFrenz, who became director of Aramco's Management Development Program in 1963, explained: "It became apparent Saudis were going to go beyond the supervisory level because it was considered a necessary move to lengthen our concession. The one thing we needed on our side was able Saudi Arabs who were in positions with considerable responsibility and felt they were treated right, and wanted to continue in this manner."

The company began to move faster than ever before to prepare Saudis for positions of greater responsibility. Between 1960 and 1963 the number of Saudis on out-of-Kingdom training assignments rose from 29 to 110. In the same three-year period, the number of Saudis at the senior-staff level increased from 50 to 110. In addition, the company began to examine its organizational structure, in order to redesign job requirements and otherwise facilitate the movement of Saudi Arab employees into higher-level jobs.

The accelerated promotion schedule benefited employees from other Arab countries as well as Saudis. In April 1961, the company suddenly elevated about a dozen teachers from Middle Eastern countries other than Saudi Arabia to senior-staff status at grade code 11. They were the first teachers of Arab extraction to reach that grade code. One of them was Khalil R. Nazzal, a native of Palestine and a graduate of Durham University in England, who began teaching English at the Dhahran ITC in 1960. Nazzal and other university graduates among the Middle Eastern teachers had been hired at the grade code 10 level. They quickly became resentful at not having the same senior-staff privileges as the grade code 11-plus, university-trained American teachers enjoyed.

"Several of us in the Dhahran ITC had applied to teach in the Sudan," Nazzal recalled. "We were about to leave when we got the surprise of our life. Frank Jungers (then acting general superintendent of Industrial Relations) came to the school and took three of us into an office. He closed the door and said, 'You three have been appointed to senior-staff status.' It was like someone coming up and saying, 'You've been made a lieutenant general.'"

Nine or 10 other Middle Easterners teaching in Abqaiq and Ras Tanura were promoted that same day in the same unceremonious manner. After that, every bilingual teacher classified as an advanced ITC teacher was recruited as a grade code 11.

One of the first things Nazzal did to take advantage of his senior-staff appointment was to find housing on the Aramco compound in Dhahran. He was never told why he and the others were elevated so abruptly to the senior-staff level. Jungers, many years later, said, "They were selected because of outstanding performances and bilingual qualities. We wanted to equalize their grades with Americans but were careful to raise grades only on those who had certain high-level qualifications."

An instructor, center, and trainees in electronics, 1964.

In 1962, a Western-educated attorney, Ahmed Zaki Yamani, was named director of Petroleum and Mineral Resources for Saudi Arabia. He adopted a moderate approach in government relations with Aramco. In a 1963 speech, Yamani said the country's economy was based on free enterprise, and therefore any forced Saudization of Aramco would upset the economic balance.

The company's worst fears were eased, but the longer-range requirement remained. Barger, in a letter to key management personnel in 1963, wrote: "Perhaps the biggest challenge we face in the next few years is in training and developing Saudi Arab employees. We have done a commendable job in training Saudis for performance-level jobs and for first-level supervision. Continuing aggressive action is required to achieve similar accomplishments in moving Saudis into professional and management-level positions."

The company set out to identify Saudis who might one day occupy top positions in Aramco. Tom Collier, director of Management Development, reported to the Executive Management Council in September 1963 on employees who seemed to have the potential to fill key management positions. His list included Saudi Arab names for the first time. Collier reported that 190 Americans and 20 Saudis had been identified as candidates for the top 210 jobs in the company.

"Currently," he said, "we have over 100 senior-staff Saudis. About one-third of them are foremen in nonprofessional areas. On average they have significantly less than a secondary-level education; however, about one-third of the group are college graduates. Careful appraisals by supervisors indicate that the most optimistic forecast would be that with additional experience and development, 20 of the group have the potential to advance into key positions."

Saudis may not yet have reached the point where they could claim the qualifications to occupy key management positions, but neither could American management any longer feel as if it had a permanent monopoly on positions at the top of the corporate pyramid.

Aramco's Relationship with the Saudi School System

In the 1960s, Aramco began receiving more and better-educated recruits from the Saudi school system than ever before. During his tenure as Saudi Arabia's first education minister from 1953 to 1960, HRH Amir Fahd ibn 'Abd al-'Aziz Al Sa'ud (who became King in 1982) reorganized and modernized the school system.

Early in his ministry Amir Fahd got Aramco to sign an agreement under which the company was obliged to build and fund the operation of public schools for the children of its Arab and Muslim employees. Amir Fahd also negotiated an agreement in which Saudi Arabia and other members of the Arab League established a standard number of years for each level of schooling, from elementary to college. The agreement called for a six-year elementary school cycle, a three-year term of intermediate schooling, three years of secondary schooling, and a four- to five-year higher-education program. During his ministry, the government initiated a scholarship program through which college degrees were earned by hundreds of bright young Saudis, including several men who went on to have distinguished careers as Aramco executives. One of them

The Aramco-Built Government School for boys at Madinat Abqaiq, 1959.

92

of them was Ali M. Dialdin, who eventually became the first Saudi general manager of Training and Career Development. Another was 'Abd Allah S. Jum'ah, later president and chief executive officer of the company. When Amir Fahd became minister in 1953, there were only 321 schools and 41,678 students in the Kingdom. By 1960, his last year at the Ministry of Education, there were 789 schools and a total enrollment of 114,176 boys. The schools were so new, and the expansion had been so rapid, that 99 percent of all students were still in elementary school.

Aramco television broadcast Hashim Budayr's chemistry lessons to Saudi secondary schools in Dammam and Hofuf, 1964.

Saudis who did not complete elementary school or high school were often referred to as "dropouts." In many cases it was an unfair label. In Western terms, "dropouts" meant persons without the desire or capability to continue in school until they earned a diploma. For many Saudi youngsters growing up in the 1950s and '60s — even the '70s in some remote areas — there simply were no schools within a reasonable distance of their home. Or, if there was a school, it might not provide all the levels of instruction required for an intermediate-school or high-school diploma. It wasn't until the 1980s that the government school system was developed enough so Aramco could require new-hires to have a high-school diploma.

In 1959 the government asked Aramco to establish intermediate schools in the Eastern Province for boys in grades seven through nine. It would be, the government argued, merely an extension of the 1953 agreement to construct and pay operating costs of elementary schools for the sons of Arab and Muslim employees. The company did not contest the point, and in a letter to the deputy minister of education, Shaykh 'Abd al-'Aziz ibn Hasan, on March 24, 1959, Aramco agreed to build and fund the operation of intermediate schools. The company's first two intermediate schools for boys, each with a capacity for 150 students, were built in Dammam and Hofuf for a total cost of $721,000 and turned over to the government in 1961.

The next step was girls' schools. The drive to open public schools for girls, a momentous step for Saudi Arabia, gathered force under Amir Fahd's tenure. As early as 1956, one of King Sa'ud's advisers, Shaykh 'Abd Allah ibn 'Adwan, was quoted as saying the country's number one social problem was the lack of education for girls. The problem was causing newly educated Saudi men to seek educated wives from other countries, he said. This feeling echoed throughout the country, especially strongly in the Eastern Province. In October 1959, King Sa'ud announced his resolve to open schools for girls. After that, it was only a matter of time before Aramco became involved. On September 3, 1961, in response to a request from the director of education for the Eastern Province,

The Aramco-Built Government School for girls in Rahimah, near Ras Tanura, 1965.

Aramco agreed to build schools for the daughters of Muslim and Arab employees "at a rate of approximately two schools per year, or a capacity of 300 pupils per year." The letter was signed by James V. Knight, then on the policy and planning staff of Government Relations.

However, due to delays in deciding which government agency the company should work with during construction of girls' schools, it was not until September 1964 that the first Aramco-built girls' school opened. In the meantime, the company made substantial donations of furniture to Dammam Government Girls' School No. 1, which opened in

Saudi students watch an educational program broadcast by Aramco TV, 1959.

late 1960 at the Workmen's Compensation Building in Dammam. In the beginning there was a critical shortage of qualified teachers for girls' schools. Although it was not part of its agreement, Aramco in 1962 financed a summer program of intensive training for government-selected teachers at Beirut College for Women. The company also provided 12 portable buildings as temporary classrooms for girls' schools in the Eastern Province. The demand for space was overwhelming. Harry Snyder reported that 900 girls attempted to enroll for 500 spaces at the Damman No. 3 government school.

College of Petroleum and Minerals

n 1963, the government established the College of Petroleum and Minerals (CPM) in Dhahran, the first facility in the Kingdom devoted to training and research in the petroleum industry. As early as 1956, when he was education minister, Amir Fahd had envisioned a technical institute to train men for the oil industry. On September 23, 1963, after years of proposal and counterproposal, the college was established by royal decree. Ahmed Zaki Yamani, the new minister of petroleum, with the support of King Sa'ud, was the moving force behind the college. Yamani chaired a four-member college planning board that included among its members Robert King Hall, the former director of training at Aramco. Another board member, Salih Ambah, holder of a Ph.D. in petroleum engineering from the University of Southern California, became the first dean of the college. Hall became the college's first senior-staff advisor.

King Faysal inaugurated the college in February 1965. The college operated out of temporary buildings. Its offices on the north side of al-Khobar-Dhahran Road were in a partially finished, four-story structure originally intended to house a local branch of the Ministry of Petroleum and Mineral Resources. Aramco supplied the land for the college and 149 aged buildings in the old Saudi camp area. Some of those buildings were converted into college dormitories.

In his inauguration speech, King Faysal said: "It is not of importance that we build institutes and celebrate their inauguration. The important thing is that we exert all our efforts to benefit from such institutes, to realize what our nation expects from us, and to find among our ambitious young men persons who would strive for a prosperous future with all their might, and with self-denial in the service of their religion, country and nation. I ask Almighty God to make this institute the beginning of other institutes, and not an end in itself."

The college adopted English as its language of instruction. It began by giving students a one-year orientation program of remedial work, mainly English-language instruction, designed to prepare them for entrance into the college's two-year program. The first class, the class of 1964, consisted of 67 students, 55 from Saudi Arabia and 12 Algerians on fellowships. The one-year remedial program was completed in 1965, and students began a two-year course in general engineering subjects. In 1967 the college sent

The University of Petroleum and Minerals campus in Dhahran, 1975.

The university's water tower was one of the first campus projects.

its first 24 students to the United States to continue their work toward a four-year degree.

An architectural firm, Caudill, Rowlett and Scott from Houston, Texas, was retained to design a campus for the college on 1,500 acres atop a hill overlooking Aramco headquarters. Their contract stipulated that the exterior design must emphasize Arab architectural traditions. Initial construction was completed in 1969. By 1970 the college, now a four-year school, had moved onto a magnificent new campus. In 1975, by royal decree, the college was given university status. The new university had a faculty of 202 and 1,810 students. It was called the world's largest university of petroleum technology. (During a royal visit on December 25, 1986, the school was renamed King Fahd University of Petroleum and Minerals.)

In addition to giving land and some buildings to the college, Aramco advanced $550,000 for engineering and architectural design work on the first buildings. In 1965 and in 1970, the company provided $14.5 million in the form of grants and donations to the college building program. The first two students sent by Aramco to CPM enrolled at the school in 1969. They were 'Abd Allah Ghaslan of the Maintenance Organization and Shalah Jarman from the South Refinery Division. It was the first time Aramco employees had been assigned to an institution of higher learning in Saudi Arabia.

The Challenge of Technical and Management Training

raining Saudis for highly skilled technical and supervisory jobs in the industrial work force presented a continuing challenge to Aramco. Such jobs as relay technician, instrument technician, laboratory technician, instrument shop foreman and supervising machinist required completion of the full eight-year ITC curriculum or its equivalent. In addition, the trainee needed well-developed analytical and problem-solving skills, and had to be capable of absorbing highly specialized training, possibly at technical institutes outside the Kingdom. A Technical and Supervisory Training Program was adopted at the start of the 1964 school year in hopes it would solve the problem. The T&S Program, as it came to be called, required that Saudis selected for technical training spend two years in full-time training followed by one year in full-time work. This cycle — two years of training, one year of work — was to continue until the employee completed the ITC curriculum. It could take up to 12 years to complete the cycle. New-hires in certain divisions were to be placed full time in ITCs until they reached the 4C level. Then they began a cycle of one year on the job and two years in full-time ITC training. The T&S requirement did not extend to clerical and administrative trainees hired under the program. They followed the usual daily schedule of four hours of work and four hours of training.

Although this enforced regimen had the potential to solve training problems, it created manpower problems. One superintendent's experience illustrates

Trainees study air conditioning and refrigeration, 1960.

95

the problem many superintendents and managers experienced while trying to balance training and work requirements. In August 1965, Joe Mahon, superintendent of plants in the Plants and Pipelines Department at Abqaiq, wrote to his supervisor complaining: "The T&S Program has very seriously limited our ability to cover the 'solid jobs' for the near future." Mahon figured he had 63 people with either the experience or the potential to be trained to fill 51 technical jobs in the division. But 26 of the 63 people were enrolled in the T&S Program. In the next two years he would get full work benefit from only five of the 26. The most severe shortage would be at the Ain Dar Gas Injection Plant, where he had seven people to fill 14 jobs.

Mahon asked for, and got, approval to modify the program so the trainees from his division would spend one full year instead of two in training and the next full year on the job. During his year on the job, the employee would be required to take two hours of "R" time (paid overtime) training after his eight-hour shift. In addition, training could be slowed down if necessary to keep the plants running. "This revised program, while not as desirable from an academic education standpoint, will allow us to get productive work out of the T&S Program candidates sooner than the current training pattern," Mahon wrote.

Technical and Supervisory Training called for above-average English-language skills. In 1965, the basic level one-through-four English program was changed in a manner that helped meet that need. For the term starting September 1, 1965, Training lengthened English-language class periods to two hours. One-hour periods had been standard since training centers were first opened. The reason for the change, according to Training's 1965 Accountability Report, was that "One-hour courses were found to be inadequate to prepare trainees for secondary-level courses." In secondary-level courses, Saudis developed the English reading and writing skills necessary for supervisory positions and for out-of-Kingdom training assignments.

O'Grady Replaces Pendleton as Director of Training

In 1966, a long-time Aramco executive, Gary Owen, vice president of Government Affairs in Washington, D.C., retired. Director of Training, John Pendleton, a former executive in Aramco's Government Relations office, was seen as the ideal replacement. Pendleton had served as director of Training for nearly four years. It was a time of steady decline in the total number of Aramco employees, in the number of Saudis in training, and in expenditures for training. In 1963, Pendleton's first year as director, more than 4,000 Saudis enrolled in training programs. Direct expenses for training during that year totaled $5.3 million. By 1965, his last full year as director, Saudi enrollment in training programs had fallen to 2,800 and net direct expenditures budgeted for training had been cut to about $3.25 million.

Pendleton, in a speech before the American Manufacturers' Association meeting in New York in 1965, summarized as well as anyone the way in which training programs evolved at Aramco. He described the company's current training programs to the audience and concluded: "Our present practices did not just happen. They are, rather, the fruit of trial and error, of much soul-searching and disputatious discussions." As to the future, he felt sure another Aramco trainer some 10 years hence would deliver a speech "which treats in a condescending fashion as something outmoded the training system I described to you in glowing terms."

The new director of the General Office Training Department, effective July 1, 1966, was William P. O'Grady, a graduate of Iona College in New Rochelle, New York, and Fordham University in the Bronx, and a former teacher in the tough inner-city New York schools. O'Grady said he was "flabbergasted" by his appointment and that it was "a bit of a shock" to other trainers as well. They

96

*William P.
O'Grady,
1966.*

had expected a veteran administrator, like Harry Snyder, would get the job. In fact, Snyder transferred that same year to the neighboring College of Petroleum and Minerals as assistant dean of academic affairs.

O'Grady's administrative abilities had been tested in a number of posts since his arrival in Saudi Arabia in September 1953. He had compiled a good record as principal of the Ras Tanura ITC. When he became principal of the Dhahran ITC in 1960, O'Grady said, he found an administrative nightmare: "The place was being run mostly by clerks." He contrived with some difficulty to get it back on an even keel. Then he took over Saudi Development and ran it within budget. Since 1961 O'Grady had been staff adviser in General Office Training, a troubleshooter-type job, and he had served as acting director while Pendleton was on vacation. He received excellent reviews as a public speaker and was reputed to have a near-perfect memory.

The department he headed was small, not much changed since its creation in 1952. The key players reporting to O'Grady were called training coordinators. They were Don Richards, Industrial Training Centers and Industrial Training Shops; Radcliffe R. Daly, Aramco senior-staff schools; Frank Jarvis, Management Training and Saudi Development; Harry Snyder (who was soon replaced by Mustafa Husam Al-Din), Aramco-Built Government Schools; and the senior Arab linguist, Merril Van Wagoner. The General Office advised the company on training policies and procedures, but had no administrative control over employees aside from the five training coordinators and 20 or so persons in the General Office support staff. The various districts had administrative authority over the remaining 250 people engaged in training.

Innovations in Management Training

Between 1965 and 1967, the Training Department produced the *Aramco Basic Management Series,* a series of 10 or more textbooks for supervisory and management training programs. The series started with a 200-plus-page manual, *Essentials of Leadership,* intended for potential supervisors. There were four separate bilingual textbooks for segment-level supervisor training programs. Five other volumes, written entirely in English, were designed to train candidates for jobs at the foreman level or beyond. Both sets dealt with the same four topics: safety, communication, organization and supervisory techniques. Each volume had a similar format. Nearly every page contained drawings to illustrate the text, plus one or more pithy sayings set off in boxes from the main text. Examples of these sayings included: "Knowing must be accompanied by doing. ... Leaders must set good examples. ... Advancement must be earned at Aramco. ... Sincere praise encourages initiative. ... For best results, lead, don't drive." The various volumes contained chapters on subjects such as grievance procedures, planning work programs, orientating a new employee, correcting mistakes, planning, delegating authority and building employee morale. Instructors in the Basic Management Series Program included Hassan Natour, Jamil Milhem, Jim Allen, Frank Weaver and Frank Jarvis.

Aramco's first full-time training course for supervisors began in the Management Training Center at Dhahran in September 1967. The course, called the Supervisory Training Program, ran eight hours a day, five days a week, for four weeks. Frank Jarvis organized the 20-day course, which he expected would become the core program of supervisory training in the future. He queried various departments about the weaknesses of their supervisors and developed the program's agenda from their responses. One-fourth of the entire 160-hour course was devoted to improving communication techniques. Other subjects included personnel practices, management techniques, safety, organization, on-the-job training, medical functions and timekeeping. About 30 supervisors took the course at a time. They were divided into teams of five or six men each for the duration of the program. There were lectures, discussions and case studies of various subjects. The course was aimed primarily at improving the skills of Saudi supervisors, but some Americans and supervisors of other nationalities enrolled as well. "We wanted to provide a good foundation for all supervisors coming up in the organization," Jarvis said. In the first three months, 59 employees, 40 of them Saudis, completed the program. One of them was a woman, Dhahran nursing supervisor Mary Ann Pettigrew.

The four-week course concluded in a unique way. The entire final day was given over to the "Oil Game," a kind of board game that simulated the real risks of capital investment in the oil industry. Teams of supervisors representing fictitious oil companies started out with a $14 million stake. As would real oil companies, the teams studied subsurface profiles and calculated the price of getting rigs to the site and of building pipelines and terminals to move their crude oil, if there was any, to market. Then they decided whether or not to risk their money by buying a concession area. The game was played on a box about 4½ feet long, 3 feet wide and several inches deep, with concession areas marked on top and spindly miniature "rigs" used as game pieces. A slender rod was inserted into predrilled holes, and if it went down far enough, oil was struck. Some teams went bankrupt; others enjoyed a bonanza. "They were faced a little with the realities of the oil industry. We wanted them to appreciate a little more about the oil industry, the finance situation risks," Jarvis said. The "Oil Game" was developed by Jan Bakkum, personnel supervisor of Shell Company of Qatar, Ltd., and evaluated by an 18-member group of Aramcons before it was approved for the training program. One of the 18 members on the evaluation team was a young Aramco geologist, Ali Al-Naimi.

Geologist Ali Al-Naimi prepares to set down a rig in his team's concession area as part of Aramco's "Oil Game" training activity, 1966.

Participants gave the Supervisory Training Course high marks. One of them, 'Abd Allah Al-Faysal, a trainer from Abqaiq, spoke at a dinner attended by the first 30 supervisors to finish the course. He said: "This course made us doubt, or question, if you like, the methods we follow in our daily activities. We need new techniques and answers to cope with the various questions employees ask. We don't pretend to be experts in benefits, for example, but surely we should know something more than to say 'It's company policy.' This course has helped us learn new things and clarify some other things."

The Oil Game proved such a powerful learning tool that it became a feature of seminars on the international oil industry conducted by Aramco in the 1960s and '70s for businessmen, journalists and other visitors. The seminars were normally conducted by Thorn Snyder, the company's economist and public speaker extraordinaire. Of the Oil Game he said: "The emotional stress inherent in this game stimulated more learning about investment risk and uncertainty than millions of words." Snyder was well known for his 60-minute orientation programs for new-hires, which he began by holding up a small bottle of crude oil and declaring, "For most of you, this is all the oil you will ever see while you are in Saudi Arabia. ..."

Muhammed R. Al-Mughamis

"If the company wants to succeed, the training effort must continue. ... Production of oil and training go hand in hand."

Muhammed R. Al-Mughamis, looking back over his childhood, considered himself "one of the lucky ones." He grew up on a small farm near Hail, in northern Saudi Arabia, in the 1950s. In those days, formal schooling was still a new concept, and illiteracy was still widespread. "Graduating from elementary school was like graduating with a Ph.D.," he said. Fathers were divided over the value of formal education. "Some had their sons work instead of going to school, and others preferred that their sons get an education to be prepared for the future."

Al-Mughamis' father needed help on the farm, but he took the latter view. "I remember vividly his telling my mother, 'Look, it's time to have this kid start school in order for him to achieve something,'" he said. His father bought him a notebook and pencil and sent him off to elementary school in 1951. Al-Mughamis graduated in the summer of 1957.

Because he had relatives working for Aramco in the far-off Eastern Province, employment with the company was a logical next step. "The incentive basically was to assist my family," Al-Mughamis explained. Family bonds were strong, and a young teen — even a boy of 10 or 11 — was expected to "behave like a man" and help support his family. "In those years the words 'the company' meant only Aramco. There were no other companies, only little shops here and there. So most people just talked about 'the company,' as there was little else to talk about. Radios were just beginning to appear."

To reach Aramco's Abqaiq employment office, Al-Mughamis rode in the back of an open truck across open desert at night. The journey from Hail to Riyadh took two weeks. In the capital, he boarded one of the old "green trains" for the Eastern Province.

Al-Mughamis' first job with Aramco in 1957 was as a clerk at the Abqaiq Clinic. By 1960 he had arranged for a transfer to Training, where he became an ITC trainee/teacher studying advanced subjects in the morning and teaching lower-level English, math and Arabic in the afternoon.

Al-Mughamis enjoyed teaching and was good at it. In 1964, the company sent him abroad for the first of several out-of-Kingdom educational assignments. He earned an associate degree from Wesley College in Dover, Delaware, in 1966 and a bachelor of arts degree from Hofstra University in Hempstead, New York, in 1970.

When he finally returned to Abqaiq in the early 1970s, it was not as a teacher — Training had closed its operations there — but rather as a Government Affairs representative. "I found it a very satisfying, very interesting job, hectic and frustrating, dealing with people of different levels and interests," he recalled. Most of his career with Aramco and Saudi Aramco focused on Government Affairs and Public Relations. In 1988 he was named general manager of Government Affairs.

Looking at the company's training strategy over the years, Al-Mughamis said Saudi Aramco is on the right track and should continue building on its success. Employees should take advantage of the educational and training opportunities available to them, he said.

"Without education and training, I don't think anyone would have a successful career in the company. Training is a prerequisite to advancement in the company, and the competition is great. I'm advising everybody to take the opportunity while it exists. This is the foundation for building higher management," Al-Mughamis said.

Training Activity Slows in 1960s

raining activity gradually slowed during the 1960s, as the number of Aramco employees continued to decline. Between September 1965 and September 1967, assigned-time enrollment at the company's ITCs and ITSs fell by 29 percent and voluntary enrollment dropped by 60 percent. The Training Department's annual Accountability Report listed three main causes of the decline: 1) more stringent requirements for enrolling employees in training during work hours; 2) a declining work force; 3) a change from the trimester to the semester system at training facilities. In that same period there was a 25 percent reduction in the ITC staff. Some of the staff reduction was due to the dismissal of 14 foreign-contract teachers following the outbreak of the June 1967 Arab-Israeli war.

For a change, the Training Department had the luxury of extra space. For example, the Dhahran ITC had 24 classrooms. Training didn't need all those classrooms, so about half of the top floor of the ITC was converted into a Management Training Center, which included one large conference room and several smaller rooms.

O'Grady felt that the quality of the teaching staff had improved during the persistent drawdown. "We trimmed down the teaching staff because of the reduced load on us in terms of classes," O'Grady explained. "We just kept the very, very best teachers and let the weaker teachers go. As a result, by the time we were reaching the period of 1970, we had probably the best ITC and ITS staff that we were ever going to have."

Ali Twairqi could attest to the quality of the teaching staff. In 1966 Twairqi was a free-spirited, 16-year-old eighth-grade dropout, who had just landed a job with Aramco. He was working as a ticket taker and broom pusher at the Abqaiq movie theater. Twairqi was glad to be an Aramco employee, no matter how menial the job. He had prestige. He was someone who got free medical treatment and might someday own a home through the Home Ownership Program. But training was not on the list of things he wanted to do. He was happy just to get to see a free movie and earn a little money.

Trainers were sent to remote drilling sites, such as this one in the Rub' al-Khali, 1967.

Twairqi might have remained an unskilled laborer, had it not been for a supervisor who insisted he take training. "He told me they didn't have room for people (at Aramco) who didn't want to train." So Twairqi went back to the classroom. It was a tough, mind-opening experience. Like so many other Saudi trainees of the time, Twairqi vividly remembers the burly ITC principal, Larry Emigh, "who would stand by the door of the ITC every morning and lock the doors at exactly 7 a.m." Trainees who came late didn't get in. Emigh kept close tabs on his teachers, too. "He had this huge box in his office," Twairqi said. "It was an intercom, and it was connected to the different classrooms so he could listen in on classes to make sure they were running, there was dialogue, and nobody was asleep at the wheel.

"We had instructors then who were very professional and hard working, and they made sure we knew immediately who was boss. They were paid well to train young Saudi employees and they did it very, very

hard and very, very well. They made no bones about it. They told you, 'You want to learn, you stay in my class; you don't want to learn, you get out now, because we don't tolerate nonsense.' They were people who had commitment." Other instructors he remembered especially well included Hashim Budayr, a science teacher, and Ibraham Shihabi, a teacher of English. Thirty years later Twairqi, the eighth-grade dropout who didn't want to go back to school, became head of the company's Training Organization.

Milestones of the 1960s

Many milestones were passed along the road to Saudization during the decade of the 1960s. In 1960 Sa'id Al-'Awwami graduated from the American University of Beirut to become the first Saudi to earn a degree in pharmacy through Aramco's Saudi Development Program.

In October 1961, Aramco trainers opened the first ITC training sites deep inside the Rub' al-Khali. Hassan Sa'di held classes for Saudi workmen at Structure Drill Site No. 5, about 400 miles southwest of Dhahran. 'Abd Allah Hamdan opened a classroom at Structure Drill Site No. 2, some 300 miles southwest of Dhahran. Once a month the trainers were relieved for one week by Sa'ad 'Abd Al-Rahman from the Dhahran ITC. They taught Arabic, arithmetic and English to Saudis during classes that met in air-conditioned trailers after working hours.

In the 1950s and until the department reorganization in 1968, the Abqaiq district was responsible for training outside of Dhahran and Ras Tanura. Abqaiq operated small satellite ITCs on offshore drilling rigs and at onshore sites as distant from Abqaiq as Safaniya, 185 miles away. In 1965 enrollment at these satellite ITCs totaled about 200. Abqaiq principals and assistant principals logged many miles on the road and in planes trying to keep up with these operations.

The company adopted the standard 40-hour, five-day work week in 1961. The work week had been cut from 48 hours to 44½ hours in 1954, and to 42 hours in 1958.

Women participated for the first time in Aramco's 1962 summer work program for Arab college students. Forty-nine Saudi students were employed in the three districts that summer, four of them women.

An all-Saudi crew ran a full shift at the fluid hydroformer at Ras Tanura for the first time in March 1962. The seven-man crew, under supervisor Ahmed R. Al-Magbil, had trained and worked together for more than three years. The 10-story-high fluid hydroformer was a complicated and relatively rare installation used to make high-octane gasoline from low-octane feedstock. At the time there were fewer than 10 other hydroformers of the same size and type in the world.

The first all-Saudi crew to run a shift at the fluid hydroformer in Ras Tanura, 1962.

Salih Al-Sowayigh, 1964.

An all-Saudi crew also ran a shift at the new Abqaiq Liquefied Petroleum Gas (LPG) Plant for the first time in 1962. The plant, built to remove raw LPG from natural-gas liquids and send it by pipeline to Ras Tanura, opened in 1961. The Saudi crew, under supervisor Yousif Mohammad Gasham, took over some shift operations at the plant only eight months after its opening.

Two of the company's first senior-staff Saudi employees retired during the decade. Ahmed Rashid Al-Muhtasib, deputy company representative in Jiddah, retired October 1, 1962, after nearly 30 years with the company, and Salih Al-Sowayigh, deputy company representative in Government Relations, Dammam, retired July 15, 1964. Both men had reached senior-staff level in 1950.

Saudi student checks for results of qualitative analysis in a Training Department chemistry lab.

By the 1963-64 academic year, the company had established 60 scholarships for nonemployees in agriculture, the sciences, engineering, teacher education, medicine (including two in nursing) and business management. In addition, the company continued to support the Arab Refugee Scholarship administered through the United Nations Relief and Works Agency. The scholarship funds were not related to the company's program of sending qualified employees to colleges and universities.

In 1963, Aramco's senior-staff schools started compulsory Arabic-language courses for students from kindergarten through third grade. Four new teachers were hired for the courses. Up to that time, the only language course other than English had been French, an elective in grades seven to nine. Arabic courses were later added through grade six.

Training produced new aptitude and mental-abilities tests to challenge the better-educated Saudi recruits of the 1960s. The old Saudi Test for Job Assignments (STJA), an oral examination created in 1953, when almost all job applicants were illiterate, was simply too easy for the new recruits. A new multiple-choice test, called the Job Aptitude Test and written in Arabic, was devised and first administered in 1961. Training used this test as a hiring and placement tool for more than 15 years.

Among publications issued by the Training Organization during the 1960s was the *Aramco English Series*, the first company-produced English-language textbooks. Ron Goodison, a linguist with a Ph.D. from Cornell University, and Jim Sitar, an ITC English-language teacher, worked on the series for three years. The 12-volume series, containing a vocabulary of more than 3,500 words, was published between 1960 and 1962. Each volume was designed to cover one trimester during the first four years of ITC English-language training. Although some of books in the series were soon replaced with off-the-shelf materials, *Aramco English Series* books continued to be the standard texts in English levels one and two for 15 years.

In 1963, the Saudi government greatly expanded its interest in the oil industry and its support of projects that gave Saudis training and experience in the petroleum industry. The creation of a Saudi government Ministry of Petroleum and Mineral Resources was followed by the formation of the General Petroleum and Minerals Organization, better known as Petromin, to supervise oil activities in the country. Between 1964 and 1967, Petromin purchased all Aramco product distribution facilities that served the Kingdom. Saudis employed at Petromin gained valuable experience in the marketing and refining of petroleum products for Saudi Arabia.

In 1963 Petromin purchased all Aramco distribution facilities in Saudi Arabia.

In May 1964, with the approval of Aramco, the director of education for the Eastern Province recruited seven British instructors to teach English at the Aramco-built intermediate schools at Dammam and al-Khobar. They were the first British language teachers in the government school system. The results of the "experiment" were successful, and within a year, the

Ministry of Education was attempting to recruit more than 100 more British instructors.

Because of a shortage of trained intermediate school teachers, Aramco undertook a one-year program in 1964-65 of telecasting physics and chemistry courses given by its ITC teachers to the intermediate schools in the Eastern Province. The company provided TV sets for the government schools.

The first refinery simulator was installed at the Ras Tanura Refinery in the summer of 1965. The $23,000 Carmody University process training simulator featured a display board, eight feet wide and seven feet high, on which operating procedures and conditions in a refinery could be simulated. About 340 employees from the North and South Refinery Divisions trained on the simulator during the first four months of its operation.

On August 1, 1965, Zafir H. Husseini became the first Saudi department manager in Aramco. Husseini was appointed manager of the Products Distribution

Zafer H. Husseini, 1967

Department, headquartered in Dhahran. He was the first Saudi to reach the executive level at Aramco. Until all distribution became Petromin's responsibility, his department supervised the distribution of petroleum products to the central, eastern and northern areas of the Kingdom. It also operated 11 distribution stations in the eastern part of Saudi Arabia. As a department manager, Husseini had broad administrative powers. He was responsible for long-range planning and coordination of services with other departments. Husseini had joined Aramco in 1952, after earning a degree in economics and history at Ripon College in Ripon, Wisconsin, through the government scholarship program.

In 1967, Aramco ITCs acquired a new training aid called the English Language Laboratory System. Students at individual carrels put on earphones to hear tape-recorded instruction. The system was first used to teach English with audio tapes based on the *Aramco English Series.* Two instructors seated at a console controlled tapes heard at the stations. The instructors could talk to individual students, but the students could talk only to the instructors, and only after requesting to do so electronically.

In 1969, Mustafa al-Khan Buahmad became the second Saudi department manager at Aramco. He was appointed manager of the Public Relations Department on September 1st of that year. Buahmad joined Aramco as a filing clerk

A Learning Laboratory instructor could contact one student or the entire class.

trainee in 1944, attended the Jabal School, taught Arabic to new American employees at the Sidon training center, and took a business degree from Antioch College in the United States in 1963. He was superintendent of Saudi Riyal Personnel in Abqaiq prior to becoming manager of Public Relations.

'Abd Allah Ali Al-Zayer became the first Saudi Arab to become a principal of an ITC, in 1969. He had served for two years as assistant principal at the Dhahran ITC before being moved up to principal. His promotion was made without any great fanfare. "He was experienced, reliable and competent," O'Grady said. He seemed "the logical choice." The son of a Qatif-area farmer, Ali Al-Zayer had joined Aramco in 1950 at age 12 as an office boy in Abqaiq. A year later he moved to Dhahran as a kitchen worker in Community Services and attended ITC classes on his own time in the evening. He transferred to Training in the early 1950s and became a trainee teacher.

In the late 1950s he was sent to a small New England school, Lyndon State College at Lyndonville, Vermont. He was the only Saudi in the college, in the town and perhaps in the state. He earned a degree in education and English and returned to Dhahran in 1964 for a career as a teacher and administrator in Training and Career Development.

Disaster Strikes

Shortly after midnight on the morning of Saturday, April 18, 1964, the Aramco community suffered one of the worst tragedies in its history. Middle East Airlines flight 444 from Beirut disappeared while making its approach to Dhahran International Airport during a raging sandstorm. The next afternoon a rescue helicopter spotted a wing of the jetliner riding just above the water in the Arabian Gulf about 15 miles south of Dhahran. There were no survivors. Forty-nine people, including 22 Aramco employees, died. One of the victims was Rashid Al-Rashid, acting assistant principal of the Abqaiq ITC, and a rising star within the Training Organization. Harry Snyder, coordinator of SAG-Aramco educational services, was scheduled to be on the flight with Al-Rashid, but shortly before departure time, Snyder chanced to meet Oil Minister Ahmed Zaki Yamani and Yamani asked him to remain in Beirut that evening for a business conference.

Continuing Adjustments

The number of Aramco employees declined steadily throughout the 1960s. By 1965 the employment had fallen to 13,100, the lowest in 20 years. In contrast to cutbacks in employment and other areas, production continued to grow at a brisk pace. On February 14, 1962, the one billionth barrel of oil was processed at the Ras Tanura Refinery. In June of 1962, total production throughout company history reached five billion barrels. The giant Ghawar field yielded its two billionth barrel of oil on October 16, 1963. Production averaged more than two million barrels a day during 1965. On December 22, 1966, a record 3.1 million barrels were produced in 24 hours.

Aramco's first shipment of liquefied petroleum gas (LPG), the equivalent of 75 million cubic feet of gas, was loaded onto a specially designed tanker at Ras Tanura in December 1961. It was the first step in a process that by 1979 would make Aramco the world's largest producer of natural-gas liquids, including LPG and natural gasoline.

In March 1963, after months of negotiations with the government, the company agreed to relinquish about 75 percent of its original concession area and committed itself to further relinquishments that would, by 1993, reduce the company's concession to 20,000 square miles, or less than three percent of the original and preferential areas once held by the company.

In 1968, Aramco became the first operating company in history to produce more than a billion barrels of oil in less than a year. That same year marked the 30th anniversary of the year in which well No. 7 came in, and confirmed the existence of oil in commercial quantities in Saudi Arabia. During those 30 years, the company produced more than 10 billion barrels of crude oil, most of which was consumed in Europe and Asia. In 1969 nearly 50 percent of Aramco's exports went to Europe, 36 percent to Asia, and less than three percent to the United States.

The company's Home Ownership Program was another exception to the general slowdown in activity. By 1966, when Home Ownership celebrated its 15th anniversary, more than 6,000 Saudi employees had acquired homes through the program.

Ali Dialdin

n April of 1968 a young geologist and former elementary school teacher, Ali Dialdin, joined Aramco. Dialdin had first visited the company in 1960, but was not favorably impressed. That year Aramco recruited 116 outstanding eighth- and ninth-grade students from various parts of the Kingdom for the company's Secondary Student Employment Program. Dialdin was recruited out of a school in his hometown, Madinah. The program was supposed to help youngsters develop good work habits and familiarize them with the oil company, which was still relatively unknown outside the Eastern Province. The program was "a major disaster," Dialdin later recalled. The students were housed, two in a room, in uncomfortable, in-line dormitories in the old Saudi camp. Dialdin's roommate was a childhood friend, Ibrahim Qabbani, later deputy ambassador of Saudi Arabia in Morocco. They were not allowed to use company facilities. They had to cook for themselves or eat at specialty shops. "We didn't find any program, or anybody to take care of us," he said. They felt like they were "dumped" into jobs in various departments. Dialdin, who had some math training, helped Saudi employees fill out forms for withdrawal of funds from the company's Thrift Plan and worked in the Cash Office during the payday rush. He also attended English-language classes at the Dhahran ITC. The program lasted three months. Dialdin left with "a bad impression" of Americans who ran the program, and "a very bad impression" of Aramco in general, although he had a "very kind, very understanding" boss in the Thrift Plan office, 'Abd Allah Al-Jame'.

Ali Dialdin, 1969.

Dialdin attended San Diego State University on a government scholarship and received a bachelor's degree in geology. He wanted to go after a master's degree, but it was 1967, the year of the first Arab-Israeli conflict, and "the government was not in a mood to send anybody anymore" to the States. He was offered a chance to go to Aramco, but turned it down. Instead, he went to work for the Ministry of Petroleum and Mineral Resources in Jiddah. Dialdin was assigned as a liaison person with the United States Geological Survey group in Jiddah. He was the only Saudi in the group. He was told there was no office space for him, so he sat outside the group's offices with the coffee vendors. After three weeks, he asked for a new assignment. Aramco was suggested once more. With some hesitation he agreed to join Aramco. He was assigned to Industrial Engineering in Ras Tanura, but several months later his boss, Bill Griffin, was transferred to Training and asked Dialdin to join him. In this roundabout way, Ali Dialdin, who would eventually become head of the Training Organization, found his way to Aramco and a career as an educator.

Aramco Reorganization Affects Training

n January 1, 1968, the company underwent a general reorganization. Management was centralized in Dhahran at the expense of top-level management jobs in the districts. The company eliminated the jobs of the district manager, assistant district manager and five general superintendents in each of the three districts. No one's pay was cut, but men who thought they were in position to move up in the organization suddenly found themselves in the painful position of having lost their job titles, and having to choose whether or not to accept a lower-level job with less authority and prestige.

Additions to the Training director's administrative domain due to reorganization included 123 people employed in the new Industrial Training Division, which encompassed the ITCs and ITSs in all three districts, 121 employees of Aramco's senior-staff schools in the three districts, and 23 people in the Saudi Development/Management Training Division. For the first time since training

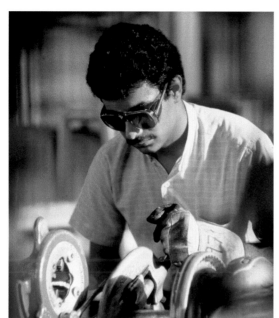

Job skills training.

was decentralized in 1948, the director of Training, Bill O'Grady, had administrative as well as functional authority over training activities in all districts. From an administrator in charge of a General Office staff of 23 employees, with a payroll of several hundred thousand dollars, O'Grady became, overnight, head of a Training Department staff of 283 people with a budget of millions of dollars. It was done "with no change in pay grade," O'Grady said. The Training Department and the people working in the Aramco Schools reported administratively to the vice president of Industrial Relations, George F. Larsen. Larsen's authority also extended over the Employee Relations Department, Industrial Relations Policy and Planning, the Medical Department and the Safety Department.

The company had more changes in mind for Training. Training was centralized by subject — academic training in Dhahran, job skills training in Ras Tanura. Abqaiq training was to be shut down. Management, believing it could find Saudis to fill key positions among graduates from the new College of Petroleum and Minerals, also decided to cut back on Saudi development programs. The Dhahran ITS was closed in 1968, as was the Central Welder Training facility at Ras Tanura, and the outlying ITCs at Udhailiyah and Nariya. The four-week, 160-hour Supervisory Training Course was suspended. Dr. Charley Johnson, arriving in Saudi Arabia to teach English at the Ras Tanura ITC, was informed that future prospects looked dim. The first two levels of English instruction had already been dropped for lack of trainees; levels three and four would probably follow soon. Then the Ras Tanura ITC would be closed. "You'll be the one to turn out the lights for the last time," Johnson was told.

The reorganization caused some hard feelings. Frank Jarvis, then coordinator of Management Training and Saudi Development, decided to resign. "I just couldn't stand to see the organization's capability destroyed," he said. "Why change something that has been working so well?" Warren Hodges, transferred from Ras Tanura to Dhahran to become the first superintendent of Industrial Training, lasted just one year in the job before he, too, decided to leave.

By 1970, both the Abqaiq ITC and ITS had been closed and the Abqaiq junior high school shut down. The junior-high students and their parents were transferred to Dhahran as part of a plan to phase out Abqaiq as a family camp and make it a bachelor and commuter facility. The only training facilities in Abqaiq were five classrooms in the Main Gate Building. At one point the Abqaiq ITC had only 52 part-time students.

The 1969-70 training year enrollment totals included 777 in assigned ITC training, compared to 1,200 the previous year; 153 in assigned ITS training, compared to 231 a year earlier; and 285 employees registered for voluntary-time classes, compared to 389 the year before. Fewer than 14 percent of the Saudi employees attended ITC or ITS classes. Total enrollment was about one-quarter of what it had been when the decade began.

By 1970, the reduction of the work force had been going on for 18 years. Aramco employed 10,606 people in Saudi Arabia, compared to a high point of 24,120 employees in 1952. In the 18 years since 1952, the Saudi payroll was down by 41 percent, the number of U.S. employees had dropped by 72 percent and the total number of other foreign workers fell by 87 percent. The company work force was now more than 80 percent Saudi.

The decade of the 1960s saw a changing of the guard in Training. Many of the pioneers who had devoted the largest part of their adult lives to Saudi training left Aramco during the decade. Paul Case, who planned the first drivers' training school and launched the Saudi Development section of Training, left in 1961 after 17 years with the Training Organization. Reg Strange, industrial shop supervisor and an artist of note, resigned in 1963 after 12 years. George Trial, supervisor of Dhahran District Training, retired in 1964. Don Richards, coordinator of Industrial Training in Dhahran, retired in 1965. Vince James, the former Jabal School teacher and first superintendent of senior-staff schools, retired in 1965 after 20 years on the job. Other trainers who retired or quit during the 1960s included George Larsen, Ed Thompson and James Mileham. Warren Hodges, the superintendent of Industrial Training, former head of training at Ras Tanura, industrial-arts teacher, woodworker and a founder of the Ras Tanura yacht club, retired in 1969 after 25 years in Saudi Arabia. "I felt," he said, "that I had done my job. I decided I would build a sailboat and just go off over the horizon, which I did." So did many others.

Warren Hodges teaching in a classroom adjoining Ras Tanura's first Industrial Training Shops.

6

An Unexpected Expansion 1970-1975

"We were competing to get high grades so we could join the company. We knew also that getting high grades would qualify us for a scholarship after we joined Aramco."
– Khalid Abubshait

ramco's 1970 annual report told a familiar story: crude oil production increased by 19 percent during the year; the Kingdom's stockpile of proved petroleum reserves, already the largest in the world, went up by several million barrels; and the company's payroll was down by nearly 400 persons. It was the 29th consecutive year of increases in both production and petroleum reserves, and the 18th straight year of decreased employment at Aramco.

Those who helped the company plan for its future needs had become accustomed to this pattern of increased production and decreased employment. Oil industry experts, both inside and outside Aramco, saw the world market beginning to open up, indicating more demand for Saudi Arabian crude in the near future. Company planners, however, did not expect an increased demand for oil to halt the downward trend in Aramco employment. Quite the contrary, forecasts presented to executive management in 1970 estimated that Aramco would

Opposite: An Aramco exploration team navigates the Rub' al-Khali. Below: Saudi drillers training on a working Aramco rig.

be able to cut its work force by at least 40 percent over the next 10 to 15 years. This would be accomplished through the twin benefits of improved efficiency due to manpower training programs, and the development of automated equipment. One report predicted that the number of Aramco employees would fall from 10,000 in 1970 to about 6,000 by 1980, including a cut in the American work force from 880 to 400. Another forecaster saw the company payroll, not including upper management, falling as low as 5,000 by 1986, with only 250 Americans left in Aramco.

The company's 1970 Operating Plan, in step with these prognoses, told Training personnel to develop a long-range plan to cope with "a rapidly decreasing work force, while keeping the cost per training hour as low as compatible with high training standards. ..."

Training activity was already at a low ebb. During the previous two decades, Aramco had

An Aramco apprentice adjusts the hand wheel on a valve, 1973.

developed a work force of veteran, tested Saudis whose further training needs were minimal. Nearly 80 percent of the Saudi employees had been with the company for 15 years or more, and more than 98 percent had been with Aramco for at least five years. Only 1,100 Saudis were enrolled in industrial training programs at the start of 1970, compared to an enrollment of 8,200 Saudis in 1953, when many of the men in the 1970 work force were just learning their jobs.

To Bill O'Grady, director of Training, the late 1960s and early 1970s were idyllic days when "We [Training] were not being pushed to do anything of a major nature. We were just going ahead and doing a real good job with what we had.

"We were able to run what had been a good program that was now refined and developed, and there were some good people heading up the program. They were halcyon days in that sense."

Key Training personnel included Bill Griffin, described by O'Grady as "a very accomplished and dedicated superintendent" of the Industrial Training Division; Mustafa H. Al-Din, head of Aramco-Built Government Schools; Radcliffe R. Daly, superintendent of Aramco Schools; and Ellis "Ed" Hill and Jim Allen in Saudi Development.

Training's big job at the time was developing Saudi managerial talent. Over the years the company experimented with a number of programs designed to improve managerial efficiency and performance. A popular five-day course in problem analysis and decision making was presented periodically to small groups of supervisors in Dhahran and Ras Tanura by the Kepner-Tregoe firm. Candidates for key executive slots were assigned to advanced management courses at prestigious universities such as Harvard, Stanford and Carnegie Mellon. The company also invested in programs concerned with individual and group psychology. A program based on transactional analysis was brought over from the United States. Transactional analysis sessions were held at two levels, one for upper management and another for supervisors. The company also sent several managers to the U.S. for sensitivity training, a group program designed to increase self-awareness.

Managerial Grid Training

But these programs were minor in scope compared to a management training program called Managerial Grid. Between 1968 and 1970, nearly all managerial, supervisory and professional employees in Aramco — about 1,400 people in all — attended one of the week-long Managerial Grid seminars presented by the Training Department. The total cost to the company was about $1 million. Aramco had never before invested so much money and effort in a single training program.

The company hoped these seminars would improve communication and working relationships in the aftermath of the 1968 reorganization and centralization of management in Dhahran. There were also lingering tensions associated with the 1967 Arab-Israeli conflict. It was believed the seminars would help identify and overcome cultural barriers to performance in Aramco's multinational work force.

110

Salim S. Al-Aydh

"You feel that this training you went through — these programs that were given to you — helped you to become what you are today. I became obligated to this company, and I feel that no matter what I do, it's not enough."

alim Al-Aydh grew up in Ras Tanura, a community filled with Aramcons. His father worked for Aramco, and the company was the subject of daily conversation.

His earliest dream was to be an officer or pilot in the military, but when he finished ninth grade, it was decision time. A military career would take him to Riyadh, which at that time seemed far away due to transportation difficulties. His family urged him to stay in Ras Tanura and seek a career with Aramco instead.

The timing was advantageous because the company had recently created the apprenticeship program. Al-Aydh was accepted into the program in the summer of 1970 and was among the first group of approximately 100 apprentices.

Al-Aydh was impressed with the efficiency and organization of the Aramco training programs. Ali Al-Naimi, a superintendent in the Southern Area at the time, welcomed the young crop of apprentices. Al-Aydh was offered the chance to work as an operator trainee and go to school after hours. After two months, the supervising operator designated him as outside operator to replace an employee on vacation. Al-Aydh still cites this as one of his first achievements.

Al-Aydh became a control-room operator and went to school for three hours after work. There were times when work took up to 12 hours, and Al-Aydh still had to go to school on top of that. His grades were very good, however, and in 1974 he was offered the chance to attend school full time as a regular employee. He finished school and the TOEFL (Test of English as a Foreign Language) examination in 1976, after which he was selected for the company's prestigious Out-of-Kingdom Academic Training Program.

The original plan was to earn an associate degree at Kilgore College in Kilgore, Texas, in two years and then go on to finish a bachelor's degree elsewhere. However, after only three semesters at Kilgore, it was clear Al-Aydh could handle more advanced academic work and was transferred to the University of Tulsa, where he graduated with a bachelor of science degree in mechanical engineering in 1979.

He was eager to get back to Saudi Arabia and Aramco. When told he would have to go through the company's Professional Development Program (PDP) for three years, Al-Aydh felt that his career was on hold. Now, however, he sees the merits of the program as a further chance to ripen his skills and critical judgment.

Al-Aydh progressed through positions of increasing executive responsibility, including executive management positions in Downstream Development and Coordination and Saudi Aramco Affairs.

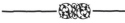

Dr. Robert Blake, Managerial Grid coordinator, opens a training session in Ras Tanura, 1968.

Managerial Grid was a six-phase program developed by two University of Texas faculty members, Drs. Robert R. Blake and Jane S. Mouton. At the time of their association with Aramco, Blake and Mouton were president and vice president, respectively, of Scientific Methods, Inc., a firm that conducted Managerial Grid seminars for industries across the U.S. and in Europe.

Aramco sent a number of people to Managerial Grid seminars in the United States and Europe before the program came to Saudi Arabia. Mel LaFrenz, director of Management Development, attended Managerial Grid seminars in Hawaii during the 1960s. He and others who had experience with the seminars gave the Grid program favorable reviews.

The company planned to present only the program's first three phases, which focused on improving communications within an organization. The last three phases concentrated on long-range business planning. Phase-one participants were encouraged to speak candidly, to say what was on their minds, even if it resulted in conflict. Conflict was seen as a welcome part of the search for solutions. Phase one promoted the use of critique — the process of looking at what had gone wrong in the past, not to place blame, but to learn how to do better in the future.

The program took its name from a depiction of a grid. The grid was composed of nine columns, representing degrees of concern for people, and nine rows, representing concerns for production. A major aim was to demonstrate that these two concerns could and should coincide.

A managerial style could be illustrated on the grid. The ideal picture would be a diagonal line connecting the outermost numbers on the horizontal and vertical lines. That manager would have a 9-9 rating. He would be enthusiastic about getting the job done and concerned about the people working on the job. The worst would be a one-one rating, indicating a manager who cared little about either the job or the people.

A series of five one-week Grid seminars began March 28, 1968, in Ras Tanura. The seminars were held in a newly renovated apartment building known as the Management Training Center, the first building to be so designated by the company. Group meetings were held in the Training Center's conference room, which occupied the west wing of the U-shaped building. Eight to 10 smaller rooms were used for meetings by teams of Managerial Grid participants.

Between 50 and 55 people participated in each seminar. They were divided into teams of six to eight people each. There was one American and one Saudi instructor leading each seminar. The American instructors were at the department-head level or higher. Nearly all top management, from Thomas C. Barger, chairman of the board and chief executive officer, on down, were instructors for one or more of these sessions. The Saudis selected as instructors held supervisory or professional jobs. Ali Al-Naimi, then manager of Abqaiq Producing, was one of the Saudi instructors.

The Managerial Grid teams were composed of people from different organizations and different organizational levels. Each team included Saudi and American employees, as well as personnel of other nationalities. No person was on a team with his boss.

Joe Mahon, manager of the Organization and Industrial Engineering Department, was reassigned as organization development coordinator in charge of the program. For the next year his full-time job was to identify and train instructors, line up participants, and see to it that each weekly session went smoothly. Every Friday he attended a kickoff dinner at the Surf House Community Center in Ras Tanura and gave an introductory speech to members of that week's group.

Training Department personnel under Al Lampman and Jamil Milhem handled the day-to-day chores. They kept the records, provided participants with textbooks and other materials, arranged for the participants' meals and sleeping quarters and saw to it that the conference center was cleaned and ready for each day's session.

The Grid system used intensive group discussion and interaction techniques similar to those used in group-therapy sessions. The candid interchange of opinions encouraged during the program caused considerable anxiety. The most dreaded moment came on the next to last day of the seminar, when team members appraised each other. The individual being appraised stood at a flip chart, marker in hand, to record the comments while other team members discussed how to describe him, and gave examples to support their statements.

"There was a great deal of apprehension about these sessions in the beginning," Mahon said. "People were afraid of the personal appraisal activity — having team members comment on their performance. They were concerned about the discussion on intercultural barriers between Saudis and Americans. We had several people who we had to send to the hospital for treatment of anxiety symptoms. In time, these situations disappeared as people learned that their coworkers had attended the Grid and reported they liked it. The foreboding and apprehension declined."

More than 90 percent of the participants polled at the end of the weekly sessions said they felt what they learned would be useful to them on the job. The last day and a half of the seminar was devoted to a discussion of relationships between various nationalities in the work force. Many participants found this part especially interesting.

One Saudi said afterward: "This is probably the first time in Aramco's history that Saudis and Americans held work discussions in an atmosphere of sincerity and candor. I think the mixing of the two cultures was long overdue, and I strongly believe that it has been in the best interests of Aramco."

An American stated: "I have a warm, good feeling for some Saudis that I otherwise would not have known, and this washes over to other Saudis."

Another Saudi concluded: "I think both groups generally thought better of each other after the Grid experience."

By October 1970, a total of 27 phase-one seminars had been held. Some 1,400 people had attended the seminars, among them about 50 women, mostly from the Medical Department. Participants included 760 Americans, 430 Saudis and 160 employees of other nationalities. The large number of American participants reflected the preponderance of Americans in managerial and supervisory positions. Despite this disparity in numbers, the main purpose of the Managerial Grid, according to Mel LaFrenz, director of Management Development, was to benefit Saudi Arabs.

"The objective was to improve Saudi Arabs so they could move into positions of greater responsibility," LaFrenz said. "Most of the Americans attending there were going nowhere but eventually home. This was aimed at Saudi management development."

A team of Aramco employees consults on a Managerial Grid problem.

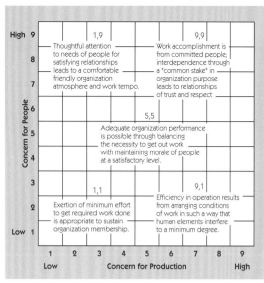

The Managerial Grid.

Phase two of the Grid explored relationships among groups that worked together daily. Teams were composed of a manager and his subordinates. This phase included discussion of what the group's ideal working relationship would be, and how this ideal compared to the actual relationship. It also involved appraisal by the team of the manager's managerial style, and agreement among team members on improvement projects for individual members and the team as a whole.

The opening phase-two seminar was held May 1-6, 1969, at Guest House 1256 in Dhahran. Participants included Tom Barger, Aramco board chairman and chief executive officer; Aramco President Robert Brougham; Senior Vice President Liston Hills; General Counsel William L. Owen; six vice presidents of the company, and Robert W. Ryrholm, president of Aramco Overseas Company. Tony Pearson, general manager of Scientific Methods, Inc., served as monitor of the session.

Management's initial reaction to phase two was favorable, but their enthusiasm gradually waned. Mahon, in a report prepared for Aramco's managers in October 1969, five months after the first phase two session, wrote: "I think it's fair to say that most of the members of the Executive Management group feel that they have not made any outstanding breakthroughs as a result of phase two activity to date. The reaction of the group at the time the seminar was completed was generally favorable. ... I think the general feeling of the group at the moment, however, is that our visible, tangible progress has been disappointing."

During the next year, only 15 organizations with a total of 95 participants conducted phase-two seminars. Shortly thereafter, the Ras Tanura series of Managerial Grid seminars was discontinued. They were dropped despite the caution by its developers, Drs. Blake and Mouton, that there would not be significant changes in performance from the first phase of the program alone. The learning of phase one must be reinforced by phases two and three, where the ideas of phase one are applied to the work situation, Blake and Mouton contended.

LaFrenz said he tried and failed to sell management on continuing all phases of the program on a regular basis. "The question they asked was, 'What's the bottom line? Are we really going to improve our ability to be effective?'

"They felt Managerial Grid was an interesting personal experience, but no one was able to point out to them how it had benefited Aramco. When you have a group of men whose dedication is to the production of oil, the shipping of oil and the refining of oil, the things that the human-relations people are interested in doing sometimes get slight consideration."

A few more Managerial Grid seminars were held in Dhahran at long intervals between 1970 and 1973. After that no more Grid sessions were held for four years. Then, in early 1977, the program was revived for two sessions. Bill O'Grady, director of Training, and Hassan Natour, a Training Department staff advisor, were the instructors. But that was the end of Managerial Grid programs at Aramco. O'Grady lamented the program's passing:

"It was a very intensive program. Many times participants, who started with scheduled courses early in the morning and then turned to team activities, would still be working at two or three o'clock in the morning. People became deeply enmeshed in the activity. Many people think they benefited from it. Aramco made a mistake in not carrying it through."

Despite its premature end, however, Mahon thought the Grid program had a lasting impact on employee relationships. "The intercultural activity provided major benefits," he said. "I think people came to see that the major barriers to performance were barriers between people and not barriers between cultures. These discussions let off a lot of steam, and people came to see that intercultural differences weren't a big deal."

Other, less ambitious programs concerned with improving communication in the ranks followed in the wake of the Managerial Grid. In January 1970, the company initiated a 32-hour communications seminar for managers and supervisors based on a program developed by Standard Oil Company of New Jersey for use at Standard's South American and European subsidiaries. The seminars, four hours daily for eight working days, were held at the ITC in Dhahran. Administrators for the program included Ellis "Ed" Hill, John Kaharl, Hassan Natour and Jamil Milhem, all of the Training Department.

In addition, the Supervisory Training Program, better known as the "160-hour course," begun in 1967 and shut down in 1968, was revived and modified in 1970 to put more emphasis on the "why" of Aramco's policies and procedures. These changes were made in line with the Managerial Grid philosophy that a fully informed employee can better align his or her personal goals with those of the company.

Hiring Remains Restricted

For nearly two decades prior to 1970, Aramco took in very few new employees. In the last five years of that period, the company averaged fewer than 60 new-hires a year. A "New-Hire Study" presented by the Training Department to the Executive Advisory Committee on February 3, 1970, recommended a moratorium on the hiring of Saudis during the coming year. The only exceptions, the study said, "should be those cases where a Saudi applies for employment who essentially possesses the education and skills to fill an open position. In no case should Saudis be hired who require additional long-term training." The recommendation was based on a survey that showed there was an adequate number of Saudis in training to cover openings for technical jobs in the near future.

The hiring moratorium could not be extended indefinitely. The Saudi work force was made up mostly of long-term employees who fit into a relatively narrow age range. Many of the employees would be reaching normal retirement age about the same time. The company needed to develop and train a number of younger men who would be ready to step in when the Saudi veterans retired.

Aramco's official policy was to preferentially hire Saudis for all positions for which they were qualified, as long as Americans continued to run the company. It was, after all, the U.S. partners' money that was financing the company and taking the risks.

A policy statement issued February 1, 1971, and signed by Liston Hills, the new chairman of the board and chief executive officer of Aramco, declared: "The company will utilize Saudi Arab nationals at all levels for which they are qualified, including jobs in the management group, except that sufficient Americans will be employed to maintain American direction of the enterprise." The statement also said that Americans were to be employed to direct the company and "on other work, when qualified Saudi Arabs are not available, and other nationals do not meet the conditions of employment."

Expectations for Saudi Vocational Schools

Aramco had hoped to find a source of fresh talent in the new system of Saudi vocational schools. When the Hofuf Intermediate Vocational School opened in 1968, operated by German educators who were contracted to the Saudi government, the headmaster predicted that within five to seven years the school would enroll 400 to 500 students. "The graduates will be nearly equivalent in skills to European vocational school graduates," he said. Similar vocational secondary schools were scheduled to come on line soon in Riyadh, Jiddah, Makkah, Madinah and Dammam.

The 1969 Operating Plan for Industrial Training confidently predicted: "Saudis who will be hired by the company during the next 10 years will have acquired their basic training through the government school system. Job-related training for these employees will be limited to some special language training, and job-skill training will be limited to specialized high-skill-level courses."

But it didn't happen that way. There were not nearly enough vocational school graduates to satisfy the needs of Aramco, as well as the needs of the Saudi government and of the Kingdom's growing industrial sector. Skilled and semiskilled workers were needed to help implement the Saudi government's ambitious new five-year plan, which focused on improving the country's defense capabilities and the diversification of its economy. The vocational training centers might eventually supply 1,600 graduates a year, but that was a mere trickle compared to the estimated need for 61,000 skilled and semiskilled workmen in the Kingdom.

The company looked for other sources of new talent. It considered recruiting students from the College of Petroleum and Minerals (CPM). Under one plan, selected CPM students who had completed the first year of the Applied Engineering Program would attend advanced industrial training courses at Aramco for half a day and perform on-the-job training assignments the rest of each day. The college was receptive to the idea. However, an Aramco delegation that visited the school wasn't encouraging about the project. Bill O'Grady, director of Training, reported to executive management: "Our enthusiasm for this project and for what it offers has been slightly dampened by what our people saw at CPM. They [the Aramco delegation] now feel that it might take all of three years to bring the CPM students to the point where they could complete our advanced industrial training courses."

The Apprenticeship Program

Finally, as it had at other times in the past, Aramco decided it would have to develop its own new talent. An apprenticeship program was adopted. It was not a new idea. The company first considered such a program in 1964, but shelved the idea because the Saudi government had no labor laws covering apprenticeship programs. That roadblock was removed in November 1969 when the government adopted new Labor and Workmen Regulations that included a definition of apprenticeship programs. Shortly thereafter, the company organized an Apprenticeship Task Force to develop specific recommendations for an apprenticeship program. The seven-member task force, under the chairmanship of Training's Bill Griffin, met between March 1 and April 25, 1970. Task force members in addition to Griffin were Neville Robinson, O.C.P. Camp, 'Abd Al-Rahman Bubshayt, Ali I. Aburayyan, James A. Bowler and James E. Jay. Alternates were R.F. Mount and John C. Tarvin. Its recommendations were presented to the Executive Advisory Council on May 26, 1970, and adopted with minor changes. The plan was also approved by the Saudi Ministry of Labor and Social Affairs.

An apprentice checks a pump and motor as part of his on-the-job training.

The company's apprenticeship program was to be a five-year program run by the Training Department. During the first three years, apprentices would take basic core classes in English and math, and learn the use of hand tools and simple machine tools. In the final two years, they would be placed in specialized programs and allocated to line organizations for training as craftsmen, plant operators or technicians. Only those who successfully completed the program would be offered jobs as regular Aramco employees.

Griffin called the apprenticeship program "an economical and logical approach — the best system yet developed for screening individuals prior to regular employment."

The company, relying on forecasts that Aramco's manpower needs would continue to decline, tentatively limited apprenticeship classes to 100 candidates per year. The assumption was that 50 to 60 out of each class of 100 apprentices would complete the five-year program. That would be enough to offset the annual attrition that the company expected in the veteran work force.

One of Aramco's first apprenticeship classes.

"Those apprentices who complete the program," O'Grady told executive management in September 1970, "will be an extremely attractive pool of talent from which to hire for company needs. But the company would not be under any obligation to offer a job to an apprentice," he said, "nor would the apprentice be under any obligation to accept a job with Aramco."

Although Aramco did not advertise for apprentices, word got around, and more than 400 applications were received during the summer of 1970. Candidates for the program had to be at least 16 years old, and graduates of an intermediate school (nine years) or higher. At first, Training wanted to require a high-school degree (12 years), but investigation revealed the pool of candidates would be too small. Fewer than 2,000 students graduated from high schools in the Kingdom in 1970 compared to more than 10,000 graduates from intermediate schools. Applicants also had to score better than 50 percent on the company's General Training Aptitude Test, obtain at least English-three and math-four levels on placement tests and pass medical and vision tests.

The company's Employment and Placement Office tested the applicants and carefully selected the top 100 candidates based on test scores. Of the 100 candidates, 20 scored high enough on tests to cause trainers to believe they might complete the program in four years instead of five.

At the last minute, the Comptroller's Office requested an apprenticeship program for office workers. A three-year program for 15 additional apprentices per year was devised. Office-worker trainees would take a year of English, math and science, then go into a program of half-day training and half-day clerical work. With this addition, net intake for the first apprenticeship class was 115 young Saudis.

The apprenticeship program began September 1, 1970. Apprentices were enrolled in classes at the Ras Tanura ITC and ITS. They were housed in company quarters at Radhwa Camp on a double-occupancy basis, and received a stipend of SR410 (about $120) a month to start — to be raised in steps to SR630 (about $170) a month by the final year of the program. They were also entitled to free medical care and had use of all the facilities available at Radhwa Camp.

Salim Al-Aydh was one of the original 100 apprentices. He grew up in Ras Tanura, son of an Aramco employee, and dreamed of becoming a military officer or a pilot. When he completed ninth grade, however, his family encouraged him to apply for Aramco's apprenticeship program rather than join the military. Al-Aydh was both surprised and apprehensive during his first days as an apprentice. He remembered going to Dhahran for testing: "We were amazed, at least I was amazed, by the people who talked to us. They were a good example to me. I thought I would like to work hard and become like the guys who were giving us the tests. They sent us to the clinic, and later on we visited the ITS and ITC facilities in Dhahran. It was a good start."

When classes actually began at the ITC in Ras Tanura, Al-Aydh wondered whether he had made the right decision. "I came from a government school, and I think the environment [in the ITC], the building, the bulletin boards, the way they were put together, the teachers — I think it was a completely different environment. That's how I saw it. I felt like a school dropout because I did not continue high school after the ninth grade. The high school is where all of my friends were going, and here I was turning to a different program."

Yacoub I. Nasr, vocational instructor, demonstrates engine-disassembly techniques to apprentices in Dhahran, 1973.

Those fears proved groundless. Al-Aydh not only completed high school, he entered the company's out-of-Kingdom Training Program and earned a bachelor's degree in mechanical engineering from the University of Tulsa. Returning to Aramco, he advanced through a series of supervisory positions in oil and terminal operations, and was later selected as a vice president of the company.

Khalid Abubshait, son of a long-time Aramco employee, was another of the original 100 apprentices. He joined the apprenticeship program because it was the fastest and easiest way to join Aramco.

"The competition among my group was very intense. We knew from the beginning that the program was for nonemployees, which meant that if we didn't do well, we could be dropped from the program at any time. We were competing to get high grades so we could join the company. We knew also that getting high grades would qualify us for a possible scholarship after we joined Aramco," he said.

Abubshait got his scholarship and graduated in 1980 from Hamline University, St. Paul, Minnesota, with a degree in business administration. He returned to Dhahran and joined Training as an academic advisor for out-of-Kingdom Training and then went back to the States as administrator of Career Development in Houston. Abubshait became a general manager and later, director general in the office of the Minister of Petroleum and Mineral Resources of Saudi Arabia.

Employees demonstrate a turbine-balancing machine to visiting Saudi students in Dhahran Shops, 1970.

He recalled some of the apprenticeship program trainers, men like Khalil Nazzal and Wadie Abdelmalek, English teachers; Tony Shehadah, English and geography teacher; Eliyas Madouk, English teacher; Nizar Dindo, science teacher; and Ali Humaidah, math teacher.

In March 1971, six months after the start of the program, Bill Griffin, in a memorandum for office files, wrote of the original apprentices: "The general opinion of the ITC teachers and ITS instructors is that the group is, on average, the most capable group of young Saudi Arabs yet encountered at Aramco."

They were indeed a special group. More than 70 percent of them went on to pursue degrees in higher education. Many others beside Al-Aydh and Abubshait had distinguished careers with Aramco. One of them, Ali A. Al-Ajmi, became president of Petron Corp., a Saudi Aramco downstream affiliate and the largest refiner and marketer of petroleum products in the Philippines. Others who reached high-level positions in the company included Zaki M. Al-Jishi of Southern Area Producing and Fahad S. Al-Aboudi and Saeed A. Al-Khabaz, both prominent in Training.

Unexpected Expansion

By early 1971, it was apparent that the rapidly increasing demand for oil in Western Europe, Japan and the United States called for production increases and expansion at Aramco beyond anything anticipated. Estimates of future employment needs made only a few months earlier now seemed greatly understated. That year, Saudi Arabia was recognized by the international petroleum press as the leading oil-exporting nation in the world, with Aramco supplying about 98 percent of the Kingdom's petroleum exports.

In response to the changing picture, the size of the apprentice class was tripled and the five-year program was modified. Instead of 100 apprentices, as originally planned, the company took in 339 during 1971. The majority of new apprentices were assigned to Dhahran, where more living space and more classroom space was available. The program was modified so that about two-thirds of the apprentices would follow a three-year regime, while the other one-third, the most academically able, would continue in the five-year pattern. In addition, 20 Dammam Vocational School graduates entered a special one-year apprenticeship program for the Producing and Terminal organizations.

December 1971 saw the first direct "raids," as the Training Department called them, on the apprenticeship program. Eager for full-time help immediately to meet increased production quotas, the Refinery and Terminal Organization offered the 20 apprentices from the Dammam Vocational School jobs as regular Aramco employees. Almost all the apprentices accepted. They were the first Saudis, other than college graduates, hired for full-time jobs by Aramco in several years.

The apprentices who accepted full-time employment with Aramco took lower-level positions than they had been targeted for following their apprenticeship, but, as regular employees, they were paid much better than apprentices. In addition, their dependents became eligible for free medical care and other company

Clerical apprentices practice typing, 1973.

benefits. Many of these young men, it later developed, supported their parents or were already married and were parents themselves. The free medical care was more important to them than the salary increase they received.

The apprenticeship program continued to expand and change during 1972. The company took on 788 apprentices, 449 more than the previous year. The original five-year program was abandoned altogether in favor of the three-year program introduced a year earlier. Qualifications were relaxed so that candidates who tested at English level two and math three, instead of English three and math four, were accepted as apprentices. Seventy more apprentices were hired away from the program by line organizations.

Trainers urged the line organizations to wait for apprentices to complete the program before offering them jobs. "What would you rather have on your payroll, a dummy or a trained man?" they asked. The answer was obviously "a trained man," but severely understaffed organizations continued "raiding" anyway.

Line organizations took 347 more apprentices during 1973. Producing took the most, 155, followed by Plants and Pipelines with 97. By year's end, less than half of the 1,912 apprentices recruited since 1970 remained in the program. "We have written our apprenticeship program in disappearing ink," Jim Knight remarked. "We are eating our seed corn," Jim Ehl declared.

In 1973, for the first time since the apprenticeship program started, there was a large-scale addition of new personnel outside that program. The company hired nearly 1,800 new employees during the year, the majority of them unskilled workers who did not meet apprenticeship specifications. In November of that year, the apprenticeship program was split into two parts. There was an "A" program for ninth-grade graduates or higher, and a "B" program for those 16 years old or older who had completed the sixth, seventh or eighth year in government schools. The "B" program apprentices would attend ITC or ITS classes for one year and then complete their training by spending one more year in on-the-job training. "A"-level apprentices continued to follow a three-year pattern. The company recruited 478 apprentices for the "A" program and 135 for the "B" program during 1973.

Apprenticeship Program Reevaluated

Aramco's executive management, facing an increasingly severe labor shortage, created a two-man committee in early 1974 to analyze the effectiveness of the apprenticeship program. The committee, composed of John I. Tucker of the Training Department and Khalid A. Mulhim, supervisor of Saudi Employment, recommended to executive management on October 5, 1974, that the apprenticeship program be discontinued.

Instructor demonstrates a chemistry experiment for apprentices in an advanced ITC science class, 1973.

The two-man committee found an "alarming" attrition rate in both the "A" and "B" programs — nearly 50 percent a year and rising. Most of the intermediate-school graduates who quit the "A" program left to take higher-paying jobs

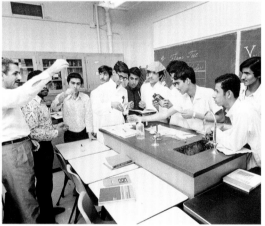

elsewhere. Competition within the Kingdom for the relatively small pool of intermediate-school graduates had become so intense, the committee said, it was unlikely that the company could recruit even 50 percent of the total "A" program apprentices called for in the long-range manpower program. As to "B"-class apprentices, nearly 75 percent of those who quit the "B" program did so to take jobs as regular Aramco employees.

"The Saudi manpower picture in 1974 is very much different from that which existed in 1970 [when the apprenticeship program began]. If the situation that Aramco faced in 1970 can be characterized as a buyer's market, with a surplus of employment candidates competing for substantially fewer job opportunities, then 1974 is a seller's market as far as Saudi Arab intermediate graduates are concerned," the report said.

In conclusion, the committee recommended that "Aramco convert the present apprenticeship program to industrial training programs and bring Saudis

Khalid Abubshait

"Young Saudis should take advantage of every opportunity available to them anywhere in the company. Opportunity only knocks once."

or Khalid Abubshait, working for Aramco "was an ambition that I wanted to achieve. My father worked for the company for nearly 40 years before his retirement in 1988. So I consider myself a second-generation Aramcon."

At first, it appeared that young Abubshait might not have an opportunity to realize his aspiration. "Unfortunately, at that time, Aramco was hiring only college graduates or job-qualified applicants," he recalled. "However, in the middle of 1970, Aramco's first apprenticeship program for nonemployees was introduced. It was the fastest and easiest way to join Aramco, so I decided to join this program."

The camaraderie of apprenticeship training made a lasting impression on Abubshait. "I remember most of my classmates," he said. "We even knew each other's Aramco ID numbers. Over the years we developed a very close relationship, and we keep in touch with most of our classmates who are still working with the company."

Abubshait set lofty goals for himself during his period as an apprentice. "My hope was to get high grades, join the company and get a scholarship," he remembered. He achieved all of that and more. In 1980, Abubshait was awarded a bachelor's degree in business administration from Hamline University in St. Paul, Minnesota. In 1995 he completed the Executive Program at the University of Virginia.

"At the beginning, I was very proud to get a scholarship from the company,"

he said. "But the turning point in my career was after graduation, when I joined my sponsoring organization, Training & Career Development.

"Being the product of the Out-of-Kingdom Training Program and working as an academic advisor in Dhahran and Houston — and later as administrator of Career Development at ASC — gave me an opportunity to understand the training process. I was able to serve the company and its sponsored trainees in the best way possible. I am very proud and honored to have been a part of this organization, whose mission is to train Saudi nationals."

In 1989, after more than a decade with Training, Abubshait was named manager of Saudi Aramco Affairs, first in Riyadh, and later in the Eastern Province. In 1995 he was selected by HE Ali Al-Naimi to serve as the Director General of the Private Office of HE the Minister of Petroleum and Mineral Resources. He was promoted in December 1996 to General Manager/Special Assignment.

Abubshait's advice for young Saudis enrolling in today's training courses is simple: "My advice to them is to work hard to achieve excellence in their individual training programs and personal development. They should remember that success always means more when it happens to someone who has truly earned it."

into these programs as regular employees with equal compensation and benefits. We believe that by offering regular employee status Aramco will substantially improve its ability to attract intermediate-school graduates to join the work force and reduce training program attrition."

The recommendations were approved by management, and the company's first apprenticeship program was officially terminated as of September 30, 1974. Aramco had accepted 2,763 apprentices during the four years of the program. Of these, 930 quit the program, 525 became regular Aramco employees, and as of September 30, 1974, there were 1,308 apprentices converted to employees and classified as Industrial Trainees.

The apprenticeship program was designed to satisfy the company's needs under conditions that had prevailed for nearly two decades prior to 1970, but the world and the oil industry had changed in ways and to an extent largely unforeseen when the apprenticeship program was created.

Growing Pains

ramco strained to take advantage of these changes — most notably a sharp increase in demand for crude oil from the West and a nearly 400-percent rise in the price of oil. Between 1970 and 1974, the company's oil exports to North America jumped from 13 million barrels a year to 150 million barrels a year. During those same four years, the Western world enacted strict environmental laws, encountered an "oil embargo" related to the fourth Arab-Israeli war, and felt the pangs of a fuel shortage. OPEC imposed production cutbacks just at the moment when demand from Western Europe, Japan and the United States was skyrocketing. This drove prices up rapidly. In a span of three months, from October 1, 1973, to January 1, 1974, the price of a barrel of Saudi Arabian light crude jumped from $2.80 to $10.84. The $8-a-barrel increase was unprecedented. For nearly 30 years prior to 1973, oil prices had hovered around the $2 level, and fluctuations of $1 or more a barrel had been considered major adjustments.

The company's annual crude oil production more than doubled in the first four years of the decade, rising from 1.3 billion barrels in 1970 to 2.99 billion a year in 1974. Average daily production of crude oil jumped from 3.5 million barrels to 8.2 million barrels in those same four years. Between 1972 and 1974,

Much of the material for expansion came through Dammam Port.

Aramco built more than 800 miles of major pipelines, drilled some 1,000 wells, built 24 GOSPs and added more than $1 billion worth of new turbines, generators and stabilizing columns to its physical plant.

In August 1972, during this period of hectic activity, the company experienced its worst industrial accident to date when a fire during pipeline repairs near Abqaiq injured 26 employees, 13 of them fatally.

The company's outlay for materials, about $65 million in 1970, shot up to nearly $450 million in 1974. In 1974, the year the new Ju'aymah Terminal opened, the company loaded 4,470 oil tankers with the aid of a modern new crude-loading system and record-sized pumps.

Far from a reduction in the number of employees, as had been forecast at the start of the decade, the Aramco work force had nearly doubled in five years — going from 10,778 workers in 1970 to 19,467 in 1975. Aramco built and operated construction-worker camps at eight sites in the Eastern Province. The overflow people were housed on barges that were stacked up to five stories high with portable buildings and anchored in the Arabian Gulf.

Saudi employees at an Aramco welding shop.

During the same five-year span, Ras Tanura grew into one of the largest oil ports in the world. Abqaiq, targeted a few years earlier as a bachelor-only community, resembled a boom town. The junior high school had reopened, housing on camp was full and landlords in Madinat Abqaiq demanded high rents from contractor employees. In Abqaiq, Dhahran and Ras Tanura, two-man trailers, prefabricated efficiency apartments, and new family and bachelor houses were added by the hundreds, along with new recreation, shopping and medical facilities. The 'Udhailiyah camp, mothballed several years earlier, was reopened and expanded to provide working and living space for some 1,500 employees involved in the projected growth of oil development, water injection and gas facilities in the Ghawar field.

The so-called "halcyon days" in the Training Department were over. In just two years, from 1969 to 1971, the number of hours of assigned ITC instruction went up by 76 percent, and the hours of instruction at Industrial Training Shops increased by 230 percent. Between 1970 and 1975, enrollment in the company's Industrial Training Centers and Shops jumped from about 1,100 to 5,500 trainees. Because hiring standards had been lowered, the department found it necessary in 1974 to add an English-one course to the ITC curriculum for the first time in eight years.

Curriculum and Test Development

Curriculum and Test Development Unit (CTDU) was formed within the Training Department in 1973. Its goal was to standardize the ITC curriculum at all levels in all districts. The unit was also responsible for standardizing tests and scoring test results. CTDU let the instructors know what material had to be taught in order for students to pass the tests. No longer were instructors able to manipulate test scores, either to satisfy the students or to make the instructor's record look good.

When CTDU's first supervisor, Dr. Charlie Johnson, arrived in Saudi Arabia in 1966, he found training districts operating more or less independently of each other. "Each ITC had its own curriculum. Although some of the same materials were taught in each district, they were not necessarily taught in the same order, and although some direction was provided by the administration in Dhahran, teachers pretty much determined what they taught in their classrooms, and they prepared, administered and scored their own tests." CTDU had the delicate task of bringing districts long accustomed to operating independently in line with each other.

"We had to create more alternates for tests and maintain security; oversee administration of tests, and even administer tests in some cases; observe and evaluate teachers in the classroom; provide orientation and training for administrators and teachers in U.S. contractor-operated training facilities; provide supplementary materials that were more culturally accessible than those available commercially off-the-shelf, and materials that expanded and/or reinforced thinking and reading and writing skills. ...

"All things considered, and given the tremendous pressures placed on both CTDU and the ITCs, I think it remarkable there was not more friction between the two groups," Johnson concluded.

Aramco firemen in training.

Between 1974 and 1976, a CTDU vocational analyst, Khaled M. Shehabi, produced the first set of ITS training manuals written in Arabic. The seven manuals covered the basic courses for shop trainees, starting with a volume on using hand tools and concluding with instruction in welding. They replaced a series of loose-leaf instruction sheets that had been created some 15 years earlier, when most industrial trainees could not read Arabic or English. The skimpy nature of the old sheets irritated Shehabi. "How," he wondered, "could Saudis be expected to learn a task from these sheets — just a picture and a couple of words in English or Arabic at the bottom? The answer was, they couldn't. They waited for an instructor to demonstrate the task and then followed his moves." That time had passed. Now, most new Saudi employees could read Arabic.

Meeting the Challenges of Growth

By the end of 1973, the wave of new trainees was challenging the department's ability to find space for training. A section of the TV building in Dhahran was turned into an ITS, the former Supervisory Training area in the Dhahran ITC was converted into classrooms, and two portable buildings were appropriated for ITC classrooms. Dormitories in a former construction camp at Ras Tanura had been converted into extra ITC classrooms, and the community's ITS was expanded into the Recreation Annex. Classes were held in makeshift facilities at the two newest Aramco communities — Safaniya to the north and 'Udhailiyah in the south. Training occupied three rooms in a portable building at Safaniya and one room in a former dormitory at 'Udhailiyah. Most of those attending classes at Safaniya or 'Udhailiyah did so on their own time, as volunteer students in English or math. A few employees were sent to the ITCs on overtime, but as yet there was no allowance for regular assigned-time classroom training at these two bachelor camps.

The space shortage was especially acute at Abqaiq. At one point the Abqaiq ITC was running from 7 a.m. to 9 p.m. to accommodate shift workers and make maximum use of classroom space. The Abqaiq Junior High School had reopened, but only eight small classrooms were available for industrial training, six in the Main Gate Building and two in the Preventive Maintenance Building. The department considered, but finally rejected, a proposal to convert barges anchored in the Arabian Gulf into floating classrooms. The space crisis eased somewhat in 1975, when the old Abqaiq ITC building, taken away from Training for use as housing during the slowdown in 1970, was returned to the Training Department.

For the first time, the company located training facilities several miles distant from an Aramco compound. In September 1972, Management Training operations moved from Dhahran and Ras Tanura to leased space on the top floor of the Saudi Cement building in Dammam. Instructors and students who could not commute to their homes lived either in Dhahran or in bachelor quarters leased by the company in the Bin Ali and Lulu Palace buildings in Dammam. About 350 Saudis completed Management Training courses during the first year of operations in Dammam. The center offered eight separate courses, the longest being the 160-hour Supervisory Training Program. Four staff members, Fahmi Basrawi, Ellis "Ed" Hill, John Kaharl and Jim Sitar, manned the center.

That same year Aramco leased space for industrial training classes in a building under construction near the Dammam railroad station and in a former

hotel in the al-Hasa oasis town of Mubarraz, near Hofuf. The two leased facilities,
when finally completed and renovated,
would have room for up to 1,000 trainees.
At first, contractors were hired by Aramco
to teach classes at these centers, but within a year the company decided the contractors' training operation was not up
to company standards, and the Training
Department took over the centers.

Ali Dialdin, then a staff advisor for out-
of-Kingdom Training, traveled as far as Mal-
ta, five hours' flying time from Dhahran, in
search of additional training space. In 1974, he found an abandoned refinery in
Egypt, near the city of Suez, at the north end of the Gulf of Suez. The former
Suez Oil Company refinery was a shambles after a battering by Israeli artillery
during the 1973 war, but a training center adjacent to the refinery had not been
hit. The building was leased by Aramco and renovated into a facility with room
for about 165 trainees. The training center was staffed largely by Egyptians,
using training programs supplied by Aramco. It offered special craft training in
areas such as machine operation, basic electrical work, the basics of instrumentation and air conditioning and refrigeration. Trainees were assigned to Suez for
terms of nine months to two years. The Suez center had a unique advantage
over other training centers. "The refinery was left just as it was when the Israelis
destroyed it," Dialdin recalled. "So our trainees used to go and pick up pieces of
equipment from the refinery for use in training. You need a pump? Go find one
in the refinery and bring it back to the shop for training."

Saudi trainees examine a model of Qurayyah Sea-water Treatment Plant.

Later on, Dialdin located classroom space for another 100 trainees at the
American University in Cairo. Dialdin hired an Egyptian travel agent to find
housing for the students and arrange bus transportation for them to and from
the university.

Training Department manpower nearly doubled between 1970 and 1975,
increasing from 258 to 591, but it was not enough to completely absorb the
increased workload. By 1975, the number of trainees in the company's Industrial
Training Centers and Shops was four-and-one-half times more than it had been in
1970, rising from 1,000 to 5,500. At some locations, classes were held in double
shifts — one full day of classes from 7 a.m. to 1 p.m. followed by another from
1:30 p.m. to 8 p.m. Some teachers were on their feet in front of classes for 10

Geography was a basic course at Industrial Training Centers.

hours a day. The department struggled to find
instructors qualified to teach English as a second
language. In most cases, Aramco recruiters could
offer teaching prospects only bachelor-status
housing, although the salaries offered were very
good by Middle East standards. The company had
limited success recruiting in Jordan, and civil strife
in Beirut ruled out trips to Lebanon, so the recruiting effort centered on Egypt and the Sudan.

"We began to find out that even in Egypt
there was a limit to the number of people who
could meet our standards (for teachers)," Bill
O'Grady said. "To match our teacher load and
staffing standards, we had to spend a great deal
of time and many, many recruiting trips to
get sufficiently well qualified instructors to meet
our needs."

Change and Turmoil in the Aramco Schools

Just as the expansion of manpower and facilities was picking up steam, a major problem arose in an area of the Training Department that had always been a point of pride: the Aramco schools operated for the children of senior-staff employees. The problem created an uproar in the expatriate community, added a new "four-letter word" to Bill O'Grady's vocabulary and resulted in a rapid turnover of Aramco school superintendents.

When Vince James, the first superintendent of Aramco Schools, retired in 1965, he was succeeded by Radcliffe R. Daly, an ex-World War II fighter pilot and former superintendent of Lincoln Park, New Jersey, schools. At the time, Aramco's schools were considered to be as well equipped as any schools in the United States, and the teachers, each with at least five years of classroom experience, were thought to be the equal of any staff in the U.S. But all sorts of new educational theories and programs had been developed in the 1960s. The notion grew that the traditional teaching system used by Aramco Schools needed to be updated.

In 1969, Daly was succeeded by Dr. Owen C. Geer, who had recently introduced a "modern K-9" program to expatriate schools at the Volta Aluminum Company in Tema, Ghana. Geer had previously held a number of teaching and administrative jobs in California, one of the first states to experiment with "individualized, team-teaching, open-classroom" educational programs. He became the chief advocate for adoption of an "open-classroom" system in Aramco Schools.

In 1970, with the ardent support of Geer and a favorable recommendation by the Training Department, the company decided to introduce an instructional system developed by the Westinghouse Learning Corp. and known as PLAN, an acronym for Program for Learning in Accordance with Needs. The new program was to be put into effect in all of Aramco's schools at all grade levels at the same time. Westinghouse people remarked that Aramco's was the first school system to convert all schools and all grades to the program at one time. Other school systems introduced the program gradually, a few classrooms at a time, they said.

Aramco's senior-staff schools introduced the controversial "open classroom" in 1970.

"In hindsight," O'Grady said, "two major mistakes had already been made that laid the groundwork for the turmoil that followed. First, we should not have taken for granted the information we were getting from the people who were proponents of the program — our own people who desperately wanted it, who saw it as the cutting edge of education. Second, we should have sent some people out to schools using PLAN, and not just to those that were recommended by the proponents or the company. I think if we had done that we would have been told that the program did have promise, but it had to be taken in very small bites and handled very, very slowly."

PLAN was to be introduced in four stages. The first stage went into effect September 1, 1970, the opening day of the new school semester. Geer sought to both explain and promote the new system through a series of articles in the company paper. "The arrival of PLAN," Geer wrote, "and adoption of an open-classroom teaching philosophy is spelling the end of the traditional classroom ... like the buggy whip holder that appeared on the first automobiles, the egg crate school with its isolation booth classrooms will disappear. ..."

"They will be replaced by a new type of school," Geer wrote. "The heart of the new school is the learning center. It is more than a library of books and audio-visual materials; it is an area where perhaps 25 percent of the student body can be found at any one time, working independently or in small groups on instructional units and projects prescribed by teaching specialists on the basis of diagnosed needs."

In fact, according to observers, it was more like bedlam. The "learning centers" were created by combining from three to five classrooms into one large room. More than 100 youngsters might be in the learning center at any one time. The noise level was too high for normal conversation. Students sat wherever they felt like sitting, came and went almost as they pleased and studied or read pretty much whatever they wanted. Other youngsters passed the time socializing with classmates in the hallway outside the center.

"The kids were somehow lost," Nimr Atiyeh, a teacher of Arabic in the Dhahran school, recalled. "They had too much freedom. The change was too much. The teachers themselves didn't know what to do. Those [teachers] who claimed to know what to do were just afraid of being called traditional, or old-fashioned or something like that."

Owen Geer, 1969.

What's more, there were problems with PLAN's computer system. The computer was supposed to score the student's tests, feed back information to the teacher and the student on future study assignments and record the student's academic history on a computer punch card. But many data entry errors were made at the schools, and there were frequent breakdowns in the system itself.

"There was tremendous unhappiness with PLAN in a relatively short period of time," O'Grady remembers. "I, as director [of Training], was receiving calls from executives, from people in the districts, from wives — it was a case of an unmitigated disaster. We were trying to cope with an extremely unhappy community."

Geer and Aramco parted ways in early 1971, only four months after the PLAN system's first stage was introduced. Geer was succeeded as superintendent of schools in February 1971, by Dr. Harlan DeBoer, the Aramco school psychologist. Under DeBoer the final three stages of the PLAN program were implemented, in the face of rising criticism from the Aramco community.

"It is apparent," the Training Department's 1971 Accountability Report acknowledged, that in implementing the PLAN system "Aramco Schools have tried to go too far too fast."

A group calling itself "Concerned Mothers" and claiming to represent a majority of the Aramco expatriate community, wrote to Liston Hills, chairman of the board and chief executive officer, to complain about PLAN.

O'Grady was called before the Executive Advisory Committee to explain the problem with Aramco Schools. "Fortunately," O'Grady said, "I was dealing with an executive management group that was going to be sensible and cool headed. I don't remember what I told them about what was going on and what we were going to do to correct the situation, but I was able to get out of that room with my head still on my shoulders. I think that was the critical point in my career."

Aramco management decided an elected school board should be established to guide the Aramco Schools, just as is done in most school districts in the United States. A divisive election battle shaped up between those who wanted to go back to the traditional, more structured classroom system and those who wanted to continue the PLAN system.

Those favoring a return to the old system alleged that Aramco community schools were not only in a state of general disorder and failing to teach students good study habits, but that the PLAN instructional material was unacceptably "liberal." Frank Fugate, manager of construction for Aramco, decided to run for the school board as an anti-PLAN candidate after reading some PLAN material brought home by his daughter.

"The school board election campaign was a hot one," Fugate said. "I was labeled as a guy who wanted to return to the one-room school."

On election day, July 1, 1972, Fugate and other opponents of PLAN won six of the seven elected school board seats. Three persons were elected to the school board from Dhahran and two each from Abqaiq and Ras Tanura. They were: from Abqaiq, Harold Streaker, general manager of Oil Processing & Movement, and Mrs. Pat Morris, an Aramco Schools teacher; from Dhahran, Fugate, along with William Wallace, a superintendent in Construction, and Mrs. Ellen Speers, a housewife; from Ras Tanura, Richard Gardner, a refinery plant operator, and Warren Otter, an engineer. The company appointed two additional board members. One of them, Richard Lawton, head of Employment Policy and Planning, had served at one time on the Hillsborough, New Jersey, school board. He was named president of the new Aramco school board. The other appointed member, John Kelberer, then general manager of Government Relations, was named vice president of the board.

The board promptly voted to abolish PLAN in all Aramco schools by the end of the 1972-73 school year. The schools would return to the traditional, single-teacher classroom system. In September 1972, Jack DeWaard, former assistant headmaster of the American Community School in Beirut, replaced DeBoer as superintendent of schools. His job was to reinstitute the old system. He was the fourth Aramco Schools superintendent in three years.

Jack DeWaard, 1972.

The relationship between Aramco and PLAN ended in a fittingly unusual way. The company could not simply stop using PLAN. All PLAN materials had to be disposed of in a manner prescribed by contractual agreements between Aramco and Westinghouse. A representative of Westinghouse came to Dhahran to witness the execution. DeWaard spent half a day at the computer calling up and deleting each individual unit of the PLAN program while the Westinghouse man noted each deletion on a form. In addition, two truckloads of printed PLAN materials were burned in an incinerator at the company's garbage dump before witnesses. Hal Fogelquist, vice president of Corporate Planning and Administration; DeWaard, the new school superintendent; and the Westinghouse representative watched the materials burn. Some goats that were hanging around the dump chewed up any papers that missed going into the incinerator. Afterward, Fogelquist formally notified Westinghouse that all PLAN materials in Aramco's possession had been destroyed or deleted and the company was no longer using the PLAN program.

The PLAN controversy drove some parents to seek schooling for their children outside Aramco. One option was the Dhahran Academy, a school opened for the children of the American Consulate staff in the early 1960s. The consulate employed eight Americans on its staff in 1972, and they had just two children of school age among them. But enrollment at the Dhahran Academy in January of that year was 600 students, indicating that 598 students came from outside the consulate, many from Aramco.

The Aramco school board handled many controversial issues in the next few years, including whether corporal punishment should be permitted in Aramco's schools, but none so hotly debated as the PLAN issue.

PLAN was not easily forgotten by those who experienced it. When the acronym "PLAN" was mentioned to Bill O'Grady some 20 years later, he promptly snapped, "That's a four-letter word!"

By October 1973, parent-teacher associations (PTAs) were formed in Abqaiq, Dhahran and Ras Tanura to provide a channel of communication between the board and parents. The PTAs were also made responsible for supervision of school-board elections. The Parents' Advisory Committees, forerunners of the PTA, had been abolished when the school board was established.

Changes in Aramco's Relationship with the Government

In the early 1970s, the Saudi Arabian government took two giant steps toward a complete buyout of Aramco. First, on February 21, 1973, the Saudi government acquired a 25-percent interest in Aramco's crude oil concession rights, crude oil production and production facilities. Government participation in the ownership of Aramco, long a matter of discussion, had become a reality. Then, in 1974, the government's interest was increased to 60 percent, and negotiations continued toward 100-percent ownership by the government. From that point on, it was largely Saudi Arabia's money that was being used to run the company and Saudi Arabia's prerogative to decide where the money would be spent, how it would be spent, and who would be in charge of company affairs.

Frank Jungers, newly elected chief executive officer of Aramco, said the participation agreement gave further impetus to management training for Saudis.

"This was not so much because percentage increases in Saudi government ownership demanded automatic corresponding increases in Saudi management," he explained, "but rather, because we wanted to be sure that Saudis would be ready to move up. Realistically, we had to expect that participation could accelerate demands for Saudi management. Perhaps our good record in this area was the reason that Saudi utilization never became an issue in participation discussions."

In 1972, shortly before the first participation agreement was announced, Aramco's Management Development Committee ordered a study to forecast what percentage of Aramco's managerial and professional work force would be Saudi by 1990. The study found that if Saudization continued at its current pace, Saudis would occupy just 28 percent of the executive and department-head positions by 1990. It further predicted that, unless the pace of Saudization increased, Saudis would hold only 54 percent of the managerial and professional jobs in the company by 1990.

Frank Jungers, right, congratulates 'Abd al-'Aziz Mohammad Al-Shalfan, employee No. 4, on the receipt of his 40-year service award.

When this study was presented to the Executive Advisory Council, it was obvious that management expected Saudis to move into the company's top jobs at a much more rapid pace than projected by the study. Joe Mahon, one of the authors of the task force report, recalled: "The consensus in management was that the figures may be right, but the answer is wrong."

At the time, Saudis held only 12 percent of the company's management jobs. Four of the company's 44 departments were headed by Saudis — Mustafa al-Khan Buahmad, director of Employee Relations; Faysal M. Al-Bassam, manager of Public Relations; Ali I. Al-Naimi, manager of Southern Area Producing; and Said M. Tahir, manager of Local Industrial Development. Saudis also held 22 division-head and key staff posts. The top-ranking Saudi in the Training Department was Mustafa Husam Al-Din, manager of Aramco-Built Government Schools.

Executive Management's determination to prepare more Saudis for jobs in the company's upper ranks was reflected in training statistics. Saudis generally received professional and management training through courses at the company's Management Training Center and by way of company scholarships to colleges and universities in Saudi Arabia and overseas. In 1972, there were 71 Saudis attending colleges and universities on full-time Aramco scholarships. That same year, a total of 174 people, expatriates as well as Saudis, attended management training courses. Within three years these figures doubled and tripled. By 1975, the company had 166 Saudis on full-time scholarships in colleges and universities and 580 persons enrolled in courses at the Management Training Center, then newly relocated in portable buildings alongside the school on Third Street in Dhahran.

Ali Dialdin became superintendent of the Training Department's Advanced Training Division in 1975. In this capacity he was responsible for the Saudi Development Unit, out-of-Kingdom Training, Continuing Education, Management Training, the Arabic Language Program, and the Cooperative Program with the University of Petroleum and Minerals (UPM). These were not entirely new responsibilities for him. Dialdin's first job after he joined Training in 1969 had been as a staff advisor for out-of-Kingdom Training. Early in his training career, Dialdin had initiated the Cooperative Program with the University of Petroleum and Minerals. In this program, applied-engineering students who had completed their junior year joined Aramco for six months to gain work experience and earn college credit. The program started in 1969 with six students, and over the years grew to as many as 85 students a year. Later on, Dialdin served as assistant superintendent of the Industrial Training Division and as acting superintendent of the Advanced Training Division.

Training's Ali Dialdin, standing, fifth from right, with 1972 Aramco scholarship students, including Ali Twairqi, seated, fourth from right, who later succeeded Dialdin as head of Training.

Dialdin was not easily deterred when he wanted to get a job done. For example, he told how, as a relative newcomer to the Training Department, he took it on himself to do something about the overcrowding of training facilities in Dhahran, despite his bosses' claims that everything possible had already been done. To the surprise of his superiors, Dialdin managed to locate and secure for the department's use several portable buildings, a rare commodity in those days. Then there was the story of his plan to change Aramco's policy of not paying for wives to accompany their husbands on out-of-Kingdom training assignments. In this endeavor he was aided and abetted by Mel LaFrenz, director of Management Development.

In the late 1940s and early 1950s, when Aramco began sending Saudi employees to colleges and universities in Lebanon and Egypt, the company gave married men an allowance so they could take along their wives and their children, if any. But the family allowance was discontinued when the company began sending sizable numbers of Saudis to colleges in the States. A few wives still accompanied their husbands, at the husbands' expense, but most women stayed behind. Students normally returned home just once a year, resulting in lengthy separations from wife and family. This caused serious problems, problems that grew worse as the period of separation grew. A husband, returning home after five years at a college or university overseas might feel he had little left in common with the woman who stayed behind. The situation too often led to divorce.

Dialdin submitted a proposal in 1971 to reverse the policy and permit wives to join their husbands on out-of-Kingdom assignments at company expense. The proposal was rejected as being too expensive, but that didn't end the matter. Sometime in 1975, Dialdin and LaFrenz devised a plan to bring the problem before Frank Jungers in a dramatic way.

Jungers, the company's board chairman and chief executive officer, and a man with a reputation for supporting Saudis and Saudization, was invited to dinner at LaFrenz's home. "Across the table from Frank we seated this Saudi lady and her husband," Dialdin said. "We made it a point to explain to Frank that this lady went with her husband, at his expense, when the husband was sent to college in the States."

The lady took it from there. She described ways in which she and her husband both benefited from the out-of-Kingdom training experience. She said how unfortunate it was that other couples could not enjoy similar benefits. She spoke of the strain that long separations during out-of-Kingdom training put on the husband, his wife and their children. She described the sense of estrangement that developed as a result of these long separations and its possible consequences.

"The lady was very impressive," Dialdin said. "She was able to carry on a sophisticated conversation in a social setting, and do it in English. She was an excellent example of the benefits of having a wife accompany her husband on out-of-Kingdom training assignments.

"Jungers got the message," Dialdin said. "The very next day we submitted a proposal. We told O'Grady, 'Look, we're sure if you take this thing to the Management Committee the committee will approve it.'" O'Grady agreed, and the committee approved the proposal. Starting in 1976, the company paid for wives and children of Saudi employees to accompany their husbands on out-of-Kingdom training assignments.

Years later Jungers remembered the woman and the conversation, but did not recall that it was directly tied to the policy decision on wives accompanying their husbands on out-of-Kingdom assignments. "However," Jungers said, "it may well have heightened my belief that wives should be with husbands on long assignments. Otherwise, we would have been confronted with many family problems and with the difficulty of wives not 'growing' with their husbands."

Ali Dialdin credits Jungers with instituting other changes that marked the real beginning of Saudization at Aramco. One of those changes cleared the way for more Saudis with proven ability to earn a four-year degree.

Saudi employees attending colleges and universities were divided into two groups, A and B. The group-A students were in their junior or senior years and working toward a four-year degree. The group-B students were in their freshman or sophomore years. After group-B people completed two years of college, they came up for review before the Saudi Development Committee. The committee decided whether or not they should continue their studies toward a four-year degree or leave school and return to Aramco. The committee customarily followed the desire of the line organizations that had sponsored the employee's education. More often than not, the line organization wanted the man back after two years, even though he had shown himself capable of earning a four-year degree. A few of the men who were denied a scholarship for the final two years quit Aramco and earned a four-year degree on their own, but most returned to work. Dialdin estimated that 60 to 70 percent of the Saudis who successfully completed two years of college were refused the chance to earn a bachelor's degree. These people were "victims," Dialdin said, "because without a four-year degree they had no hope of advancing into management positions."

It was an old conflict — a tug-of-war between those who would train Saudis only as far as needed to qualify them for a particular job and those who thought Saudis should be trained to the maximum of their potential. Jungers came down on the side of the latter. Things began to change, Dialdin said, after Jungers became chairman of the board and chief executive officer of Aramco in late 1973. With Jungers' support, the Saudi Development Committee became more independent. If a man had demonstrated the capability to finish college, the committee would help him do it, whether or not the sponsoring line organization approved. Often this was a matter of finding a new organization to sponsor the final years of the man's college education.

"The foresight was not there until Jungers became chairman," Dialdin said. "True Saudization began to take place after that."

Saudization Progresses

t the end of 1974, Jungers' first full year as CEO and board chairman, Saudis held 366 of the 820 supervisory positions in Aramco, an increase of 23 percent over the previous year. Faysal M. Al-Bassam was elected vice president of Public Affairs in August 1974. He was the first Saudi Arab elected to the post of vice president at Aramco. A month later, Ali Al-Naimi, manager of Southern Area Producing, was promoted to manager of Northern Area Producing, a department that administered 11 of the company's 15 producing oil fields. He was succeeded as manager of Southern Area Producing by Abdelaziz Al-Hokail.

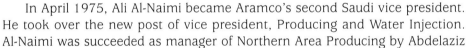

Saudis held about 20 percent of the top management jobs in Aramco in 1974. Two years earlier, Saudis had occupied only 12 percent of these key positions. In 1974, there were 12 Saudi Arab department heads and 39 Saudi division heads, an increase of eight department heads and 17 division heads in two years. Both the Northern and Southern Area Producing Departments were under Saudi supervision for the first time, as were all the major operating divisions within the Ras Tanura Refinery. Saudis were moving swiftly into the professional ranks as well. There were 197 Saudis classified as professionals at Aramco in 1974, an increase of 60 professional employees over 1973.

Faysal M. Al-Bassam, 1972.

In April 1975, Ali Al-Naimi became Aramco's second Saudi vice president. He took over the new post of vice president, Producing and Water Injection. Al-Naimi was succeeded as manager of Northern Area Producing by Abdelaziz Al-Hokail, the same man who had taken over Southern Area Producing from Al-Naimi only seven months earlier. Al-Hokail was replaced as manager of Southern Area Producing by Sulaiman Al-Kadi.

Saudi trainee practices soldering in an electronics class, 1976.

Aramco relied more and more on contractors as the rate of expansion increased during the 1970s. Aramco payments to contracting firms topped $500 million by 1975, a more than 800 percent increase over 1970 payments. Contractors built pipelines, GOSPs, stabilizers, boilers, electrical facilities, roads and other oil-field necessities. They constructed homes and offices in the fast-growing Aramco communities. They provided services such as drilling-rig moves, equipment repair, warehousing and carpentry. One contractor owned two of the largest tractor trucks ever built, each capable of carrying a complete drilling rig.

Contractors had a problem finding skilled workers, just as Aramco did. To deal with the problem, some contractors established training programs of their own, usually in cooperation with Aramco. The first such school was the welding training school opened at Thugbah in March 1972 by Sa'ad Qahtani. A former Aramcon, Jim Ripley, came out of retirement to become the chief instructor. In order to work on projects for Aramco, graduates had to pass a certification test administered by a company representative. Out of 40 students in the first welding-school class, 16 passed the Aramco test and were certified as pipeline welders.

The number of contractor schools grew rapidly. There were training and retraining schools for positions such as heavy-equipment operator, sheet-metal worker, crane operator and rigger and pipe fitter. Some of these schools enrolled more than 100 students at a time. In mid-1973, a Contractor Training Group was formed in the Industrial Training Division of the Training Department. A handful of regular Training Department employees, supervised by George Papp, were responsible for certifying that graduates of the schools were qualified to work as contractor employees on Aramco projects. In 1975, they administered more than 500 final or certification tests at these schools. The percentage of passing grades that year ranged from 100 percent at the school for crane operators to 49 percent at the welders' school.

Further Manpower Growth Is Anticipated

ramco's streak of 34 consecutive years of increasing crude oil production ended in 1975. Aramco produced 16 percent less crude oil during that year than it did in 1974. The company averaged 6.8 million barrels a day in 1975, compared to 8.2 million barrels daily in 1974. It was the first such decline since 1941, in the dark days of World War II, and it was only a one-year phenomenon — a sudden, temporary slump in world demand for crude oil — but it showed that nothing can be taken for granted, including yearly increases in crude oil production by Aramco.

Despite the drop in production, the company remained the largest corporate producer and exporter of crude oil in the world, accounting for about 13 percent of the world's total oil production in 1975. As it had every year since commercial production began, Aramco added to its known petroleum reserves in 1975. The discovery of three new oil fields, Lawhah and Ribyan, offshore in the Arabian Gulf, and Dibdibah, in the northwest corner of the country, plus the development of other fields, added 4.4 billion barrels to the company's proved reserves, which now totaled 108 billion barrels.

Far from a rapidly decreasing Training workload as forecast at the start of the decade, enrollment at company training centers increased by more than 500 percent between 1970 and 1975. Training Department manpower more than doubled in the same five years, and the Training budget increased by some 450 percent ($38.8 million, compared to $8.6 million in 1970). During those same five years, Aramco built and turned over to the government 15 more schools. In the 21 years of the program, 52 schools had been constructed at a total cost of $79 million.

Behind this growth was a soaring worldwide demand for oil, coupled with increased oil prices that boosted Saudi Arabia's earnings from $2.7 billion in 1972 to $27.8 billion in 1974. The income provided fuel for an extraordinary government program that launched the company into a period of even faster and steeper growth.

A Time of Great Projects 1975-1980

"…this is like an international school except that the teachers are the ones coming from different countries, not the students."
– Demetrio Ignacio Asistio

In February 1975, the Saudi government unveiled a 1975-80 Development Plan calling for expenditures estimated at $143 billion. It was one of the most ambitious modernization programs ever put forward by a developing country. The five-year plan assigned top priority to construction of a multi-billion-dollar, Kingdom-wide network of diversified industries. Aramco was asked to design, develop and operate a gas-gathering and -processing system to supply fuel for this vast new industrial network.

Like everything else in the Development Plan, the size of the new gas system, known as the Master Gas System, was staggering. The project would cost an estimated $10 to $15 billion and require a total of more than 30,000 men to build. It would be one of the largest engineering and construction projects, if not the largest, ever tackled by an oil company. The system would make use of associated gas, the gas that often comes to the surface with oil. Instead of being flared off as most of it was in the past, this gas would be treated at new facilities and pipelined to industrial centers where it would fuel aluminum smelters, steel mills, petrochemical plants, water-desalination plants and electrical generators, as well as provide liquid gas for export. When completed, the Master Gas System was expected to have the capacity to harness about 4.5 billion cubic feet of gas per day — enough to supply nearly half of all the gas-fueled homes in the United States. Frank Jungers, Aramco's chairman and chief executive officer, called the company's role as a major player in this immense development project "both an honor and a challenge."

The 1975-80 Development Plan proved to be the final contribution of HM King Faysal for the benefit of the Kingdom. He died in March 1975, just a month after the five-year plan was announced. The Crown Prince, HRH Amir Khalid ibn 'Abd al-'Aziz Al Sa'ud, immediately assumed royal responsibilities for the Kingdom, and continued to support and direct the Development Plan along the lines indicated by King Faysal.

Throughout the decade of the 1970s the tempo of Aramco's unprecedented growth and expansion continued to increase. The busiest year yet was 1977. The company had three of the largest projects in the world going at one time — namely,

Opposite: Aramco's Yanbu' plant at night. Below: Saudi operator monitors gas-plant systems.

135

The Berri Gas Plant, first plant completed in the Master Gas System.

the Master Gas System; the offshore Zuluf-2 GOSP, completed in late 1977; and the Qurayyah Seawater Treatment Plant. Aramco added about $1.7 billion worth of properties, plants and equipment for its own operations in 1977 and another $2 billion in connection with the government programs in which it was involved. The number of employees reached 26,517 during 1977, eclipsing the previous record high set 25 years earlier. The average annual salary of Saudi workers topped $10,000 for the first time. In November 1977, HM King Khalid commissioned the new Berri-Jubail Gas Plant, the first plant in the Master Gas System to come on line. Site work began during the year for two giants of the system, the Shedgum and 'Uthmaniyah gas plants. Crude oil production set another record, averaging more than nine million barrels per day. In the summer of 1977 the company opened a new, larger office complex in Riyadh and completed three-story additions to the administration buildings at Abqaiq and Ras Tanura.

A Review of Saudi Development

learly, undertaking the Master Gas System project, in addition to continued expansion of the company's own oil production facilities, would require large increases in the work force. This, in turn, would further accelerate expansion of Aramco's training programs, which were already straining under enrollment increases totaling more than 500 percent over the previous five years. Given such prospects, it was no wonder that members of the Executive Committee, meeting in Dhahran on April 26, 1975, reacted with surprise and alarm to rumors of deficiencies in the Saudi Development Program and widespread confusion over the company's training goals. At first they were skeptical of such talk. Joe Mahon, then general manager of Project Planning and Construction, remembered: "Many of the people sitting around the table in Conference Room A found this incredible. They said, 'It couldn't be. We've had training for years and years. Everyone understands what we're supposed to do.'"

But when the rumors persisted, the Executive Committee decided to have Mahon survey the company's Saudi Development Program. For the next nine months he was an observer at Saudi Development meetings. During these meetings, trainers and representatives of various departments reviewed the progress of high-potential Saudi trainees and drew up future training and development schedules for them.

Mahon reported his findings to the Executive Committee on February 11, 1976. He concluded that what executive management had been hearing was true. His main points were: (1) Training practices varied widely depending on each department manager's interpretation of the company's intent in training and developing Saudis; (2) In many cases, Saudi employees were not receiving training opportunities to develop their full potential; (3) Some promising new Saudi employees were never enrolled in training programs, and no one could explain why.

Mahon came across the most intractable of all training problems in the course of his survey — whether to train a man to do a job or to continue training him until the man had realized his highest potential. More than a decade earlier, top management had chosen the option of training Saudis to their highest potential, but to some men this remained an ideal and not a practical goal. For example, Mahon reported, one of the managers was reluctant to send his best people away on college scholarships or long-term supervisory training programs. This manager saw his job as "running the plants properly" and didn't feel he should be concerned with "making engineers." Mahon admitted the man made some good points. Trainees were filling key jobs. They were needed on the job, and they needed to get operating experience to prepare them for working in the new plants coming on line. What's more, the manager said, too many candidates sent for engineering training in U.S. engineering schools failed to complete their courses and became unhappy, unproductive workers upon their return to Aramco.

Building a 700-mile-long NGL pipeline across the Arabian Peninsula.

In some areas, Mahon concluded, the best that could be hoped for was that talented Saudis would be identified and programmed for more intensive training in the future. As one supervisor said, "We must keep our operation running, so some people aren't going to go to school. That is the long and the short of it."

The problems outlined in Mahon's report were not new by any means; however, they were management problems, not Training Department problems. Mahon found that the Training Organization "had done an excellent job of working within the policies and directives given to them" by management. "Department managers, in general, appeared to be genuinely trying to carry out management's intent — they just have different ideas of what that intent is," Mahon said.

In a cover letter accompanying his report, Mahon wrote: "It may be significant that only three vice presidents/general managers participated in the Saudi Development meetings" that he attended. It seemed that executive management was not much involved in Saudi Development anymore. Saudi Development had been pushed down to another level, so upper management was not likely to be aware of what was going on, in contrast to the way it had been before the expansion of the 1970s, when top-level management was nearly always present at Saudi Development meetings.

"Some improvement in this situation can and should be made," Mahon wrote. "A starting point might be for executive management to articulate its intent on Saudi Training and Development in terms that are specific enough to reduce the current ambiguities and multiple interpretations. This may be all that is necessary."

To the best of Mahon's knowledge, management did not respond to his suggestion. No statement on training goals and policies was forthcoming. The lack of response was understandable, Mahon said, "given the context of the time." It was an inherent fault of an overworked system. "When you see the magnitude of what we were doing at the time, the fact that we didn't have everything in Training neatly tied up isn't at all surprising. We were just overwhelmed." As a consequence, several years would pass before these problems were addressed by management, and then only on an emergency basis.

Large Projects Dominate Times

n 1976 and for several years beyond, attention was focused on planning for and building a maze of high-priority projects. The engineering or preparation phase of the Master Gas System alone employed some 2,500 engineers and craftsmen and would take nearly 200 million man-hours to complete. In addition, the company was rushing to complete contractor camps, now being expanded to a capacity of more than 40,000 men. There was also a

Fast growth required the use of portable houses for new residents at 'Udhailiyah, 1977.

pressing need for more employee housing and facilities in Aramco communities. Within five years the combined population of Abqaiq, Dhahran and Ras Tanura was expected to increase by 70 percent. In addition to enlarging those communities, the company needed to establish family housing and expand bachelor accommodations at 'Udhailiyah and, possibly, at Safaniya. Along with housing would come a need for new commissaries, dining halls, recreation facilities, sanitation works, roadways, and so on. At the same time, Aramco faced a critical company-wide shortage of office space. Two or three men were often crowded into an office designed for one man. At Dhahran, temporary offices had been set up in what had been an industrial cafeteria, in buildings originally intended as nurses' quarters, and even in some family housing. More than $700 million was allocated in 1976 for new housing and new industrial support facilities such as offices, medical facilities, materials supply depots, and training facilities. Major office projects included 90,000-square-foot additions to the administration buildings at Abqaiq and Ras Tanura, and a 10-story office tower in Dhahran.

In the midst of these preparations, the company accepted yet another large and challenging project. In August 1976, after several months of negotiations between the government and Aramco, a royal decree was issued creating the Saudi Consolidated Electric Company (SCECO). The company was formed by consolidating Aramco's electrical network and the bulk of the company's generating facilities with 26 private power companies in the Eastern Province. Aramco agreed to run the new company during its first five years of operations. Because of Aramco's adherence to U.S. standards and specifications, it was decided to adopt the 60-cycle electrical standard used in the United States rather than the 50-cycle system then in regional use.

The immediate aim of SCECO was to provide the quantity of electrical power required for the industrial areas being developed at Jubail and elsewhere in the Eastern Province. SCECO was also required to furnish electrical service to some 100,000 people in 200 widely scattered communities in the province,

many of which were being served by small, unreliable local power companies. SCECO's original service area was about 110,000 square miles, roughly the size of the state of Nevada or the country of New Zealand. Its planned capacity was 5,800 megawatts — nearly three times the average daily power consumption of the city of Los Angeles.

Hassan M. Natour, O'Grady's staff advisor and an Aramco trainer since 1950, was asked to establish a training program for SCECO's Saudi recruits. He began by hiring about 25 vocational trainers, instructors, program designers and writers, primarily through Aramco recruiting offices in Jordan and Egypt. He borrowed training materials from Aramco and used Aramco classrooms in Dammam for SCECO's training. The ties to Aramco training soon loosened. SCECO designed its own training materials, launched its own apprenticeship program using training facilities at the government's vocational school in Dammam, and began awarding trainees scholarships to colleges in the United States and Lebanon. By 1980, more than 300 trainees had gone through SCECO training programs. Natour remained involved in SCECO training for the remainder of his career.

The Aramco Training Department was hungry for space: classroom and living space for trainers and trainees alike. Plans were being made to build some new training facilities and enlarge others, but most of these plans were in the early stages of development. A few projects were completed in 1976: three portable buildings were annexed to the ITC in Ras Tanura, an ITC operation began in temporary classrooms at 'Udhailiyah, and industrial training shops were nearing completion at the Dammam and Mubarraz Training Centers. Construction began during the year on a new ITS unit in Abqaiq. A proposal had been submitted for construction of a multipurpose Management Training/Continuing Education Center in Abqaiq. On the drawing board were plans for two new industrial training buildings in Dhahran, an ITC and an ITS, and the design for a new Management Training/Continuing Education Center in Dhahran. The Facilities Planning people indicated that a building occupied by a contractor in Ras Tanura could be renovated and converted into a training facility after employees of the contracting firm vacated the building.

So Much Work, So Little Space

The expanding work force strained available living space and created what the Training Department's 1976 Accountability Report called "an acute housing shortage." The report said 463 fewer trainees were taken in during 1976 than called for in the budget, partly because there was no place for the trainees to live. The housing shortage also delayed the arrival in Saudi Arabia of 33 advanced ITC trainers.

Management Training sessions had to be reduced and seven Kepner-Tregoe training sessions were canceled for lack of instructors. In order to cover ITC classes, Training hired "part-time teachers," employees who taught a few hours a day for extra pay after finishing their regular jobs. Fred Scofield, then at the Abqaiq ITC, remembered that the ranks of part-timers included medical doctors, engineers, agriculturists, and teachers from the senior-staff schools for children of Aramco employees. They taught English and math classes, mostly, with results as good as could be expected from people who, in many cases, had never before stood in front of a class.

Saudi maintenance crew in Abqaiq.

Lack of family housing made it difficult to attract and retain qualified instructors. The Accountability Report said 10 instructors resigned during 1976 due to housing problems and five others threatened to leave. The situation was aggravated by a company policy that did not permit foreign contract employees hired after February 1, 1976, to bring their families to Saudi Arabia, even if the employee agreed to accept financial responsibility for the family's housing.

The Accountability Report also noted complaints about living quarters provided by the company outside of the Aramco compound. It cited conditions at two buildings in Dammam that housed young trainees as well as members of the Dammam Training Center staff. "The lack of facilities usually found at other Aramco living areas, recreation, commissary, movies and a library, in these buildings, combined with employee control problems typical in outside facilities, have been a major problem in 1976 …," the report said. It blamed the high attrition among trainees at the Dammam Training Center, at least in part, on living conditions in these two buildings.

Chemistry classes included individual lab work.

Trained Saudi crews helped with the relocation of derricks.

As an instructor looks on, a student demonstrates a chemical experiment.

Trainees rack pipe on an Abqaiq training rig.

An electrician trainee tracks a connection in a web of wires.

Cleaning up after a project was part of industrial training.

Skilled technicians were needed to maintain instruments in working order.

140

Education Department Separated from Training

At the time, Director of Training Bill O'Grady recalled, "I was up to my knees in alligators," trying to cope with growing pains in the training area. He was spending most of his time simply trying to find living space for trainees and trainers, and classrooms for them to use, and had precious little time left over for training itself. The Training Department had grown so rapidly in complexity and size that it had become unwieldy. There were more than six times the number of trainees and four times the number of Training Department employees as there had been six years earlier. It was too much for one man to administer. It was time, management decided, to split up the department. The Aramco senior-staff schools and the Aramco-Built Government Schools program were moved from the Training Department's jurisdiction, where both had been since their inception, to the new Education Department, which assumed responsibility for the two school systems. For nearly 25 years the Aramco-Built Government Schools program had been a part of the company's cooperative relationship with the Saudi government. The senior-staff schools had played an important role in Aramco operations for more than 30 years, providing well-regulated schools that helped keep families happy and helped the company retain good employees.

A search for a director of the new Education Department began in the spring of 1976. Someone asked Gene Jacobsen, an educator working at the U.S. Consulate's school, the Dhahran Academy, if he knew of any likely candidates for the job. Jacobsen thought a former colleague, Edwin "Ed" Read, a professor and chairman of the Graduate School of Education at the University of Utah, might know of someone. Read was contacted and agreed to help in the search. He lined up 10 men he considered to be qualified candidates for the post. Bill O'Grady went to Salt Lake City in August of 1976 to interview the 10 candidates. When the interviewing was over, O'Grady returned to Aramco Services Company offices in Houston convinced that Read, rather than the men he had interviewed, would be the best man for the job. Sometime later, Read got a telephone call from Fritz Taylor, vice president of Industrial Relations in Dhahran, inviting him and his wife to come to Saudi Arabia and look the place over. Read agreed to do so.

Read attended several social events in Dhahran, where he met members of the school board, some Aramco school employees and several company executives. He was impressed. They "were really fine people," he said. He looked at Aramco's school facilities. One of these, the ultra-modern Dhahran Hills School for kindergarten through sixth grade, had opened only a year earlier. Read concluded that the facilities were better than those he was working with back home. "Frankly," he later said, "after working in public schools and public universities [in the States] as long as I had, this looked to me like it would be a great experience. ..." So Read accepted the position. He felt obliged to finish out the 1976 autumn semester at the University of Utah rather than leave the university on short notice. In the meantime, Bill Griffin, former superintendent of Industrial Training, acted as director of the Education Department. The new department had two divisions. Jack DeWaard, superintendent of

Senior-staff elementary school in Dhahran, 1975.

Aramco Schools since 1972, headed the Aramco Schools Division. Ibrahim 'Akif was the first manager of the department's Aramco-Built Government Schools Division.

Read returned to Saudi Arabia and began his Aramco career in December 1976. He originally planned to stay at Aramco for 2½ years. Instead, he stayed for almost 10 years, heading the Education Department from 1977 to 1986. During these years, enrollment at Aramco schools for expatriates soared from about 1,400 students in 1975 to a peak of more than 3,600 students in 1983. He presided over a staff that grew from about 215 people in 1976 to as many as 570 in 1982. The annual operating budget for the Education Department increased to about $25 million, larger than the operating budget of Read's former school, the University of Utah. When Read took over the department, there was only one expatriate school in each district. By 1980 there were three expatriate schools in Dhahran, two in Ras Tanura, two in Abqaiq and one in 'Udhailiyah.

"But the biggest project I ever handled in my life," Read said, "was the planning and writing of specifications for a whole new generation of schools for Saudi children in the Eastern Province." Read, in cooperation with the Ministry of Education, participated in the design of schools to be constructed, equipped and maintained by the company under the Aramco-Built Government Schools program. The new design incorporated "a lot of modern ideas that were the top ideas for America at that time," Read said. Some of the ideas were akin to those in the controversial PLAN system that had failed to win acceptance at Aramco senior-staff schools during the early 1970s. The new Aramco-built schools had clusters of four to six classrooms around open areas where students could pursue individual studies or participate in group discussions. Each school was equipped with a library, a media center, a science lab, industrial arts or home economics classrooms, a canteen, dining and activity space, and a large prayer room. In the next decade, Aramco built and turned over to the government 24 schools of this modern design. They included the first company-built high schools. These high schools, both those built for boys and separate schools for girls, came equipped with computer and office-machine centers. Aramco also renovated and improved more than 50 older company-built schools during that time.

Aramco Schools for Saudi Children Started

solution was found during Read's tenure to the longstanding conflict between the educational needs of Western children and those of Saudi children at Aramco's schools. The Aramco schools provided a high-quality Western-style curriculum and school environment for the children of expatriate employees, but they did not satisfy the educational needs of Saudi children. In the Western-oriented Aramco schools, Saudi youngsters often became more adept at reading and writing English than Arabic. As a consequence, they were unable to continue their education at Middle Eastern schools without first taking time to improve their Arabic language skills and learn more about their own heritage.

During the early 1980s, the company decided it would be best if Saudi youngsters and expatriate children attended separate schools. Aramco offered to build a private school using the very best design and materials for Saudi students in a residential area near Dhahran known as Doha. (The decision also applied to Saudi children attending senior-staff schools in

Students at a government elementary school line up to select books from Aramco's mobile library.

Abqaiq, Ras Tanura and 'Udhailiyah, but the company did not offer to build private schools for them.) The school would be an alternative to Aramco staff schools. It would provide a modern, high-quality education for children of Saudi employees using essentially the same curriculum as the Saudi government schools. Read took a personal interest in the project and deferred his retirement in order to see it through.

Read chose Dhahran Ahliyya (Private) Schools to run the Doha school. These schools had been set up by a prominent Saudi businessman, Khalid Alturki, and his wife, Sally, as nonprofit, Arabic schools with separate sections for boys and girls. Sally Alturki's first school was opened in small, renovated houses at the outskirts of al-Khobar in 1977. "I visited her school," Read recalled, "and I could see she was very capable. She had some other teachers she had selected rather carefully. Later on, I traveled all over Saudi Arabia, all the way to the west coast and even to Dubai, interviewing people who ran private schools for Arab children. After carefully and honestly looking for a better team to run the Doha school, I still came back to Khalid and Sally Alturki."

The attractive new school opened in 1985. In lieu of lease payments to Aramco, Dhahran Ahliyya Schools reserved one-half of their spaces for dependents of Aramco's Saudi employees and charged them 65 percent of the regular tuition for students. In 1995, Dhahran Ahliyya Schools had about 1,600 students, nearly half of them dependents of Aramco employees. Aramco began to phase Saudi students out of its senior-staff schools in 1984. No new Saudi students were accepted in the 1984-85 school year, and Saudis in the fourth grade or below could not continue in Aramco schools after the 1985-86 school year. The ruling affected about 40 students in Dhahran senior-staff schools, and three or fewer Saudi students in Abqaiq and Ras Tanura. Since Dhahran Ahliyya did not offer kindergarten classes at that time, Saudi children were allowed to continue attending kindergarten at Aramco senior-staff schools.

Training for Saudi Women

For Aramco, 1976 was a memorable year. During the year, Aramco became the first company ever to produce more than three billion barrels of crude oil in a single calendar year. Production averaged about 8.3 million barrels per day during 1976, a 22.6 percent increase over 1975. During 1976, the company invested a record $2 billion in properties, plants and equipment in order to carry out its own projects plus the two major development programs assigned by the Saudi government. The year also saw the start of a remarkable new episode in training history.

On January 24, 1976, Aramco opened a training center for Saudi women. The Special Clerical Training Center, as it was known, began with a student body of 10 young women and a staff of three instructors, plus a supervisor who was also a part-time instructor. The center was housed in a small, one-story brick building across the street from the Administration Building in Dhahran.

The training center had been a long time in coming. In 1971, when the first apprenticeship program for men was gearing up, Ali Dialdin proposed a similar apprenticeship program to train Saudi women for clerical work. Bill O'Grady, the director of Training, and Bill Griffin, head of Industrial Training, enthusiastically agreed. The proposal went up the chain of command. Hal Fogelquist, vice president of Corporate Planning and Administration, supported the plan, as did Frank Jungers, Aramco's chairman and chief executive officer. The proposal was submitted to the Saudi government. The Minister of Labor in Riyadh pronounced it an excellent idea. Ahmed Zaki Yamani, the Minister of Petroleum and Mineral Resources, fully supported the proposal. But the idea faced a logistical problem due to a shortage of office space. Article 160 of the Saudi labor law declared,

143

"In no case may men and women comingle in the places of work or in the accessory facilities or other appurtenances thereto." The Special Clerical Training Center program was intended to prepare Saudi women to take over office-type jobs. The company had compiled a list of 231 clerical positions, such as correspondence classifier, typist, receptionist, medical clerk, and library assistant, for which Saudi women could be trained at the center.

Circumstances were not yet right for such a program. In a memorandum to Ali Dialdin and Ed Hill in August 1973, Bill O'Grady advised: "Because of lack of space, I'm afraid we will have to forget about trying to get girls' training in operation this year." "Not only was there a space problem," O'Grady added, "but the rapidly expanding male apprenticeship program was occupying so much of the Training Department's attention there was no time to spare for starting up a women's program."

The issue was not taken up again until 1974, when the expansion of Aramco's work force had begun to outrun the number of Saudi men available for hire. In the spring of that year, in response to a letter from Jungers, the government and Aramco opened discussions on an Employment Agreement covering Saudi women. The talks culminated in a letter from Yamani in August 1975, outlining the conditions under which the government would permit a women's training program to go forward at Aramco. Foremost among these conditions was that the program be under the control and supervision of the Office of the Chief Qadi (Judge) for the Eastern Province, the Office of the Head of the Public Morality Committee in the province, and the Ministry of Petroleum and Mineral Resources. Yamani stressed that the young women must be kept separate from men during training, and any Saudi woman employed by Aramco must work only with other females under a female supervisor. After a few more meetings and a few revisions, an Employment Agreement was signed giving approval for a Special Clerical Training Program.

While awaiting approval of the program, the Training Department assembled a faculty for the training center. The department selected Anne Tandlich to head the Special Clerical Training Program. Tandlich was supervisor of the Stenograph/Vari-Type and Graphic Arts Unit in Dhahran, responsible for recruiting, training, orienting and placing all secretaries, stenographers, graphic artists and varitypists for Aramco organizations in the Dhahran area. A versatile woman, she also wrote an astronomy column for the company paper, judged the horse shows at Dhahran's Hobby Farm, collected coins and stamps and was an avid amateur photographer. However, she said she would always consider that being asked to organize and operate what was, in effect, a women's school, her most rewarding and challenging activity.

Alice Sealy reviews an English lesson.

Members of the center's first teaching staff were Selma Obeid, an experienced English teacher from Syria; Alice Sealy, an American casual employee who taught English, math and typing courses; and Khalida Al-Khayyal, an English and math teacher and the only Saudi on the staff. Al-Khayyal had no teaching experience, but she had been raised in the United States, the daughter of a Saudi diplomat, and she had a degree in psychology. Tandlich, in addition to supervising the training center, was scheduled to teach a daily one-hour office practices course once the center was up and running. All those working in the Aramco women's training program were keenly aware of the need to steer a cautious course between the company's requirements and respect for the values and practices of a Muslim society.

Douhan Al-Douhan

"The mission of training and development will grow even more complex. The maintenance technician can no longer be nomadic; he has to be someone who is comfortable with the computer."

As a teenager in the 1950s, Douhan Al-Douhan traveled across the Saudi desert from Najran to Abqaiq in the back of a truck to seek work with Aramco. His journey across the dunes — exposed to the sun and wind, packed shoulder to shoulder with more than 20 other passengers — took three days. "It was rough, rough travel, no roads to speak of," Al-Douhan said.

In those days, the two big employers were the government and Aramco. "Aramco had programs and vision," and people were talking about it, Al-Douhan recalled. He was "looking for a future," for a way to better his life. "In the United States, people said, 'Go west, young man, go west.' Here they said, 'Go east.'"

Several relatives and others of his tribe worked for Aramco, but family connections didn't get him a job. In fact, he wasn't hired that first time because he was too young. He returned to Najran, and, when he was older, headed off to Abqaiq once again. After a wait of several months, during which he worked for a local contractor as a welder's helper, his opportunity at Aramco finally came. He was given a battery of tests — arithmetic, English, Arabic — and was selected as a trainee. Al-Douhan was sent to the Industrial Training Center, where he studied English, mathematics and sciences. Then he was assigned to the Transportation Department, where he trained as an auto mechanic and worked as a shop clerk while continuing his ITC studies.

Al-Douhan decided he would pursue a college education, but college was not a requirement in Transportation. With the cooperation and encouragement of his superiors, he transferred to the Safety Department, later known as Loss Prevention.

In 1965, Al-Douhan was selected for out-of-Kingdom training as part of a group sent to Temple University High School in Philadelphia to earn secondary-school diplomas. He was the first member of his tribe ever to travel abroad for education. Al-Douhan went on to receive a bachelor of science degree in engineering technology at Memphis State University in Tennessee.

His career began a steady climb, from safety advisor to Loss Prevention engineer, to superintendent of 'Udhailiyah Producing in 1979, to manager of 'Udhailiyah Producing Operations in 1983, to director of Participation and Environmental Affairs in 1986 and to executive director of Management Services in 1988, with various posts in between.

Al-Douhan, who retired in 1995, considered Saudi Aramco's emphasis on training essential to the company's mission. Training will be even more complex in the future due to the requirements of science and technology, he predicted. But he said the trainees will be up to the task. Unlike past generations, today's young people are familiar from an early age with computers and other high-tech gear, Al-Douhan observed. In the age of digital controls, the new maintenance technician cannot simply have a superficial knowledge of the hardware. He must be interested in the systems, enjoy working with them and be willing to investigate, Al-Douhan said.

Queen Elizabeth of Great Britain arrives at Dhahran on the Concorde in 1979.

Tandlich was named head of the Special Clerical Training Center on March 1, 1974, but she had to wait until May of that year to move into the new training center building. She remembers going frequently to the construction site on Eastern Avenue to check on the progress of work on the building. After moving into the building, several construction faults were discovered, further delaying the start of training programs.

Nineteen months would pass before the Employment Agreement with the government was signed and the Special Clerical Training Center could begin its primary mission. In the meantime, Tandlich and her staff ran informal tutoring programs for female employees, most of them for Medical staff members who needed to learn more English in order to improve their job efficiency. They also started an English-language program for female residents of the main camp facilities. The classes included women from India, Pakistan, Spain, Brazil, Egypt, Lebanon and Jordan, in addition to Saudi Arabia. In 1978 this informal program developed into the Women's English as a Foreign Language (WEFL) program, covering English levels two through seven. This program was open only to Saudi women. Ali Al-Naimi's wife was in the first WEFL graduating class.

In November 1975, Tandlich and her staff conducted orientation sessions for 15 girls, all of them high school graduates, who comprised the first group of prospective students at the Special Clerical Training Center. When the center opened for business on Saturday, January 24, 1976, 10 of the 15 students who had attended the orientation sessions enrolled as students.

The 10 original students were unmarried Saudi women between the ages of 17 and 20. They attended classes for seven hours a day, five days a week. They had four hours a day of English, one hour of math, one hour of typing in English and a one-hour office practices class. The women followed the same two-year curriculum and used many of the same textbooks as the men in clerical training at the ITCs. But unlike the men, once the women arrived at the center in the morning, they were not supposed to leave the building until it was time to go home — a rule that Tandlich strictly enforced. Inside the yellow brick training center building there were four classrooms with five to seven desks each, a supervisor's office, a small teachers' lounge and a second small lounge where the students could relax and have lunch.

Tandlich described her workload during those early years as "tremendous." In addition to her administrative and teaching duties, she counseled the students and did much of the clerical work required for running the program. She got to know her young students on both a professional level and a social level. Management, however, soon came to think it would be preferable if an Arab woman supervised the center. Accordingly, Tandlich was transferred in 1977 to the Dhahran ITC, where she taught typing and headed a testing unit. A new supervisor, Samia Al-Idrissi, was named to head the Special Clerical Training Center.

Al-Idrissi had been a translation specialist in Government Affairs. She graduated from the American University in Cairo with a major in political science and economics, and joined Aramco in hopes of doing research on Middle East affairs. Al-Idrissi was selected by management to head the center because she was a Saudi, she was fluent in both Arabic and English, and she was judged to be someone under whose direction the center would expand.

They were right about the expansion. During Al-Idrissi's tenure as supervisor from 1977 to 1981, enrollment at the center grew from 40 to nearly 150 students. The size of the staff increased from three to about 12, including a counselor and a full-time secretary. The center had been enlarged in 1976 to include a second, single-story structure just north of the original building. A third building, also a one-story modular, was added in 1980. The curriculum was expanded to include chemistry and science. Mona Yamak, a new-hire from Lebanon with a bachelor's degree in chemistry, started the center's first science program. A special science room and laboratory was created and furnished with some hand-me-down equipment from the ITC. With typical esprit de corps, the supervisor and her teaching staff put the finishing touches on the science room themselves rather than wait for maintenance to complete the job.

When the first group of women graduated in the spring of 1979, the students invited Tandlich to the ceremony, gave her a front-row seat and honored her by calling her their "second mother." Tandlich retired from Aramco a few weeks later.

Mona Attiya joined the center staff in 1980 as a guidance counselor. In a few years she would become supervisor and guide the women's training center during some of its most productive years as well as some of its lean times. Attiya, wife of Ali Dialdin, graduated from San Diego State University and King Saud University in Riyadh with degrees in history and sociology. In those early days, Attiya remembers, there was a lot of pressure on students to leave school and return to the traditional life as a homemaker. She counseled between three and five young women a day. "Coming to Aramco, and being so young, was a real challenge for them. They were under pressure — peer pressure and family pressure — so they used to come to me with their problems. Some I couldn't handle. I sent them to special programs in the Aramco clinic."

Aramco Scholarships for Women

fter a few months of classroom training, female students began a regimen of a half day at work and a half day in training at the center. Most of them worked in the Administration Building, just across the street from the center, or in the Materials Supply Building a few blocks north. They worked as typists, keypunch operators or clerks. Before long, Al-Idrissi realized she had some intellectually gifted young women in her care. They routinely scored higher on tests than the average male ITC students, and were eager to find an outlet for their talents. Al-Idrissi began encouraging Aramco to provide college scholarships for high achievers among women students, just as the company did for the bright young male employees. She pushed, argued and reasoned until finally Aramco approved college scholarships for selected female students. "It was the most exciting moment of my whole career," Al-Idrissi said.

The first three Saudi women accepted for college scholarships in the United States were Nadia Al-Shihabi, Badria Al-Sindi and Haifa Al-Taifi. All three scored above 500 on the TOEFL (Test of English as a Foreign Language) and were well above average on the Michigan Proficiency Test. Badria Al-Sindi had already enrolled in university courses in Riyadh, and chose not to go to the States. Nadia Al-Shihabi and Haifa Al-Taifi enrolled in 1980 in the computer science program of California State University at Fresno. Haifa Al-Taifi returned to Saudi Arabia for family reasons after a year and seven months in Fresno. Nadia Al-Shihabi transferred to a college in the Boston, Massachusetts, area but returned to Saudi Arabia, also for family reasons, before graduating.

Building 1450

he outlook in the late 1970s was for continued strong growth at Aramco. Management called for stepped-up training and increased hiring of Saudi female employees. The number of young women in training was expected to increase beyond the limits of the already crowded Special Clerical Training Center. The company decided to build a large, new women's training center, with amenities that even the men's ITCs would envy.

The new two-story center would contain 24 large classrooms, about double the number of rooms in the old facility, and accommodate up to 500 students.

It would have an administrative suite, a library, a cafeteria where hot and cold meals would be served, a student lounge, an exercise room, an audio-visual room, prayer areas, locker and storage rooms, and three restrooms. At first, Al-Idrissi remembers, the plans called for a building without windows. She argued until windows using one-way glass, so passersby could not see in, were added to the plan.

Other aesthetic touches were added. The second floor overlooked an atrium filled with plants and tended by the Gardening Unit. The complex included private, landscaped outdoor courtyards and a private rooftop terrace where the women could get some fresh air and still have the required privacy.

Al-Idrissi left the clerical training program in 1981 and was replaced as supervisor by Hilda Hamayan, an English teacher from Lebanon. During Hamayan's brief tenure as supervisor, and before the new women's training center was completed, tall plywood fences were put up to shield the old school buildings from outside view, and the student's comings and goings were more limited. The sleek new women's training center building opened in Dhahran in May 1982. It was known simply as Building 1450.

Building 1450 in Dhahran.

Men's Training Facilities Expanded

wo new industrial training buildings opened in Dhahran in September 1977, providing about 47,000 square feet of much-needed additional space for training activities. One building, a 27,000-square-foot, two-story ITC, contained 20 classrooms, three science labs, a library and an audio-visual center, plus a separate administration section with a principal's office, a teachers' lounge, a workroom and a supply room. It was adjacent to, and about 20 percent larger than, the old two-story Dhahran ITC built in 1955. The old and new ITCs, both under Principal Ibrahim Isa, had a combined capacity of about 2,000 students. The other new building, an industrial training shop, offered 19,800 square feet of space, nearly three times the area of the old Dhahran ITS, and room for about 200 trainees. The ITS, supervised by Hamad Al-Refai, had 10 classrooms and three shops, complete with equipment for hands-on instruction in subjects such as welding, electrical and mechanical work, plumbing, carpentry, and air conditioning and refrigeration.

For the first time, all industrial training facilities in Dhahran were located adjacent to one another. The buildings had been constructed so that the three structures formed a

The central courtyard of Dhahran's new training complex, 1977.

148

U-shaped training complex. A landscaped central mall with a small snack shop and shaded, outdoor tables gave the area a campus-type atmosphere. The training complex was located just outside the main gate to the administration area, near the former Saudi camp where the first training center in Dhahran opened in a *barasti* 37 years earlier.

Out-of-Kingdom Training

n 1977, Bob Brautovich became head of Industrial Relations in the company's Houston office, leaving his post as coordinator of out-of-Kingdom Training in the U.S. For nearly 18 years he had watched over Saudis who came to the U.S. on company scholarships or for short-term training assignments. He was one of the first Aramco recruiters and remembered fellow-recruiters going out during the rapid expansion after World War II on a hay wagon with a Dixieland band to attract potential recruits.

Brautovich was said to know every college admissions officer in the United States and nearly everything that happened, good and bad, to the Saudis who came to the U.S. on training assignments. The only memories he shared were positive ones.

"I have the greatest respect in the world for those Saudis who came to school here (in the U.S.). I don't think I could go to Saudi Arabia and do what they did when they came here," he said.

"We had guys who came to college in the States and got the highest grades in their English class. Some of them spoke better English than we [native speakers] did."

Aramco scholarship students compiled impressive records at schools in the U.S. Jaber S. Jum'ah, for instance, was awarded the International Students Award in each of the four years he attended Youngstown State University in Ohio. He ranked in the top one percent of his class during all four years, was elected to the national Phi Kappa Phi honor society, and graduated summa cum laude in 1973 with a degree in business administration.

Ibrahim Al-Khabour was elected vice president of the student body, president of the International Club, president of the Accounting Club, and 1967 student of the year at Armstrong College, a business school in Berkeley, California.

"We had one guy at a university in Ohio who was so gifted, so articulate, that he used to appear on a television program with the university president and they would discuss the political situation in the Middle East," Brautovich recalled.

In the late 1950s, when Brautovich became out-of-Kingdom coordinator, Aramco had about 30 employees training in the United States. By 1977 the number had increased to more than 500. The first college scholarship candidates went to the American University of Beirut to improve their study skills before going to U.S. colleges. In the early 1960s, Brautovich arranged a one-year residency program with Penn Center Academy in Philadelphia, allowing Saudis to earn a high school diploma and adjust to the U.S. culture before going off to college. Later, he convinced Oklahoma State University and Fresno Pacific College (now Fresno State) in California to accept Aramco's scholarship students directly from Saudi Arabia. That opened the door for direct admission of Aramco's ITC graduates at dozens of other colleges and universities in the United States.

Adnan Niazy, future Harvard graduate, and Sadad Husseini, future Brown University Ph.D., examine a seismogram in Aramco's Exploration Department, 1970.

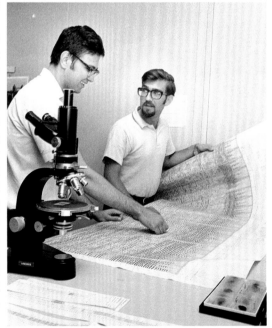

"I guess the most critical thing I did was just being there for a student who went into decline for some reason," he said. "Sometimes you'd find they were sponsored, for instance, by the Engineering Department to be an engineer, but they didn't like it or weren't cut out for it. A most challenging thing was trying to convince the Aramco sponsors who were in the operating departments to give these guys a chance at another discipline. We succeeded a number of times."

Brautovich could name 170 colleges and universities with which Aramco had some type of association over the years. He could list 228 organizations, companies or government agencies that Aramco had contacted for training. These included oil companies, drilling companies, electrical power companies, and telephone companies, plus organizations with less obvious ties to the oil industry — for example, the Brooks Institute of Photography, the Culinary Institute of America, the Pittsburgh Institute of Mortuary Science, the New York Maritime College and the Spartan School of Aeronautics.

Aramco 40 Years after Discovery

The year 1978 marked the 40th anniversary of the discovery of oil in commercial quantities in Saudi Arabia. In those 40 years, Aramco had produced more than 34 billion barrels of oil. Proven crude oil reserves, which had continued to rise in each of the 40 years, now totaled about 113 billion barrels. But, in 1978, for only the second time since World War II, crude oil production failed to exceed the previous year's output. Production was held down at government request to an average of about eight million barrels a day, a decrease of about 11 percent from 1977. The only other peacetime year of production decline was 1975.

Frank Jungers retired in 1978 amid a shower of tributes, especially from Saudi employees who considered him a champion of Saudi advancement. John J. Kelberer, vice president of Operations, was elected to succeed Jungers as chairman of the board and chief executive officer. Kelberer had joined Aramco in 1971 as general manager of Government Relations after 21 years as an engineer and executive at the Trans-Arabian Pipeline Co. That same year, 1978, Ali Al-Naimi was designated senior vice president of Oil Operations, the first Saudi to achieve that rank. At year's end, Saudis held 119 of the 805 management-level (grade 15 and higher) positions in the company. In October 1978, employees moved into the new 10-story, twin-tower office building in Dhahran. That same year the size of the work force increased by 23 percent to 34,649 employees. To fuel continuing expansion, the company imported more than 892,000 metric tons of equipment during the year. A Boeing 747 was chartered to carry cargo and personnel twice weekly from the United States to Saudi Arabia.

The busy Mubarraz Industrial Training Center, 1976.

By the end of 1978, the Training Department operated Industrial Training Centers and Industrial Training Shops at Abqaiq, Dammam, Dhahran, Mubarraz, Ras Tanura and 'Udhailiyah. Nearly 7,200 Saudi employees enrolled in training programs at those facilities. Mubarraz had the heaviest workload, with 8,700 class periods taught in 1978 compared to Dhahran, the second-busiest center, with 5,100 class periods. Eight new portable buildings with two classrooms each were added at Mubarraz in 1978, raising the total number of classrooms at the center to 51. The explanation of Mubarraz' high enrollment was simple: it was located near Hofuf, and it was the only ITC situated within daily commuting distance of the homes of almost all its trainees. The company not only saved on housing costs at Mubarraz but found that it was easier to retain young trainees when the trainees could live at home while attending ITC classes.

Management Training Centers opened in Abqaiq, Dhahran and Ras Tanura in 1978. Previously, Management Training had operated out of one district at a time, starting in Ras Tanura and moving later to Dhahran. With operations in all three districts simultaneously, there was a fivefold increase during the year in Management Training course sessions, from just 10 sessions in 1977 to 51 in 1978. About 1,000 employees attended the 1978 sessions. Abqaiq's first Management Training Center was located in portable buildings across from the community fire station. Ras Tanura operated out of a converted facility in the residential area. Dhahran's Management Training operation moved from portable buildings near the senior-staff school on Third Street to an impressive new two-story building linked to Steineke Hall, the community's largest guest facility. The Dhahran center contained four large conference rooms, each fitted with front and rear projection screens and the latest in audio-visual equipment, plus six smaller conference rooms, a reception hall and a large space for staff offices. The building was finished in stucco and fitted with recessed solar gray glass windows. When first opened, the Dhahran center had only three trainers and offered three courses, Basics of Supervision, Effective Aramco Supervisor and Effective Aramco Manager.

Aramco Recruits in Additional Countries

In 1978 the Training Department decided to look for industrial training instructors from outside the traditional recruiting areas, that is, outside of the Eastern Mediterranean, Egypt, the Sudan and the United States. During the hiring freeze of the late 1960s, the number of instructors had been reduced to a bare minimum. Recruiting of new instructors could not keep pace with the rapid expansion in Training that began in 1970. It became increasingly difficult to recruit new instructors, in part because Aramco could not offer them family housing. Wadie Abdelmalek, then supervisor of Training's Curriculum and Testing Development Unit, remembered recruiting trips to Egypt and the Sudan during which, out of some 100 candidates tested and interviewed, only five or six met Aramco's minimum standards for classroom instructors, and some of them would not agree to work at Aramco on bachelor status.

An Arabic-language typing class.

By this time, many line organizations, driven by the demands of expansion, had already sent recruiters searching for trained workers in nations never before approached by Aramco. New-hires had been pouring in from all points of the compass. According to the company's Annual Report, during 1978 alone, the number of nationalities represented in Aramco's work force increased from about 15 to 40.

The Philippines proved to be a bonanza for Aramco recruiters. In the Philippines recruiters found large numbers of well-educated professional and technical

151

Trainee applies a cutting torch to a pipe section.

people, almost all of them fluent in English, and nearly all willing to work for wages much lower than those required by their counterparts in Western countries. In just three years, Filipinos became the largest expatriate group in Aramco. The first Filipinos were hired in 1977. By the end of 1979 there were 3,820 Filipinos on the payroll, surpassing North Americans, who had been the largest expatriate group in Aramco since the founding of the company.

Great Britain was another recruiting hot spot. Between 1976 and 1979 the number of British in the work force more than tripled, going from 758 to 2,486. By 1979 they had become the third-largest expatriate contingent in the work force behind Filipinos and North Americans.

The Training Department sent its first recruiter to Great Britain in 1978, and the following year the department dispatched a recruiter to the Philippines for the first time. Khalil Nazzal, principal of the Dammam ITC, recruited in Great Britain. Nimr Atiyeh, assistant principal of the Dhahran ITC, went to the Philippines. Both men were after English-language instructors, particularly those qualified to teach intermediate and upper-level English classes. These forays went ahead despite considerable debate in training ranks about the suitability of instructors from either country. The accents associated with Filipino and British language instructors were a major concern.

Although Nazzal held a degree in English from Durham University in Great Britain and owned a holiday home in the British Isles, he was skeptical about recruiting British instructors. He had two concerns. One concern was whether the British were too traditional, not in touch with the American-style classroom methodology used by Aramco. He eventually changed his mind on this point, and even helped to bring a British-developed language training system to Aramco. His second concern was with accent, a concern that soon presented Nazzal with a dilemma.

One of the first candidates Nazzal interviewed was a Scotsman, a veteran teacher with a "wealth of knowledge" and a master's degree in languages from Nazzal's old school, Durham University. "I spent 2½ hours with him," Nazzal said. "He would start talking without a noticeable accent and then suddenly shift into a Scottish accent. He'd become conscious of it and get out of it very quickly. He'd go on all right for another 15 or 20 minutes, then revert back into a Scottish accent." As Nazzal listened, he was torn between a desire to recommend hiring a learned fellow graduate of Durham University and the disquieting thought of Saudis learning to speak English with a Scottish accent. "I had a hard time turning him down," Nazzal said. "I've always felt something like a traitor to my alma mater."

Nazzal recommended hiring three of the candidates he interviewed in Britain, but Bill O'Grady, director of Training, evidently had hoped to get more than three out of the trip. When Nazzal reported that in two days he had seen 30 candidates and recommended three, O'Grady cracked, "What was it, interviews or an inquisition?" Those three candidates became the first instructors from the British Isles hired to teach at Aramco Industrial Training Centers, but they were by no means the first British teachers in Saudi Arabia. British instructors had been teaching at Saudi public schools, including the Aramco-Built Government Schools, since 1964, with good results.

Nimr Atiyeh spent May 19-26, 1979, in Manila where he interviewed and tested 28 candidates. He laid the groundwork for these interviews carefully. Atiyeh visited the Manila Speech Clinic, where a

Nimr Atiyeh

152

Mrs. Ellen Serrano helped him compile a list of English words for candidates to read aloud as a test for accent problems. He called at Power Speech, a school specializing in communication courses for businessmen, where the director helped acquaint him with the country's education system. The next day he went to the University of the Philippines and met with the chairman of the English Department and two professors. They gave him their evaluation of the English and education departments at various universities, told him about professors' salaries, and even sent him some job candidates.

Of the 28 candidates interviewed and tested, Atiyeh found that 15 of them would have no trouble fitting into the company ITCs. In a report on his recruiting trip, Atiyeh wrote: "When compared with the average candidates hired from our traditional sources in the Middle East, the Filipinos are superior in every respect except that of accent." Not every candidate had the same problem, but all candidates had one or more problems with accent. By far the most commonly mispronounced words were words with a /zhu/ sound, like "usual" and "visual."

The final question was whether the department wanted to accept a Filipino accent in the classrooms. The reaction of ITC principals who listened to tape recordings of the interviews ranged from outright rejection to reserved acceptance. Atiyeh himself suggested hiring a few Filipinos on an experimental basis, and that's what the Training Department did.

In the summer of 1979, the first five Filipino instructors hired by the Training Department arrived in Saudi Arabia. One of them, Demetrio Ignacio "Jun" (for Junior) Asistio, a former university instructor, had been ranked No. 1 out of the 28 candidates interviewed by Atiyeh. Asistio and another Filipino, Adolfo Casis, were assigned to the Dhahran ITC. The other three Filipinos went to ITCs in separate Aramco communities.

"Jun" Asistio

To Asistio "everything at the Dhahran ITC was new and different." They had instructors from America, from various Arab and African countries, and now two Filipino instructors. "I was saying to myself, 'this is like an international school except that the teachers are the ones coming from different countries, not the students.'"

Asistio found his new job frustrating at times. At the University of the East in Manila he had taught first-, second- and third-year English classes, conducted a speech laboratory, and instructed third- and fourth-year engineering students in "technical" English. All his students had been familiar with English since childhood. The language of instruction in the Philippines at the time, from elementary school on up, was English. By contrast, in Dhahran, for a salary four to five times his Manila salary, Asistio was teaching very basic English, "almost the ABCs," to groups of older industrial trainees on shift schedules. He taught basic grammar, vocabulary and writing to classes of about 20 trainees at a time. Although the students had several semesters of English-language instruction behind them, they had yet to master some very basic vocabulary. Asistio was at a loss as to how to get across the meaning of some English words. He tried acting out word meanings, and "I found myself ridiculously in front of the class trying to demonstrate, for example, how a lame person walks, and what bald looks like."

At one point he went to Atiyeh and said, "I can't teach. I can't teach these groups." Atiyeh asked why. "I am running out of words to explain what I mean. I can't go any further back than the basic words."

Eventually, Atiyeh managed to smooth away the frustration. Asistio stayed on to teach progressively more advanced English classes. He became a senior instructor of English in the most advanced of all the company's English-language programs, the College Preparatory Program for Saudis about to leave on Aramco scholarships for colleges and universities overseas.

Competition Affects Employee Mix

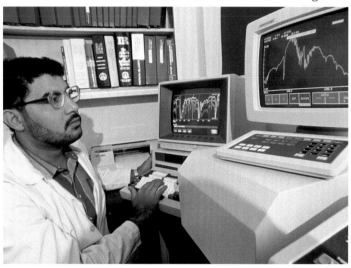

Computer use was slowly increasing in the mid-1970s.

s line organizations turned more and more to foreign job markets to fill their manpower needs, the expatriate side of the company's payroll increased at a much faster pace than the Saudi side. In 1970 the company's work force had been 80 percent Saudi. By 1975, when the Master Gas System was announced, it was 75 percent Saudi. Just three years later, in 1978, the work force was only 52 percent Saudi. Between 1975 and 1978 the net increase in the Saudi work force was about 3,360 men, while, in the same period, more than 11,800 expatriates were added to the payroll.

The decline in the percentage of Saudi employees coincided with a high turnover rate among Saudi trainees, especially in the group of trainees with the most years of schooling. Between 1973 and 1977, the attrition rate among trainees with six to 11 years of schooling, the "A" and "B" trainees, was about 40 percent, while the attrition rate among the direct hires, many of whom were dropouts with six or fewer years of schooling, was considerably lower, about 30 percent. Many advanced trainees as well as skilled workmen left Aramco during these boom years for higher-paying jobs. New or expanding companies found it easier and less expensive to offer higher wages to a trained Aramco employee than to recruit and train their own employees.

Aramco had very little success finding replacements with credentials to match those of the "A" and "B" trainees who left. Out of 1,900 Saudis hired in 1977, only two percent were college graduates and four percent were high school graduates. About 95 percent of all Saudis hired between 1973 and 1977 did not have a high school diploma. Almost 50 percent of the new Saudi employees were direct hires. Some direct hires, such as welders, were skilled workmen who required no further training to perform a job. Many others came into Aramco with few or no skills and were immediately placed in low-level jobs. Of the direct hires, 44 percent had less than six years of schooling, and 41 percent scored 25 or less out of a possible 90 on the company's job placement test, barely higher than the score expected by random marking of answers to the multiple-choice test. The average skill level in the Saudi work force suffered as a result of these trends. According to some estimates, by 1978, more than 50 percent of the Saudi employees were less than fully job qualified.

Renewed Focus on Saudi Work Force

n April 1978, the company's new chairman of the board and chief executive officer, John J. Kelberer, addressed the twin problems of upgrading the skills of the existing Saudi work force and increasing the percentage of Saudis in the work force. In a policy statement drafted for release to department managers and above, Kelberer wrote: "We are presently engaged in an extensive recruiting program to place a large number of young Saudi employees into our work force. Concomitantly, we are faced with an acute need to upgrade the skill proficiencies of our present Saudi job holders, many of whom joined the company within the past three-year period.

154

Frank Jungers

"The development of Saudis was my major accomplishment, because that had an everlasting impact."

Frank Jungers covered a lot of territory in that statement. He was chairman of the board and chief executive officer of Aramco from 1973 to 1978, five of the most difficult and productive years in company history. Jungers dealt successfully with development of the Master Gas System, creation of the Saudi Consolidated Electrical Company and negotiations leading to Saudi government ownership of Aramco — any one of which would be considered a major accomplishment. During his tenure as CEO, Aramco grew from 13,500 to nearly 32,000 employees. In that time, more Saudis than ever before were given positions of authority. In his first year as CEO there was a 23 percent increase in the number of Saudis in supervisory positions.

Under Jungers in 1974, Faysal Al-Bassam became the first Saudi Arab to be elected a vice president of Aramco. The following year Ali Al-Naimi became the second Saudi Arab vice president.

In the opinion of Ali Dialdin, who was later named general manager of Training & Career Development, "True Saudization began after Jungers became CEO. The foresight was not there until then. That's when things began happening."

Jungers had a history of making things happen. As a 26-year-old superintendent of Maintenance and Shops at Ras Tanura, Jungers realized the company was wasting money and fueling Saudi resentment by failing to utilize and promote Saudi workmen. Jungers told the craft foremen, who were mostly American men old enough to be his father, that he wanted a one-quarter reduction in foreign contract labor by the end of the year. The foremen warned that the quality of shop work would nose-dive, but they were proven wrong.

"The Saudis stepped in with a will and with surprising speed and flexibility," he said. "They had, in fact, been well trained by the supervisors who were reluctant to use them. This was really the beginning of what I myself saw as the necessity to train the Saudis, to integrate them in the work force and to devise a series of policies that kept them working and to discontinue a series of policies that were really devised for foreigners."

In 1961, as superintendent of the Maintenance and Shops Division in Dhahran, Jungers promoted the first Saudi craftsman to the level of foreman. In all, he promoted seven of the first 12 Saudis to reach the foreman level. In 1962, when Jungers was acting general superintendent of Industrial Relations, about a dozen teachers from Middle East countries were elevated to senior-staff status. They were the first instructors of Arab extraction to reach the level previously reserved for American instructors.

Jungers was born in the farming town of Regent, North Dakota. He was a teenager when the family moved to Eugene, Oregon, where his father purchased a service station. Jungers enrolled at Oregon State University as a pre-engineering student and attended college under a U.S. Navy pilot training program during WWII. He left the Navy after the war and enrolled in the University of Washington, graduating in 1947 with a degree in mechanical engineering. Soon thereafter, an Aramco recruiter came to Seattle and gave Jungers what he called "the best job offer financially that I got."

He spent the next 30 years with Aramco. Years later, reviewing his highly successful and eventful career, Jungers concluded: "The development of Saudis was my major accomplishment, because that had an everlasting impact."

"To accomplish these goals in the most effective manner and in the shortest possible time frame, we must initiate expanded and accelerated OJT [On-the-Job Training] in all of our operating areas. The results of this increased OJT activity will be a major factor in determining our ability to provide the fully qualified Saudi work force we urgently require to meet our growing manpower needs." The statement went on to say that, in the interest of a uniform approach to OJT, a General Instruction manual had been prepared for the guidance of personnel who will be implementing and operating OJT activity. (Kelberer once told an interviewer that the word "Saudization" was coined about this time by his administration. But former Aramcons say the word was in general use in the 1950s. Saudization may have had a more specific meaning in the 1970s, i.e., a 75 percent ratio of Saudis in the work force as called for in Saudi Labor Law.)

John Kelberer

The company launched an aggressive campaign to inform young Saudis about the career opportunities in Aramco. The number of Aramco recruiting offices in the Kingdom was increased from four to 20. Four green-and-white vans were purchased for recruiting trips to isolated communities. Instead of waiting for prospects to come to a recruiting office, recruiters went into communities and schools to drum up interest in Aramco. A full-length recruiting film, "The Enduring Resource," was produced along with more than 100 filmstrips portraying specific jobs in Aramco. The filmstrips were used by "career counselors" to help recruits select a career path to follow. Recruiting brochures were distributed to all secondary schools and all universities in the Kingdom. "The idea," Ali Dialdin said, "was to get people thinking of Aramco." The Saudi Employment Division was reorganized and streamlined. A new division, the Saudi Arab College Relations and Professional Development Division, was created to assist in recruiting at colleges and universities in the Kingdom.

Media Production employee edits a training film.

Through intensified OJT and beefed-up recruiting, the company sought to realize a seemingly impossible goal: to raise the percentage of Saudis in the work force to 80 percent by 1980, just two years away. To reach this goal, the company would need to replace more than 10,000 expatriates with Saudi employees in the next two years, assuming the size of the work force remained steady at its 1978 level. The company would need not only to add 10,000 new Saudi employees, but also to upgrade the training of thousands of other Saudis in order to qualify them for jobs vacated by expatriates. As a Training Department report to executive management stated: "It is probably fair to say that no private overseas industrial organization anywhere has ever before faced such a challenge. We know of no company that has ever entertained such goals of native-employee utilization in such magnitude and in such time-frame urgencies as our Aramco basic policy asks us to accomplish."

Intensive On-the-Job-Training programs had been used before to raise the skill level of employees in a hurry, but not with the speed nor in the numbers suggested by Kelberer's policy statement. During the expansion period in the early '50s, an intensive OJT

Saudi trainee aligns flanges in Machine Shop.

program succeeded in raising about 6,000 Saudi employees from unskilled to low-skilled and semiskilled workmen in five years. However, that was about as far as they were able to go. The job holders still needed classroom training to advance beyond the basic skill levels.

Expansion of On-the-Job Training

On January 1, 1972, just as a new expansion period was beginning, an On-the-Job Training Section was organized within the Industrial Training Division of the Training Department. It began modestly enough at the transportation garage in Dhahran with a half dozen vehicle maintenance trainees, one instructor and one old GMC truck. The OJT section provided course manuals and outlines for the vehicle maintenance training program. The Transportation Department provided the instructor.

A month after the OJT section was formed, a Driver Training and Testing program was added to the section's responsibilities. Then, in quick succession, came OJT programs for the Producing and Marine departments, as well as short courses in radio communications for truck drivers, gatemen and refinery operators. OJT soon became the fastest-growing training section in Aramco. It provided training materials and train-the-trainer services to line organizations. It helped line organizations establish training programs. It even operated several OJT programs on behalf of line organizations.

On January 1, 1977, the OJT section was given division status and its own superintendent, Donald L. Fink, a trainer with degrees in industrial supervision and industrial education from Western Michigan University. Fink had joined Aramco in 1966 as an instructor in Dhahran's Industrial Training Shops, and during the early 1970s was coordinator of training and employee relations for Tapline.

The Contractor Training group, formed in 1973 to test and certify graduates of private-contractor training schools, was merged with the OJT Division in 1977. By that time, the OJT Division operated 45 training and testing programs with a total enrollment of more than 1,600 employees, not counting driver training and testing. The division operated continuous training programs for Refinery and Terminal, Warehousing, Fire Protection, Process Operator, and Marine Departments. There were separate OJT units in Dhahran and Abqaiq, and units serving the Ras Tanura Refinery and the Marine Department in Ras Tanura. Among other programs were those for GOSP operators, welders, truck drivers, vehicle maintenance men, water injection plant operators, and water survival training.

OJT relied on the age-old craftsman type of instruction. It was, to use a popular phrase of the time, hands-on, at-the-plant, over-the-shoulder-type training. But the results were difficult to measure objectively until 1977, when a vocational analyst, Tom Cowie, used the Systems Engineering approach in OJT manuals for seawater treatment plant operators.

The Systems Engineering approach evolved out of military training and was popular at the time among manpower development authorities and institutions in the United States and Europe. It was based on the premise that any job could be described and analyzed as a series of discrete work units. Each job was broken down into a series of duties, each duty into two or more tasks, and each task

On-the-job instruction for crane operators.

157

into a series of steps. The Training Department published its first manuals using the Systems Engineering approach in 1978. The duties, tasks and steps required to perform a job at a specific grade level were listed in a document called an Aramco Job Training Standard (AJTS). AJTSs tended to be much more precise in describing various job duties than were earlier training manuals. For instance, where an older manual might have simply listed a housekeeper's duty as "Mop the floor," the Medical Housekeeper One AJTS listed 19 separate steps in mopping a floor, starting with "Put on blue rubber gloves" and ending with "Remove rubber gloves and wash your hands."

With strong support from James B. Morton of Training's Program Development and Evaluation Division, the Systems Engineering approach won endorsement from executive management. AJTSs became the basic document for recording and analyzing a trainee's performance. The trainee advanced one step at a time through each task, and one task at a time through each duty in the AJTS. Each time he showed he was capable of performing a task up to Aramco standards, his competence in that task was recorded in the AJTS. When a trainee successfully completed all the tasks in all the duties in an AJTS, he received final certification for a job. He was considered qualified for the job described in the AJTS at the grade code specified by the AJTS. Task certification was the key process in OJT. It was the only recognized measure of training progress.

Driver Education

The Driver Training and Testing class was a different kind of OJT program. Thousands of employees annually enrolled in one of the many driver courses. The most popular of these was the eight-hour defensive-driving course. The Training Department, in cooperation with the Security Department and Loss Prevention, first introduced the U.S. National Safety Council's Defensive Driving Course to Aramco in 1969. About 500 employees took the course in its first year, but enrollment soared to 2,300 in the following year and was still at that level when Defensive Driving and Testing was absorbed into the OJT section in 1972. Bilingual instructors went out from Dhahran to conduct defensive-driving classes at sites from Abqaiq to pump stations along Tapline, in the north. They usually ran the course for classes of 20 to 30 people, two hours a day for four days. Basically, they taught the rules of the road, what to do when approaching an intersection, procedures for making a turn, how to pass another vehicle,

Student tests his driving skills using the "Drivotron."

and so on. For four consecutive years, from 1975 through 1978, Aramco was honored by the National Safety Council for the best performance by any overseas agency conducting the council's Defensive Driving Course. Between 1969 and 1979, nearly 29,000 Aramco employees and employees of contractors working for Aramco completed the course. However, despite such training efforts, the accident rate continued to climb as the number of motor vehicles increased beyond the rated capacity of Eastern Province highways. The vehicle collision rate of 6.8 per million kilometers when the Defensive Driving course began in 1969 had more than doubled by 1976 to more than 15 collisions per million kilometers of driving. The company's 1977 and 1978 annual reports cited traffic accidents as "a major area of concern."

In 1978, the OJT Division participated in 16 motor vehicle driver, truck driver, or heavy-equipment operator training and testing programs, running the gamut from the testing of crane and bulldozer operators, to courses on desert driving for trucks. The testing and training programs, lasting from a few hours to months, enrolled a total of about 19,000 people during 1978. That same year the company built a new driver training building in Dhahran next to the driver-training range. The following year the company, in cooperation with the government, built near Dhahran the first overpass and cloverleaf highway interchange in the Eastern Province.

Odell Bratland authored numerous driver-training manuals during his 10 years with Aramco, none more personally challenging than the first such manual he wrote for the company. When he joined the Training Department in 1975, Bratland, who had been writing automotive manuals for Dunwoody Industrial Institute in Minneapolis, Minnesota, had never seen a desert, much less driven on one. Nevertheless, his first assignment after arriving in Saudi Arabia was to write a desert-driving manual. "I didn't know where to start," Bratland said. "I had never been in the desert. I went to Loss Prevention and found an old, old (desert-driving) book written years ago. My boss, Don Fink, was always out in the desert looking for artifacts, so I went out with him. We lived in the desert for a couple of days, driving and driving, looking at the color of the sand (to identify soft spots), and so forth. He was happy to show me what he knew." Armed with some firsthand experience and the old Loss Prevention manual, Bratland put together a curriculum for training operators of the huge Kenworth trucks, which often had to cross many miles of desert in order to supply work crews in remote areas. In addition to writing manuals, Bratland designed the driver-testing range at Dhahran, where, over the years, thousands of Aramco employees took a behind-the-wheel driving test.

Gas-Plant Operator Training

JT training for gas-plant personnel began gearing up in the late 1970s. The first on-site training facility for Gas Operations opened in a portable building at the Berri NGL Center in 1978. In addition to classroom training, Saudi trainees at Berri got their first taste of what it was like to work in a gas plant. American and British operators had been hired to run the plants until Saudis could be trained to replace them. These crusty veterans delighted in giving trainees a dramatic demonstration of the hidden forces contained within the workings of a gas plant. "We'd take a gauge off a bleed valve," a veteran gas engineer recalled, "and open the little ½-inch needle valve and this jet — a sonic plume of pressurized gas — would hiss out and scare the daylights out of them." The demonstration encouraged caution and respect for gas-plant safety rules. "These were mammoth plants," the engineer said, "and if you made an error it could be a mammoth error."

Saudis assigned to Berri were targeted for training in either gas-plant maintenance or gas-plant operations. In 1978 a group of 36 trainees became the first Saudis to complete phase three of the basic five-part gas-plant operator training program. Two more groups, one of 40 and another of 35 trainees, began phase-one operator training in 1979. A metals mechanic program enrolling about 30 Saudis also originated at Berri in 1979. This program began at a very basic level — how to use a file, how to operate a drill press, etc. Saudis selected for training at Berri had

Gas-plant training simulator was used to model operating problems for Saudi trainees.

completed Elementary English 4-B, meaning they'd already had about 1,500 hours of English-language training, but curriculum developers doing a target-population study of OJT training at Berri found that most trainees had insufficient vocabulary to comprehend technical information in English, "even when presented in the simplest terms."

Management estimated it would take four to five years of training to qualify the average Saudi as a gas-plant technician at Berri, and probably longer at other plants in the Master Gas System. The Berri plant used an old-style pneumatic control system with dials, gauges and other instruments actually on the control board. Later plants, at Shedgum and elsewhere, switched to a much more complex electronic, computer-controlled board system, requiring operators with extensive training in mathematics and knowledge of transistors and other electronic components. Len Wannop, who headed Aramco's gas-plant start-up team, called training Saudis "in this electronics business the biggest bugbear of the whole works because of its highly complex nature." When he left Aramco in 1981, Wannop said, "we were just really in the initial stages of trying to train in that area."

Training Goals Reviewed

A company forecast on how long it would take to replace expatriates with Saudis in the gas plants developed into a sore point between the company and the government. "Some acrimonious discussions on the subject" occurred during a Gas Task Force meeting in New York City in August 1979, according to a memorandum from Malcolm Quint, director of the Participation Affairs Department, to his boss, Majed Elass, vice president of Government Affairs. Saudi Arabia's government representatives at the meeting demanded that a schedule be presented to the Ministry of Petroleum indicating how Aramco intended to achieve a 75 percent Saudi work force in the earliest practicable time. The representatives, Quint wrote, "took strong exception to the gas-operations manpower projections presented. They felt that these projections indicated far too great a reliance upon expatriate labor, and they used the occasion to express their serious reservations about Aramco's Saudi recruiting, Saudi personnel and Saudi Management Development programs."

It was an issue that Aramco's upper management had been grappling with for more than a year, starting with Kelberer's statement in April 1978. Kelberer had called for stepped-up OJT and aggressive recruiting of Saudi employees. In September 1978, company president Hugh Goerner and George Larsen, former vice president of Exploration and Development and a 30-year veteran of the company, were asked by executive management to review the Industrial Training Program and suggest improvements. With the help of Bill O'Grady, director of Training, and Dan Walters, superintendent of Training's Planning and Programs Division, the landmark project was completed in a month.

The Goerner-Larsen report proved to be one of the most important documents in Training history. Dan Walters, Larsen's only staff man at the time, remembered going with Larsen from district to district and department to department to collect information. "I think Larsen was close enough to the situation that he realized you needed to look at each program separately," Walters said. "He kept telling me: 'You know you can't do this with pencil and paper from Dhahran.' At each stop he would bypass all levels of communication and just go straight down to the level he wanted to talk to. He's probably one of the few executives from that time who would have felt comfortable in going out and doing this."

Larsen developed a training pattern chart. He had a training person in each department use the chart to make a diagram showing how long

George F. Larsen, 1963.

A simulator imitates problems refrigeration technicians encounter in the field.

it took for Saudis to get into an entry-level job in that department. "He forced people to sit down and graphically draw out their training program," Walters said. "Nobody had ever done that before."

The Goerner-Larsen report evaluated training in 15 departments. It gave high marks to training at the Terminal Department in Ras Tanura, to Power Systems training in Dhahran, and to Southern Area Plants training at Abqaiq. The report echoed, in some respects, the report issued three years earlier by Joe Mahon on Saudi Development. Both studies found wide variations in the quality of training programs, and a failure to coordinate training programs on a company-wide basis. Both saw a reluctance in some departments to free up Saudi employees for training. Both agreed that Saudis working in remote areas were often denied effective training. The Goerner-Larsen report also reached the following conclusions:

- Many departments, trying to conform to the company goal of a 75 percent Saudi work force, hired larger number of trainees than they could handle.
- Hiring standards were too low. Most of the new trainees were school drop-outs, some with attitude problems. "It appears that in many cases new-hires are not committed to working hard or desirous of a career with Aramco."
- Due to shortages of trained manpower, many departments were unable to free qualified instructors to conduct OJT programs.
- Most OJT training facilities were substandard, especially with respect to craft training.
- Advancement rates differed. The time it took an employee to reach grade code eight from point of hire varied by as much as three years, depending on the organization.
- The most effective OJT trainers were skilled Saudis with good academic backgrounds who could teach in English and reinforce understanding by explanation or discussion in Arabic.

"The significant increase in the number of Saudis added to the payroll since 1970 [a total of more than 9,000] has placed a heavy strain on both the Training Department and the line operating organizations," the report said. "It has become particularly difficult for the line. They have had to start almost from scratch to once again build up an in-house training organization while at the same time coping with an expansion program."

The report also took special notice of the Training Department's continuing problem in hiring and retaining instructors because of the lack of family housing.

The Goerner-Larsen report differed from its predecessors by recommending that line organizations, rather than the Training Department, take the lead in Saudi development. The report made a total of 15 "suggestions for improvement." In time, almost all of them were adopted. The most far-reaching of these was put into effect within weeks, namely that an "Industrial Training Board" under the head of Oil Operations be formed "to provide company-wide industrial training guidance and coordination under the direction of Line Management."

Trainees learning to remove and replace electrical fuses.

SAMCOM

n January of 1979, in line with the Goerner-Larsen report, John Kelberer announced the formation of the Saudi Arab Manpower Committee (SAMCOM). The function of SAMCOM, as stated in the committee charter, was to: "Within the overall guidance and approval of the Management Committee, develop, recommend and approve appropriate policies and programs for the recruitment, training, assimilation and retention of Saudis into the Aramco work force. Monitor and periodically audit these policies and programs to assure that they are effectively serving corporate objectives. Review line Saudi Manpower Plans and develop a company-wide plan based on approved line plans. Report periodically to the Management Committee."

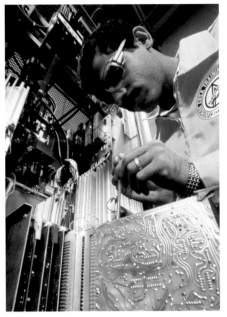

No one seeing the roster of top executives who were appointed to membership on SAMCOM could doubt the importance of this new committee. The six-member committee was composed of four senior vice presidents, one vice president and a department director, each one in charge of organizations employing hundreds of Saudi workers. The company's top-ranking Saudi executive, Ali Al-Naimi, then senior vice president of Oil Operations, was selected as the first chairman of SAMCOM. Other charter members were Hal Fogelquist, senior vice president of Industrial Relations; Frank Fugate, senior vice president of Project Management & Construction; Don Hyde, director of Corporate Planning; Joe Mahon, senior vice president of Corporate Services; and T. John McCubbin, vice president of Employee Relations and Training. The important post of committee secretary and staff leader went to another Saudi, Abdelaziz Al-Hokail, recently named assistant to McCubbin. Through Al-Naimi and Al-Hokail, Saudis had leadership roles in the training and development of their fellow countrymen for the first time.

An electrician trainee installs a printed circuit board.

Training Department Reorganized

n July 1979, the Training Department was reorganized in accordance with another Goerner-Larsen recommendation. The department, under director Bill O'Grady, was divided into Northern, Southern and Central area training divisions, each with its own superintendent. The OJT Division was abandoned, and division superintendents were made directly responsible for coordinating ITC and ITS training with OJT programs run by the line organizations in their area. In addition to the three new regional divisions, the reorganized department included a Saudi Development Division, a Curriculum and Testing Division and a Planning and Programs Division. It was the fourth time the Training Department had been administratively reshuffled. In the beginning, Training was centralized in Dhahran under one supervisor. It was decentralized in 1948 and divided into three districts, Abqaiq, Dhahran and Ras Tanura, each headed by an Education Supervisor. In 1952, Training became a full department, with three divisions and the general office in Dhahran acting in support of three training districts. In 1969, the department was centralized once again, with authority for Training activities under O'Grady in Dhahran.

Superintendents of the new divisions were Mustafa Husam Al-Din in the Central Area, Dan Walters in the Northern Area, and Zaki A. Ruhaimi in the Southern Area. Three other divisions were headquartered in Dhahran, each headed by a superintendent. Don Fink became acting superintendent of Curriculum and

Testing, Dennis Fruits headed the Planning and Programs Division and 'Abd Allah Ali Al-Zayer was superintendent of Saudi Development.

The Training Department and various business lines spent more than $68.5 million on training in 1979, including trainee salaries, about double 1976 expenditures. The department operated industrial training centers and shops with a combined total of more than 250 classrooms at eight locations: Abqaiq, Dhahran, Ras Tanura, Dammam and Mubarraz, as well as satellite programs in 'Udhailiyah, Safaniya and the Berri Gas Plant. Nearly 9,200 employees enrolled in industrial training courses in 1979 compared to an enrollment of about 1,000 in 1970 and 5,500 in 1975. About 3,000 of the trainees attended classes full time. In 1979 the Training Department had 721 employees, compared to 591 in 1975 and 258 in 1970.

New Era Starts with End of Decade

ramco's crude oil production averaged 9.25 million barrels per day in 1979 for a total of 3.37 billion barrels during the year, a record quantity. The largest percentage of Aramco's production, 41 percent, went to Europe; 33 percent was sent to Asia. Nearly 20 percent went to North America in 1979, compared to only one percent in 1970. The company's work force increased by 20 percent during the year to 42,250 employees from 45 countries. The number of employees had more than doubled in three years. The percentage of Saudis in the work force remained at 52 percent, the same record low level as in 1978. Saudis held 46 percent of the supervisory jobs, and about 19 percent of manager-and-above jobs. There were 33 Saudis in manager-and-above positions. These included one senior vice president, Ali Al-Naimi, and three vice presidents, Nassir Al-Ajmi, Faysal Al-Bassam and Majed Elass.

Rapid growth changed Aramco in many ways, some of them difficult to define. Esam Mousli, one of the first graduates from the University of Petroleum and Minerals, returned to Aramco in 1978 after earning a master's degree in business administration at a university in the United States.

"I came back," he said, "to find Aramco had suddenly taken a different shape. There were larger departments, more people, more nationalities and a different approach to handling things."

In February 1979, only a month after Kelberer announced creation of SAMCOM, a second announcement, this one from the capital in Riyadh, assured that Saudization would continue to be a high priority for Aramco. That was the official announcement from Riyadh of an agreement for full Saudi ownership of the company. In 1980, the Saudi government paid the American partners $1.5 billion for their remaining shares in Aramco. The company stepped off into a new decade under 100 percent government ownership, with Saudis now overseeing the training and development of Saudi employees.

Crude oil production set a record in 1979.

Training Goes High Tech 1980-1985

"It is essential that we improve and make more aggressive our corporate program for increasing Saudi manpower and ensure that such a program is implemented. To this end we have established the Saudi Arab Manpower Committee (SAMCOM)."
– John J. Kelberer

The Saudi Arab Manpower Committee (SAMCOM) faced some imposing obstacles in its efforts to increase the size and skills of the Saudi work force. A severe labor shortage within the Kingdom presented one major obstacle. Hundreds of new industries created during the government's multi-billion-dollar 1975-80 Development Plan were competing with each other and with the government for a limited number of Saudis entering the work force. These industries offered Saudis desirable jobs at attractive salaries, some well above what Aramco was paying. Aramco not only had difficulty recruiting enough new Saudi employees, but also was losing trained Saudi workers to private industries. Other firms found it cheaper and less time consuming to pay high wages in order to attract already trained workers from Aramco than to recruit and train employees on their own. As an Aramco executive of the period lamented, "Private industry hires them away almost as fast as we can train them."

The problem of recruiting and retaining Saudi workers was compounded by pressure to increase the size of the Aramco work force to keep pace with the escalating worldwide demand for petroleum products. According to company projections, Aramco needed to increase the size of its work force by 65 percent within five years in order to have sufficient manpower to meet anticipated production needs. By January 1979, it was already clear that the company would fall far short of its Saudi recruitment target for the year, and probably for future years as well. The Employment Department reported to SAMCOM in January 1979 that while the goal of adding 4,500 Saudi employees during the year might have seemed possible when that goal was established two years earlier, "the labor market has grown far tighter since then." Furthermore, the department predicted, about one-third of all Saudis hired during 1979 would leave Aramco within two years, most of them lured away by rival industries. The only way for Aramco to achieve a 65-percent increase in the work force, apparently, would be to hire thousands of foreign workers.

Opposite: The Aramco Training Center in Ras Tanura, 1985. Below: Trainees learn the proper techniques to lubricate machinery.

165

Rather than an increase in the ratio of Saudi workers, the number of foreign employees would continue to increase faster than the Saudi component of the work force. That trend fit the historical pattern. The percentage of Saudis employed in Aramco had always declined in periods of expansion and increased when the work force was cut back. During the company's first prolonged expansion, from 1944 to 1952, the ratio of Saudis in the work force fell from 84 to 61 percent. By 1970, after 18 years of gradual reductions in total employment, the ratio had climbed back to 82 percent Saudi. That ratio plunged again during the expansion of the 1970s. By 1978, much to the alarm of the Saudi government, almost half of the company's employees came from outside the Kingdom, the largest numbers being from the Philippines, the United States, Britain, Pakistan and India, in that order.

SAMCOM Goes to Work

When Aramco's chairman of the board and chief executive officer, John J. Kelberer, announced the formation of SAMCOM in January 1979, he stressed the need to halt the downward slide in the ratio of Saudi employees. "It is essential," Kelberer said, "that we improve and make more aggressive our corporate program for increasing Saudi manpower and ensure that such a program is implemented. To this end we have established the Saudi Arab Manpower Committee (SAMCOM), which has been charged with coordinating the plans and programs needed to achieve these objectives."

Kelberer said SAMCOM would "recommend and approve, as appropriate, policies and programs for the training, recruitment, assimilation and retention of Saudis in the work force." Acting with guidance from and approval of the Management Committee, SAMCOM would monitor and audit training policies and programs, review Saudi manpower plans, and develop a company-wide Saudi development plan. The creation of SAMCOM fundamentally changed the Training Department's role in Aramco. The department would no longer be the driving force in training. Henceforth, all training programs and policies would either be initiated by SAMCOM or brought before the committee for approval before being referred to management.

SAMCOM went to work with energy and enthusiasm. At its first formal meeting on January 13, 1979, the committee already had 16 proposed Saudi manpower development projects on its agenda. One of the original SAMCOM members, Joe Mahon, recalled that the committee usually met in a conference room on the third floor of the Administration Building in Dhahran. The meetings often began late in the day, after committee members had already put in a full day's work, and frequently continued long into the evening. "There was no doubt [SAMCOM chairman] Ali Al-Naimi was in charge," Mahon said. "There was not a great difference of opinion as to what should be done, not much serious challenge to whatever Ali pushed for. We all knew that there was a new emphasis on Saudi development."

SAMCOM did not wait for various organizations to initiate Saudization programs on their own. Instead, a five-member task force was formed under the leadership of Abdelaziz M. Al-Hokail to develop new Saudi recruiting and training activities. A statement issued by Al-Hokail on March 26, 1979,

The Dhahran administrative core area, 1982.

announced formation of the task force and explained its role. "A task force reporting to the vice president of Employee Relations and Training has been established to initiate Saudi recruiting and training activities recommended by the Saudi Arab Manpower Planning Committee. The team will be responsible for developmental initiatives until new activities are absorbed into existing Employment Department and Training Department operations."

Marine trainees examine the workings of a diesel engine, 1979.

Members of the task force, in addition to Al-Hokail, were Ali M. Dialdin, who had just completed a developmental assignment in Southern Area Oil Operations; Mike Jurlando from the Employment Department; Dan Walters, superintendent of Northern Area Training; and Zaki A. Ruhaimi, superintendent of Southern Area Training. Ruhaimi went to the Aramco Services Company offices in Houston, Texas, to develop a program for recruiting Saudis attending U.S. schools.

The task force took over a portable building adjacent to the Administration Building South in Dhahran. For the next several months, Annex 30 was a center of vigorous and creative Saudization endeavors. Each task-force member had his own specialty, but they worked as a team. They often exchanged ideas over coffee in the morning. In the late afternoon, after working all day on individual assignments, they would meet again to discuss what they had accomplished, what they felt needed to be done, and to share new ideas.

Jurlando said that the team approach was "very important" to the group's success. "We came together every day. We drank coffee together. We were often together for eight hours a day. We were devoted 100 percent to accomplishing our goals," he said.

Although the task force was in existence for only four months, from March 26 to July 31, 1979, it could claim a hand in at least 23 major accomplishments in the areas of Saudi recruitment and training. The list read like a blueprint for the Saudization of Aramco. In the margin of an old copy of the task force's final report, someone wrote a one-word description of the group's accomplishments: "Impressive."

A summary of the task force's major accomplishments should probably begin with its efforts to recruit more employees from the growing number of students attending colleges and universities within the Kingdom. The Kingdom now had six universities. Higher-education enrollment had soared from 8,500 in 1970 to more than 44,000 in 1979. In addition, it was estimated that more than 13,000 Saudis were attending colleges and universities outside the Kingdom.

Recruiting Saudi College Graduates

In early 1979 the SAMCOM Task Force appealed directly to Saudi universities for advice on getting students interested in Aramco. Al-Hokail visited the University of Petroleum and Minerals, Riyadh University (later King Sa'ud University) and King 'Abd al-'Aziz University in Jiddah, asking them frankly, "How can Aramco get your graduates, entice them to come to work for us?"

"They gave us some ideas, like Career Day," Al-Hokail said. "They were very cooperative. They gave us space to put our brochures, and we had dates for a Career Day for Aramco only at each of these universities. People from Training, people from Recruitment and someone from other business lines made presentations."

167

Saudi scholarship graduates were honored in Dhahran, 1985.

The Employment Department established a College Relations Division at its main office in Dhahran and at its branch offices in Riyadh and Jiddah. "Their main activity," Al-Hokail said, "was to contact universities and try to get us the best graduates available in certain disciplines. We looked at the universities' curriculum — what we could do to help them improve the curriculum in petroleum engineering, geology, chemical engineering, those disciplines that we desperately needed. We even helped the universities develop their labs and gave them [core] samples that they needed. We developed excellent relations with the universities."

As the Employment Department representative on the SAMCOM Task Force, Jurlando was asked to develop a strategy for recruiting graduates of Saudi colleges and universities. "We were just coming into a period when the Kingdom was generating a good portion of college graduates," he said. "We were very much in competition with the government and with other businesses" for these graduates. Jurlando began by writing recruiting brochures and testimonial advertisements in which Saudi employees told, in their own words, about their experiences and progress in Aramco. These were distributed to recruiting stations and given by recruiters to students at universities and high schools throughout the Kingdom. The advertisements, written in English and translated into Arabic, may have been the first of that type produced in Saudi Arabia. They were certainly the first of that kind put out by Aramco. "The testimonials were honest," Jurlando said, "not a fabrication of someone's story."

Aramco was at a disadvantage in the free-wheeling competition for college graduates. The Saudi government had prior claim on the services of most graduates from in-Kingdom universities. The government offered college scholarships to virtually all Saudi high school graduates who were able to gain admittance to Saudi universities. The students, in return, were obligated to work for the government after graduation from college. Hundreds of new college graduates were employed each year to staff the government's expanding network of administrative bureaus. The technical nature of the oil business also put Aramco at a disadvantage in competition with other industries. It usually required at least five years of work experience for a college graduate to become a fully job-qualified Aramco professional. Other industries could offer new college graduates immediate rewards — cars, money and supervisory titles — that Aramco could not match. Aramco's strong points were training and the possibility of future advancement.

To illustrate the Aramco recruiter's problem, Jurlando recounted a typical conversation between a company recruiter and a Saudi college graduate he was trying to hire:

Recruit: "I am trained. I have my college degree. I am an engineer. Give me a project."

Aramco Recruiter: "Oh, no. You have the book knowledge, but you don't have the experience. You don't know how to build a GOSP. You don't know how to build a bridge."

Recruit: "Well, I can go to work for an agency here in the Kingdom and they're going to make me a supervisor and give me a car. Are you going to do that for me?"

Aramco Recruiter: "No. We are going to take you into the field and give you some training."

As an instructor and fellow trainees watch, Mohammed Ali Al-Zuri adjusts a refinery-process simulator, 1979.

Sometimes the recruit had a friend who had taken a job with a company and had been given an important-sounding title with better pay than Aramco offered. In such cases, Jurlando said, recruiters might say:

"Yes, your friend, who went to work for this other company, may be called a supervisor now and may be making more money than you're being offered, but your potential with Aramco will be so much greater. We offer a lot of other things which are of substantial riyal value in addition to salary. Five years from now you will be much better off financially because of our unlimited opportunities."

"We tried to show them the growth opportunity, which was great," Jurlando said. "Some bought our message and some didn't. I think those we have in management now, those that were in the Professional Development Program, those who did buy it are glad they did, very glad they were able to see beyond next month's paycheck."

Ali Al-Naimi told an interviewer that he never tried to gloss over the stiff challenge Saudi college graduates faced when trying to make the transition from school to professional life at Aramco. The chairman of SAMCOM indicated that he spoke from personal experience.

"I always told them that the first five years of their professional life would be the most difficult. This is the time the college graduate struggles to put into practice what he has been studying since the first day of school. It is a profound test. If he can successfully master his professional tasks during this crucial period, his opportunities for further career development will be enormous. Many of us at Aramco have gone through this experience."

The Professional English-Language Program

 new program, which came to be known as the Professional English-Language Program (PELP), grew out of SAMCOM's investigation into the reasons for the company's poor record in recruiting Saudi college graduates.

"When we looked at why we were not getting more Saudi college graduates in the work force, we found out there were many obstacles for Saudis to get into the company," Al-Hokail recalled. "For instance, one of the requirements was that a person with a college degree should be proficient in English" in order to be hired. "Well, you didn't often have that, because most Saudi college graduates attended in-Kingdom schools where lessons were given in Arabic, so their English was weak.

"We said, 'OK, we need to establish English courses for new college graduates of non-English-language schools. It should not be a hindrance for people who graduate from college but don't know much English. We should accept them in Aramco.'"

The idea was not entirely new. Some in-Kingdom university graduates had received English-language training at the company's ITCs as far back as 1966. These employees, usually referred to the ITCs by their line supervisors, were offered individual or group instruction in the late 1960s and early 1970s, when the Training Department had the

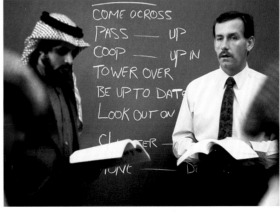

Saudi students review lessons with instructor in a Professional English-Language Program (PELP) class, 1985.

169

manpower and facilities to accommodate them. Increased demand and confusion about the schedule, goals and importance of this informal training procedure led to development of a new program, the Graduate English Course. The first session of the month-long course was held at the Dhahran ITC in late 1978. Afterward, the course director, Julia Schinnerer Erben, spoke to a joint ITC/Saudi Development meeting about ways to improve the course. During this meeting on January 29, 1979, out-of-Kingdom English-language training for graduates of Middle Eastern universities was first discussed.

After much deliberation, a recommendation for an out-of-Kingdom English-language program was made by the Training Department and endorsed by SAMCOM. A pilot program known as the Associate Professional English Program (APEP) was approved by executive management on April 29, 1979. Fred Scofield, principal of the Abqaiq ITC, was sent to the United States to check out language-training institutes. Forty-one Saudis were selected for the pilot program. They were all graduates of in-Kingdom universities. They had been hired during 1978 and 1979 and nominated for the program by their managers. Their pre-departure TOEFL (Test of English as a Foreign Language) scores ranged from 380 to 420. Members of the pilot group were assigned to three stateside institutions. The first group arrived at the American Language Academy in Tampa, Florida, on August 26, 1979. A second group reached the American Graduate School of International Management in Glendale, Arizona, in September. A third and final group arrived at the American Language Institute at San Diego State University in California in November 1979.

The stated objective of the program was to increase the TOEFL score of each trainee by at least 80 points over his pre-departure score. The Florida and California programs lasted for 10 months. The Arizona program lasted only five months and was considered the weakest of the three. The TOEFL scores of trainees in the pilot group increased by an average of 85 points during the program.

Saudi college scholarship candidates prepare for their English-language test.

The results of the pilot program were considered promising, although it needed more specific goals and objectives. In 1980 the scope of the program was defined as the "development of participants' communicative competence — the ability to get ideas across, to understand information and instructions, and to act upon them." About the same time, the program was renamed the Professional Development English Program (PDEP) to eliminate the title "associate" and any connotation that participants were not full-fledged Aramco professionals.

Khalid Nassir Al-Maghlouth, a graduate of Riyadh University with a major in economics, entered the PDEP program in 1980. He was not the typical new college-educated Aramco employee. Al-Maghlouth was a native of al-Khobar and a third-generation Aramcon. Both his father and his grandfather had worked for the company. He reimbursed the government for his college scholarship instead of working a specified time in civil service. He would have better long-range prospects at Aramco, he reasoned, and besides, Aramco ran in his family.

A sensitive and intelligent young man, he was aware of mixed feelings among older Saudi employees toward the Saudi college graduates. "Many of them had worked day and night for 20 to 25 years to reach grade code 10," he said, while Saudi university graduates entered the company at grade code 11, the professional level. For older Saudi employees, reaching grade code 11 and

Sa'ad Rowda

"A key to success, with money and health, lies in a good, solid education."

When Sa'ad Rowda came to Aramco out of the desert in the late 1950s, he went right into training as a typist, not unlike so many other Saudis in those early days. He had little formal education but was hired to work as a shop clerk with the company's Maintenance Department in Abqaiq. He was 15 years old.

Rowda learned to type up to 50 words a minute under Jerry Willard, an old Training hand, who explained to the new hires, through a translator, that they didn't need to know English in order to type. "'You guys (we didn't know what "guys" meant), you don't have to know the name of this "A" or that "B,"'" said Willard, 'Just look in the book and copy the letter,'" Rowda recalled. He remembered that Willard stressed being neat and careful with the paper. "'Never let the papers fall,' he'd say."

Through the years Rowda progressed, and he eventually received a college degree from an Arizona university during an out-of-Kingdom training assignment. When he returned to Aramco, Rowda joined Oil Supply Planning and Scheduling (OSPAS) in Dhahran, and from there went to contractor training, where he helped to oversee the company's program of bringing contractors' employees up to standard before putting them to work.

In 1979 Rowda was offered a chance to relocate to the Western Province near his hometown for a position in the Recruitment Administration Unit. Barely six years later, the company announced it would be closing down its outlying employment offices. Rowda faced the choice of returning to the Eastern Province or going to the Aramco center in Yanbu' to remain in the West. He elected to stay in the Western Province.

In recalling his training experiences, Rowda remembers Hashim Budayr, the man who made the most difference to him as a teacher. "I spent five years with him, learning Arabic. When I started I could not even write my name in Arabic. He has had a great influence on my career with Aramco. He taught me chemistry and physics. This man is an encyclopedia," Rowda said.

Rowda admits that Budayr was an especially good teacher because of his knowledge and the candid manner in which he dealt with students. For instance, if he didn't know the answer to a question, he would say so and then see if anyone in the class knew the answer. "'If I don't know it, I don't know it.' I remember Hashim saying that," Rowda said. "He has done a great work of education."

What advice does Rowda have for younger people today? "What do we need in this life other than money and health? Money will allow you to do anything, and your health will allow you to enjoy it. A key to success, with money and health, lies in a good, solid education. It is the only means that will allow you to move ahead and permit the forward movement of the country. It is education and hard work that will keep the country strong and moving ahead," Rowda said.

An instructor explains types of pipe and casing, 1982.

becoming eligible to live in Aramco's attractive senior-staff communities would be "like paradise," Al-Maghlouth said. Some old-timers resented the privileges afforded inexperienced university-educated employees and "tried to put us down," he said, while others took a nationalistic pride in the new graduates, "calling us a bright new generation" of Saudis.

Al-Maghlouth spent his first several months at Aramco doing "busy work" in the Materials Supply Department. "They had collected all these college graduates," he said, "but they didn't have jobs for us." After about six months, he boarded a chartered 747 airliner to the United States, where he was to study English for nine months as a PDEP student. The airliner, configured entirely in first-class seating, was filled to capacity with Aramco employees headed to the United States on various assignments.

He was among 17 Aramco PDEP members assigned to the language institute at the University of California-Davis, near Sacramento. They lived quite well, for students. Each Saudi had a furnished, rent-free apartment in the Sacramento area. They collected their full Aramco pay plus living expenses. Some of them indulged in expensive cars and clothes. They were conspicuously well off in comparison to the students from Asia and South America who were studying English at Cal-Davis in preparation for entering U.S. colleges.

All 17 Aramco PDEPs were in the same English class. Their instructor, a man named Bert, had a master's degree in English literature and was studying for his Ph.D. He was also learning to speak Arabic. The English class met for eight hours on Mondays and four to five hours most other days of the week. Al-Maghlouth already knew some English from school and from his father, who used it at home occasionally. "When he got irritated with my mother, he'd say something like 'leave me alone' in English, which she couldn't understand."

Al-Maghlouth seemed unsure about how successful his PDEP training had been. "Nine months was not enough time to learn a language, especially since the first two months were spent overcoming culture shock," he said. "But the experience was good. We got benefits from it, polish and experience with another culture." He may have learned more English than he realized, for his TOEFL score went from about 300 at the start of the program to 430 at its end.

In addition to English classes, there were sightseeing trips across northern California, excursions to Reno, Nevada, and to San Francisco. Al-Maghlouth took tennis lessons, got interested in meditation and attended lectures at the university's school of business.

He returned to Aramco and eventually found a rewarding job in Employee Relations for the Medical Department. For him the trip overseas had been beneficial, but too many others had treated it as a time to relax and play, one reason that PDEP programs were eventually pulled back to Saudi Arabia.

In 1982, out-of-Kingdom PDEP grew to 90 participants at eight language institutes in the United States. The participants included graduates of colleges and universities in Egypt, Kuwait and Iran in addition to Saudi Arabia. In August of that year, an in-Kingdom PDEP program with an enrollment of 68 Saudis was established at the Professional English-Language Center in the Dhahran ITC.

The PDEP was finally converted into an entirely in-Kingdom training program in 1983. The last PDEP group for out-of-Kingdom training departed from Saudi Arabia for the United States in March 1983. In 1984 the PDEP name was changed to the Professional English-Language Program (PELP). The program was expanded to include a year-long course for veteran Aramco employees without a college degree who spoke English fairly well but needed to update their reading and writing skills to qualify for promotion. The nature of the program changed somewhat. PELP became more of a business- and economics-language course. It focused on competence in letter writing and reading, in giving and following oral reports and in understanding job-related conversations and meetings. PELP proved to be a very successful program, providing thousands of Saudis with essential business-English skills.

College Recruiting Drive Expands

The company's college recruiting got another boost from SAMCOM in the spring of 1979. Al-Hokail and other Saudi executives traveled to Riyadh to ask the head of the government's General Civil Service Bureau for an allocation of college graduates from the government scholarship program. The so-called bursary students received college scholarships from the government in return for a pledge to work for the government after graduation. The Aramco executives spent almost three days in Riyadh presenting their case. Al-Hokail said one of the points made by the company during the talks was that, since Aramco was now 100 percent government-owned, it was entitled to its share of help from the government. They asked for an allocation of 1,250 bursary students over the coming five years, but the government granted Aramco only 300 to 500 graduates over five years. They were to be assigned to the company on an escalating basis, 30 graduates the first year, 50 the next year, and so on. In Dhahran it was considered a successful negotiation even though the company got less than half of what it asked for. It was the first time the government had agreed to let Aramco have any of its bursary students. The first eight college graduates were assigned by the government to Aramco in June 1979. They included four graduates with civil-engineering majors, plus individuals with majors in petroleum engineering, agricultural engineering, business administration and industrial engineering.

In 1979, SAMCOM initiated procedures to hire on a direct basis all Saudi college graduates without requisitions (without obligations to work for the government). A blanket job offer was made to all such college graduates except those who had majored in the arts, history or some field where placement in a suitable job with Aramco would be difficult. Graduates in those fields were to be handled on an individual basis.

The college recruitment drive extended to more than 1,400 Saudis attending schools in the United States. For that purpose, Zaki Ruhaimi, working out of Aramco Services Company offices in Houston, Texas, in early 1979, established the College Relations Program. All Saudi students attending colleges and universities in the United States, Saudi government-sponsored or not, were considered potential employees. The company promised it would finance the remainder of the college education of qualified candidates in return for an agreement to work for Aramco after graduation.

"We were looking mainly for skills critical for the company ... mainly engineering skills," Ruhaimi said. "We were trying to sponsor students at the time who were in their freshman or sophomore year and retarget them into some of these critical skills."

The College Relations Program hosted several weekend gatherings for Saudis from schools all over the United States. The students were invited to spend a weekend in

Aramco Services Company was located at 1100 Milam in Houston in 1983.

Houston or in Washington, D.C., at Aramco's expense, to learn about employment opportunities in Aramco. Al-Hokail went to the States two or three times to speak to such groups. "These were very successful meetings," he said, "attracting as many as 200 to 250 Saudi prospects at a time.

"We invited all government scholarship students, gave them the pitch of what Aramco was, the benefits and the kind of experience they would get. Then we asked them to sign up. Once they signed up and had graduated, we asked for their release from the [Saudi government's General] Civil Service Bureau. Then we hired them."

The college recruiting campaign was a success. The company hired 124 Saudi college graduates in 1979 and 203 in 1980, compared to 40 in 1978 and 30 in 1977. In 1979 and 1980, Aramco hired a total of 327 Saudi college graduates, three more than the total number of Saudi college graduates the company had been able to hire in the entire 20 years from 1958 to 1978.

The largest number of new Saudi graduates came from universities in Saudi Arabia. Of the 327 college graduates hired during 1979 and 1980, 122 came from schools in Saudi Arabia, an indication of the growth in size and quality of the Kingdom's higher-education system. The group included 57 graduates from the University of Petroleum and Minerals, 48 from Riyadh University, 12 from King 'Abd al-'Aziz University and five from other Saudi universities. The company also recruited 41 Saudi graduates of Middle Eastern universities outside Saudi Arabia, 35 Saudi graduates from U.S. universities, five from European schools and two from schools in Asia.

Aramco seemed to have an unlimited appetite for Saudi college graduates. Hassan Husseini, coordinator of Employment and Staff Services, was asked by an interviewer about recruiting 1980 graduates from the University of Petroleum and Minerals (UPM), located adjacent to company headquarters in Dhahran.

"Even if I could have all the graduates from UPM for the next 15 years I still couldn't fill this year's jobs," Husseini replied.

The company's top executive, John Kelberer, sent a memo to Business-Line heads on December 14, 1981, virtually prohibiting them from denying a job to any Saudi college graduate. Kelberer wrote: "Every Saudi college graduate should be hired and developed. A Saudi college graduate may not be rejected. However, exceptions must be justified in writing for review and concurrence of the functional vice president, and the endorsement of the vice president of Employee Relations and Training."

Training in oil operations included classes in drafting.

Kelberer went on to acknowledge that the company had been increasingly successful in hiring Saudi college graduates. "However, we still need to hire more graduates in order to meet our requirements." The matter "is of utmost importance," he said.

In the spring of 1979, Aramco inaugurated two training programs designed to attract outstanding high-school graduates. One was called the College Fast Track Program for Non-Dependent High School Graduates, soon shortened to College Fast Track. Under this program, Saudi high-school graduates with a grade average of 85 or better who agreed to join Aramco were sent to the U.S. for up to one year of intensive English-language training. Those who achieved a TOEFL score of 500 or better after their training qualified for an Aramco college scholarship.

SAMCOM proposed the College Fast Track program to the Management Committee on May 15, 1979. Management approved a one-year pilot program for up to 50 students. On August 8, 1979, management agreed to the Training Department's request to increase the size of the program from 50 to 75 participants. All candidates attended a three-month orientation program before going overseas, and signed a contractual agreement to work for Aramco for a specified time after returning to Saudi Arabia.

Aramco students at Arizona State University, 1985.

In September 1979, the first 35 College Fast Track participants started language classes in the United States. They were sent to three different schools: 10 participants went to the University of Denver, 11 went to Vincennes (Indiana) University, and 14 went to the University of Arizona. In December, another group of 40 participants was enrolled at three other universities: Louisiana State University, the University of Miami, and Tulsa University.

This aggressive new high-school recruiting program was also a success. In 1979 Aramco hired 796 high-school graduates, compared to only 121 the year before and to an average of just 90 a year during the previous decade. Aramco did even better in 1980, hiring 1,281 high-school graduates, more than 10 times the 1978 figure. The College Fast Track program got credit for significant recruiting help. Even though only 57 of the high-school graduates (including 13 females) recruited in 1979 qualified for the College Fast Track program, hundreds of other students had been induced to take a serious look at opportunities in Aramco, thanks to publicity about the program.

"The Fast Track program has helped to greatly increase Aramco secondary-school recruiting and has attracted high-potential candidates to the company," Bill O'Grady, director of Training, reported to management in early 1980. "Most of those with an overall grade average of 85 percent joined Aramco as a result of publicity about the College Fast Track program."

The success of College Fast Track persuaded the company to extend scholarship opportunities to the dependents of the company's Saudi employees. In July 1979, SAMCOM received management approval for the College Fast Track Training Program for Dependents of Aramco Employees. The company agreed to sponsor English-language training for up to 50 dependents who were high-school graduates and who could qualify under the rules of the College Fast Track program. Aramco also agreed to sponsor the remaining education of up to 30 dependents of Aramco employees who were already enrolled in colleges or universities in Saudi Arabia or abroad. To be eligible, students had to have a grade-point average of 2.5 or higher and be working toward a degree in scientific, technical or business fields related to jobs within Aramco. Liberal-arts majors were specifically excluded from consideration.

Placing College Graduates

iring Saudi college graduates was one thing, integrating them into the Aramco system was something else altogether. The Associate Professional Program (APP) created in 1974 was designed for that purpose. It provided for a series of rotational work assignments intended to give college graduates the experience necessary to handle an entry-level professional job. The APP was operated by the Management and Professional Development Department, which was also responsible for assuring that the company had an adequate supply of well-prepared candidates for top management positions. APP participants were assigned to a department where they were given a job target and an individual development plan. They were scheduled for a variety

175

of work assignments, including English-language training, if necessary, for a period of up to three years. This schedule exposed participants to real work situations and gave Management and Professional Development a chance to observe the person's willingness and readiness to perform on a full-job basis.

Ali Al-Naimi reviewed the program in 1978 and recommended to the Management Committee that the APP be strengthened and made more respon-

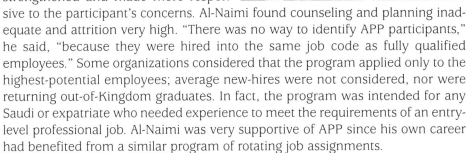

Scale models of Aramco plants, made for design purposes, were valuable training tools as well, 1982.

sive to the participant's concerns. Al-Naimi found counseling and planning inadequate and attrition very high. "There was no way to identify APP participants," he said, "because they were hired into the same job code as fully qualified employees." Some organizations considered that the program applied only to the highest-potential employees; average new-hires were not considered, nor were returning out-of-Kingdom graduates. In fact, the program was intended for any Saudi or expatriate who needed experience to meet the requirements of an entry-level professional job. Al-Naimi was very supportive of APP since his own career had benefited from a similar program of rotating job assignments.

In May 1980, the APP was renamed the Professional Development Program (PDP) and transferred from the Management and Professional Development Department to the newly created Career Development Department. At the same time as it relinquished control of the PDP, Management and Professional Development dropped the word "professional" from its title and became the Management Development Department.

Al-Naimi, in an interview with the company newspaper, spoke directly to the young Saudi college graduates who were eligible for the program:

"The PDP is not a big-brother program," he said. "It is an opportunity for an inexperienced employee to put into practice his know-how under controlled supervision. It's a unique learning opportunity.

"The program is not only a chance for Aramco to assess the potential of an employee at an early stage, but for the employee to discover what he can and likes to do best."

Measuring fluid levels precisely was a basic part of training, 1982.

With support from SAMCOM and the Management Committee, PDP enrollment steadily grew. From fewer than 400 participants in 1980, enrollment

increased to 670 in 1981 and to 750 in 1982. In deference to their education and senior-staff status, individuals enrolled in the program were called PDP "participants," not students, trainees or associates.

Under Career Development the PDP continued to be open to all college graduates, Saudis and expatriates, men and women, just as it had been when administered by Management and Professional Development. The 720 PDP participants in May 1983 included 62 women, mostly working in Industrial Relations. The sponsoring departments gave their participants individual development assignments that lasted for up to 36 months — 45 months if English-language training was necessary. These developmental plans involved a series of job assignments plus business courses at the company's Management Training Centers. Although PDP participants occupied a wide spectrum of professions, from petroleum engineers to career advisors, they followed a similar schedule of PDP classroom work. During the first 12 months of the 36-month program, participants took one of two orientation courses created by Career Development. They were enrolled in either an orientation course for directly hired professionals who knew nothing of Aramco, or in a course for Aramco-sponsored graduates with knowledge of the company. The direct-hire orientation was the only PDP course given in Arabic. In the second 12-month period, PDP participants took two courses, oral presentation and interpersonal communications. The third 12 months involved a course on effective writing and, if appropriate to the participant's career path, the Effective Aramco Supervisor course. The optional nine-month English-language training program included extra courses in business writing and time management.

Career Development Department

he Career Development Department was created in May 1980 as a direct result of the drive to recruit more Saudis into the company work force. The Training Department, with a record-high ITC enrollment of 13,600 trainees, could no longer encompass both training and developmental needs. In December 1979, T. John McCubbin, vice president of Employee Relations and Training and a member of SAMCOM, proposed formation of a new department to be responsible for monitoring and guiding the development of all Saudis below the management level. His proposal covered Saudis in grade codes three to 14, a total, at the time, of about 21,000 employees. The mission of the Career Development Department, as McCubbin chose to call it, would be to develop a pool of Saudis to fill professional and supervisory positions in the grade code 11 to 14 range. It would identify high-potential employees, coordinate their placement in appropriate development programs and generally administer to their training needs. It would be a department independent from Training, and a permanent adjunct to the work of SAMCOM. The Training Department would be freed to focus on its primary task of trainee instruction.

T. John McCubbin, 1978.

Career Development was a unique organization, formed by combining elements of the Training Department, the Management and Professional Development Department and the Employee Relations Department. From Training it assumed control over Management/Technical Continuing Education courses, consisting of 185 correspondence courses with an enrollment of more than 5,000 Saudis and expatriates. From Employee Relations it took over the Placement Unit. From Management and Professional Development it inherited the Associate Professional Program (later PDP) with 450 participants. Donald A. Rudkin was acting director of Career Development during its initial year.

On January 7, 1981, management named Ali M. Dialdin as director of Career Development. Dialdin had just returned from a four-month-long study of the corporate career-development policies used by the U.S. shareholder companies in California and New York. He was also serving as secretary of SAM-COM. The department he took over had a full-time staff of about 140, and a budget of about $50 million. At the time, some 1,150 employees on out-of-Kingdom training assignments were carried on the Career Development payroll.

Career Development comprised three divisions: the Career and Professional Development Division, headed by Jack W. DeWaard, former superintendent of the Aramco Schools; the Management and Professional Education Division, administered by Zaki Ruhaimi, former superintendent of Southern Area

Saudi machinist trainee using a drill press, 1982.

Saudi trainees and instructor disassemble a pump, 1982.

Training and a member of the SAMCOM Task Force; and the Saudi Development Division, administered by 'Abd Allah Ali Al-Zayer, who had headed Saudi Development when it was a division of the Training Department.

Career Development was responsible for making sure all Saudi employees below the management level had the opportunity to develop to their highest potential. The function of the Saudi Development Division was to identify high-potential Saudis in grade codes three to 10, provide them with developmental opportunities, both in- and out-of-Kingdom, and track their progress. The Management and Professional Education Division coordinated the placement of all company employees — expatriates as well as Saudis — in appropriate training programs. The Career and Professional Development Division administered the PDP and its companion, the Professional Development English Program (PDEP), for Saudis in the grade code 11 to 14 range.

Resetting Saudization Goals

The achievement of any training goal required cooperation from the company's various Business Lines. On recommendation of the SAMCOM Task Force, starting in 1980 management required each Business Line to report annually on its Saudization activities. All Business Lines were to include a Saudization report along with their annual business plan. This requirement spotlighted Saudization activities and sparked competition among Business Lines, just as its proponents hoped it would.

"Each Business Line had to detail its success in recruiting, training and retaining Saudis," Al-Hokail said. "Each organization was also asked for a 10-year forecast of Saudi requirements and for the percentage of Saudis that would be in the organization each year. So, you can see, when they made a presentation to the Management Committee, there was competition among Business Lines over who had the best results from their Saudization activities, and who was going to do most in the future. As a result, they became more energetic trainers."

In 1978 the Training Department began using a computer system to help organizations forecast their Saudi-development requirements. In the beginning, the system was used by Training to determine whether line-organization estimates of current training enrollments and future graduates were realistic. Later on, line organizations copied the system and modified it to fit their own needs. In the early 1980s the system was upgraded from a mainframe-computer program to a personal-computer spreadsheet program. It came to be called the Training Model, or the Crawford Model, after John P. "Crif" Crawford, the planning specialist who developed it.

Al-Hokail, when asked to select the most important results of SAMCOM's work, made several choices. "I think one major thing was the emphasis on professional training," he said, "getting college graduates in engineering, computers and management. Then there was the craft training. You go to Abqaiq, you go to the refinery, and you'll find at least 80 percent of the maintenance organization is Saudis."

Advanced wire-line skills were taught at Ras Tanura, 1981.

William P. O'Grady

"There was a great deal of satisfaction in knowing that I was involved in an operation that I could see was going to do some great things in Saudi Arabia."

illiam P. O'Grady was director of Training during some of the most eventful years in Aramco history. He became director in 1966 when the company was downsizing, led the department through the wild growth spree of the 1970s and stepped down as director in 1982 as the expansion period was ending.

There were 1,500 Saudis enrolled in the company's training classes when O'Grady became director. He had a General Office staff of 23 persons and an annual budget of several hundred thousand dollars. When he left the directorship 16 years later, Training consisted of 13,000 Saudi trainees, and the department had a full-time training staff of 2,100 persons and an annual budget of several hundred million dollars.

In 1966, Training was a General Office department of specialists who provided functional guidance to line organizations on training matters. Shortly after he became director, executive management asked O'Grady what could be done to downsize Training. Among several suggestions, O'Grady's most drastic proposal was that the department be eliminated and the specialists assigned to other organizations. The company's sudden expansion in the 1970s rendered that step unnecessary. O'Grady later came to regard the late 1960s and early 1970s as an idyllic time in Training.

The expansion period produced one headache after another for Training. O'Grady had to cope with a revolt by angry parents after senior-staff schools, then under his direction, adopted an "open classroom," individualized, computer-controlled teaching system. The system was dropped in favor of the time-tested single-room method, but not before O'Grady was called before executive management to explain the school's problems. Next, Training's Apprenticeship Program, which held great promise, had to be canceled because line organizations continually raided the program for employees to meet their critical manpower needs. In the mid-1970s, confronted by an acute shortage of space, O'Grady approved opening training centers as far away as Egypt.

O'Grady graduated from Iona College, received a master's degree in American history from Fordham University and taught at inner-city New York public schools before joining Aramco in 1953. He was assigned to the Ras Tanura ITC, where, he said, his career got "a sort of jet-assisted take-off."

"I was truly impressed with the dedication of the people in the training organization at Ras Tanura. We were in a facility that was simple and quite spartan," he said. "But there was something dynamic and exciting about that particular operation. I had come from the States, where I was teaching in an inner-city school, and I was not getting any real satisfaction. All of a sudden things turned around for me. There was a great deal of satisfaction in knowing that I was involved in an operation that I could see was going to do some great things in Saudi Arabia. The boost that I got at that time I tended to carry with me as I went on in my career with Aramco."

O'Grady earned a reputation as an accomplished public speaker and possessor of a near-perfect memory. Like most trainers, he took great pleasure in seeing those he trained advance in the company. He recalled tutoring Hamad Al-Juraifani. "I recognized that I had a student with a real potential in Hamad," he said. Al-Juraifani eventually became a Saudi Aramco vice president. Of the Saudis who worked with him, O'Grady recalled outstanding work from Ali Twairqi; Zaki Al-Ruhaimi; 'Abd Allah Ali-Zayer; Mustafa Al-Din, who "distinguished himself" as manager of the Aramco-Built Government Schools Division; and Ali Dialdin, "the best of our advisors in the Out-of-Kingdom Training Program."

"Another major accomplishment was getting a commitment to develop Saudis. It was necessary to change attitudes. It was a hard sell. A busy department head doesn't necessarily want to spend his time worrying about training. We even had to fight some of the Saudis [department heads]. They were under pressure to produce. They said, 'How can I do this if you give me [inexperienced] Saudis?' We said, 'That's the reason we give them to you, so you can develop what is required. If you don't have courses, develop courses. If you don't have special programs, develop special programs.'

A swimming survival class in Yanbu'.

"We are very, very proud of what has been accomplished. A highlight of my career was this period I spent in Training because it was a field that you could do so many things in."

A Training Building Boom

No one at Aramco in 1980 could foresee an end to the company's wild growth spree. Just the opposite, in fact. The company had experienced unparalleled growth in the 1970s, but the decade ahead appeared even more challenging. Corporate Planning estimated that more than 30,000 Saudis would be hired by Aramco between 1980 and 1985, more than doubling the size of the Saudi work force in five years. Nearly all of these new employees would have to be trained. Present Saudi employees would have to be given training to upgrade their skills so they could advance to higher positions.

At executive management's request, the Training Department compiled a report in 1980 on the status of training and the prospect for meeting the company's manpower-development goals. The report, titled *Corporate-Wide Training Activity 1980-84,* predicted increases over the period of 83 percent in ITC/ITS class periods, 106 percent in the training staff, and 131 percent in net direct expenditures for training.

"The training function in Aramco is presently facing a challenge of a magnitude never before experienced in the company's history," the report said.

The report listed three main manpower development goals facing the company:

"1. Initially meet the requirements of Article 45 of the Saudi Labor Law, which states that at least 75 percent of each employer's work force shall consist of Saudi workmen, and ultimately meet the company's goal of 100 percent Saudization.

"2. Provide training for new Saudi employees hired for anticipated company growth and replacement of workers lost due to attrition, and ...

"3. Improve the skills and abilities of those Saudis presently employed so that they may advance to higher positions in the company."

The first goal, a 75 percent Saudi work force, was so far in the future, it wasn't even discussed further in the report, except to say the company would "fall far short" of 75 percent Saudization in the 1980-84 period. Instead, the report suggested a slight increase in the ratio of Saudi employees — from the 52 percent level to 54 percent — might be possible by 1984. Progress on the second and third goals depended largely on how many additional facilities and how much more manpower the company was able to supply for Training.

A substantial improvement in the educational level of new-hires was expected. The report estimated that the average newly hired Saudi during the 1980-84 period would have about eight years of education before joining Aramco. Nearly one-third of new Saudi employees would have 11 or more years of schooling, it predicted. By comparison, in the early 1970s the average Saudi new-hire had completed six years of school, and less than five percent of new Saudi employees had had 11 years of education. The report went on to say that no matter how much schooling the recruits had, "Virtually none are expected to possess the high-level technical skills to assume craft and operator jobs at the time of hiring or soon after."

Adding thousands of inexperienced Saudis to the workforce caused further concern. In 1980, nearly 80 percent of all Saudi employees had been with the company for fewer than five years. The majority were not yet qualified to handle a full-time job. More than half of all Aramco employees were Saudi, but two-thirds of all Saudi employees were not fully job-qualified.

Marine trainees learn to operate a tanker loading arm.

High rates of attrition continued to be an obstacle to increasing the size of the Saudi work force. The task was like climbing a sand dune: two steps up, one step back. During the 1970s the company hired about 22,500 Saudi employees and lost nearly half of them to attrition. The result was a net increase of about 12,000 Saudi employees in 10 years — discouraging figures for a company that needed to add some 30,000 more employees in the next five years.

Training's report expressed an urgent need for new training facilities. Although the space devoted to training activities was about six times greater than it had been in 1970, the increase was not enough to keep up with the expanding workload. By 1985 ITC enrollment was expected to increase by 50 percent and ITS enrollment was expected to climb by 170 percent. To accommodate this growth, the Training Department report said, would require an increase of about 260 percent in the amount of space devoted to training. Training facilities were already running at or, in some cases, beyond maximum rated capacity. If additional space was not available as scheduled, the report warned, about 1,100 Saudi trainees would have to be turned away in the coming year, and the number could escalate to 10,000 by 1985.

Instructor and trainee review control-panel maintenance, 1985.

The company responded by approving a record capital budget for training. The budget included 20 major training-related construction projects at a total cost of almost $300 million. About 70 percent of the approved spending — some $210 million — was for Training Department facilities; the remainder

was for On-the-Job Training (OJT) functions. Newly approved projects would add 834,000 more square feet of training facilities by 1984, bringing the total space that was devoted to training within the company to more than 1.73 million square feet.

Fortunately, at this time Aramco was prospering as never before. Following the change in government in Iran in late 1978, the market price of Saudi light crude leaped in two years from $12.70 a barrel to $30 a barrel. Aramco production, working to take advantage of high prices, averaged a record 9.63 million barrels a day during 1980.

Saudi employee runs tests on electronic process instrumentation at Ras Tanura, 1986.

Two management decisions accounted for most of the record size of the training-related capital budget. One was the decision to build new ITC/ITS facilities to replace leased facilities in Dammam and Mubarraz. The second was the decision to build new craft-training centers. The centers were intended to answer a pressing need for highly skilled craftsmen known at Aramco as industrial maintenance men.

Aramco maintenance workers were classified as having either critical or noncritical skills, depending on how essential their work was to sustaining oil production. The critical-skill category included positions such as maintenance machinists, electricians, instrument repairmen and refrigeration mechanics. These craftsmen maintained and repaired the working parts of the company's production and distribution facilities — the pumps, valves, motors, compressors, turbines, electrical and motor-control panels and other essential paraphernalia. They also tested, replaced and calibrated the instruments that measured pressure, temperature and flow of petroleum products through the company's plants and pipelines.

The decision to build new craft-training centers was made at a time of growing concern over the shortage of Saudis with high-level technical skills. It was apparent that neither the Kingdom nor the company would be free of dependence on foreign workers until Saudis learned to maintain the sophisticated equipment used by Aramco and other industries. Outside Aramco, the only in-Kingdom vocational training effort of any size was conducted at 24 government-operated vocational and professional training centers scattered about the Kingdom. But neither the level of training nor the size of the program, a total of enrollment of 4,000, could satisfy Aramco's needs.

No firm in Saudi Arabia had more state-of-the-art equipment or required a larger number of skilled craftsmen to maintain and repair the equipment than Aramco. Nearly one-sixth of the total work force, 9,000 men, were in the maintenance field. About half of the maintenance work force was Saudi, but fewer than 10 percent of Saudi craftsmen were fully qualified for jobs in the critical maintenance fields.

Job Skills Training Revamped

arry Tanner, then vice president of Southern Area Oil Operations, was one of the Aramco executives who felt pressure to speed up Saudization of the maintenance work force, but had no hope that government schools or the Training Department's regular ITS programs could provide the necessary technical training.

"All training basically had been directed more toward academics," he said. "Training had an organization to do shop training, but it was low-level shop training, like woodworking and painting and similar skills. When we started to get into mechanical and electrical instrumentation, those high-level skills, Training just had nobody ready for or capable of performing that type of training. And those were the skills we were supposed to be primarily pushing in order to get more higher-level skilled Saudis."

Consequently, in 1979, Southern Area Oil Operations in Abqaiq started its own maintenance training program in facilities of its own, independent of the Training Department. Maintenance craftsmen were reassigned as instructors and curriculum writers. The first maintenance training instructors in the Southern Area were bilingual craftsmen from various nations — "Jordanians, Pakistanis and Indians, for the most part," Tanner recalled. "We tried to use the [craftsmen] with intermediate skills because they were closer to the employees we were trying to train."

Training manuals were compiled by pulling together information from various sources. "We took a lot of data from manufacturers' manuals, for example," Tanner said. "We would get all the information the company had as to their equipment, how to check it out, how to calibrate, etc. We would write it over into our books. As I recall, we had it written in English on one side of the page, and Arabic on the other."

Line organizations had always operated shops and classrooms for job-specific training. Maintenance training, however, was a much more ambitious program than what went before, and it was spread across several organizations in different districts.

The Mechanical Services Organization started another independent maintenance training program in 1979 using facilities at the Mechanical Services Shops Department in Dhahran. Mechanical Services was headed by Robert S. Luttrell, a company vice president with a reputation for pushing Saudi training programs. A department within Luttrell's organization, Maintenance Resources Planning Department (MRPD), was responsible for coordinating maintenance organization activities such as planning, record keeping and materials ordering. SAMCOM asked MRPD to take on a similar coordinating role over maintenance training programs. The Training Department watched while MRPD tried to organize and standardize maintenance training and SAMCOM pushed to get more Saudis trained and into maintenance jobs.

Fred Goff, then manager of MRPD, remembered early maintenance training as "makeshift efforts using every kind of shed and lean-to" as training facilities. Maintenance training, however, progressed more rapidly than many people expected. By early 1980, enrollment in the Southern Area Maintenance Training Division's program was more than 600 and growing. The Mechanical Services Shops program, started in February 1979 with two maintenance trainers and 14 trainees, grew to 130 trainees by year's end. Northern Area Maintenance Training began in September 1980 with more than 100 trainees. Jim Nolan, who followed Goff as manager of MRPD, recalled the rapid development of maintenance training: "It evolved from a kind of fly-by-night training to formal, hands-on classroom work. We came up with a 10- or 11-year training program. It took about 10 years to get a person from newly hired to grade code 11 [fully job-qualified maintenance man]. There was a small group of people in MRPD, mostly vocational trainers with Ph.D.s, who did reviews of training programs and audited them. The Northern Area had a group of people under Howard Jensen who had worked in the terminal for many, many years. A big shop building — the old machine shop — was taken over for training. It had individual bays for welding, machinists, fabrication and a little bit of electrical, but not too sophisticated. In Abqaiq, what was known as the old "Green Building" was converted into a training center and Mal [Malcolm] Werner headed that up."

Line organization craft training grew rapidly, but not fast enough to satisfy the anticipated need for maintenance men. According to projections made for the company's 1980-85 Business Plan, an additional 4,000 skilled maintenance personnel would be required by 1985, just five years away, and 4,500 by 1990. The company hoped to increase the ratio of fully job-qualified Saudi maintenance

The Materials Supply Training Center was located in a 20,000-square-foot inflatable building in 1980.

men to 55 percent by 1985 and to 69 percent by 1990, and to reach full Saudization by 1997. That would require a much larger training program than line organizations had put together so far.

"Bob Luttrell felt we [Mechanical Services] had all we could handle without getting into a mass training program," Nolan said. "So he suggested bringing in outside experts to help us with our training programs."

Dunwoody Industrial Institute

everal contractors were approached about taking over the company's maintenance training programs. Only one of them, Dunwoody Industrial Institute of Minneapolis, Minnesota, had sufficient numbers of trained craftsmen with experience in teaching high-tech maintenance courses. Aramco Training had a longstanding relationship with Dunwoody going back to the Ras Tanura Refinery upgrade in the 1950s. Aramco employees were frequently sent to Dunwoody for craft training, and Dunwoody consultants came to Aramco as special instructors and training advisors. Furthermore, in the previous two years,

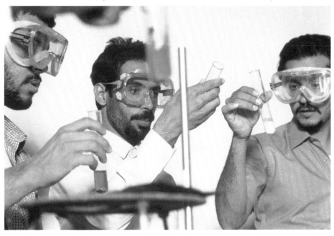

Students in chemistry lab at Ras Tanura ITC, 1986.

Dunwoody had done two studies of the company's industrial training program. One of them, issued in July 1977, recommended a restructuring of the ITS curriculum, an improved instructional staff and an out-of-Kingdom program for advanced craft training. The second study, presented by Dr. D.F. Peters in February 1979, called for development of a Maintenance Training Master Plan and creation of a series of industrial-maintenance training centers. On August 4, 1979, the Management Committee approved bringing a team of Dunwoody consultants to Saudi Arabia to develop, in cooperation with Aramco personnel, a proposal for implementing Peters' study recommendations.

A three-man team of Dunwoody consultants arrived at Aramco in November 1979. On February 25, 1980, an Aramco/Dunwoody Project Team presented a report with five major recommendations. They proposed creation of a master plan for maintenance training; construction of self-contained craft training centers; development of a special vocational English program; creation of an instructional program for the Aramco maintenance training staff; and a multiphased maintenance training program geared to instruction of Saudis with fewer than eight years of prior schooling.

New Training Campuses

he craft training centers concept was approved by the Management Committee on April 8, 1980. A company planning specialist, John P. "Crif" Crawford, worked with the Dunwoody team on sizing the proposed new craft centers.

"That was one of the first uses of computers to integrate training program throughput and attrition rates," Crawford said. "We identified all job titles, then got accounts of the number of 'expats' and Saudis in these jobs by grade codes. So we had a breakdown of four or five major craft skills and the number of Saudis and 'expats' in them. That led to forecasting how many people we needed. Company plans called for a production target of up to 16 million barrels [of oil]

per day, so we had to increase our maintenance work force by some ratio related to that. The size of the maintenance training facilities was predicated on the 16 million barrels of crude a day figure."

Sites for the new training centers were chosen in Abqaiq, Ras Tanura and Mubarraz, near the city of Hofuf. These three sites were convenient to the company's major production and refining facilities. The training centers were to be built adjacent to or interconnected with ITC/ITS facilities so as to form regional training "campuses." Two Houston, Texas, firms, Blom Engineering and Daverman Associates, designed the campuses for all three areas. Each campus included a two-story Area Maintenance Training Center (AMTC) with separate shops and classrooms for each craft specialty. The building plan called for landscaped inner courtyards and walkways, all protected by walls from sun and wind. Two of the areas, Mubarraz and Ras Tanura, would also have Vestibule Maintenance Training Centers (VMTC) for specialized training leading to entry-level maintenance jobs. In addition to offering basic maintenance training, all three areas would provide training to upgrade and recertify the skills of regular craft employees.

A class in safety procedures on a training rig in Abqaiq, 1986.

The AMTCs would occupy more than 710,000 square feet and cost an estimated $200 million to build, by far the largest and most costly training facilities yet conceived by Aramco. In combined square feet, the three centers would be almost four times larger than the two seven-story office buildings, the Exploration and Petroleum Engineering Center (EXPEC) and the Engineering Office Building, then under construction in Dhahran's core area. A single, turnkey contract for construction of all three AMTCs was awarded on October 6, 1981, to a Korean firm, Suwaiket Miryung Co. Ltd.

In early 1980, in accord with another one of Dunwoody's recommendations, the Training Department requested, and management approved, a $2 million appropriation for development of a special English-language curriculum tailored to line-organization training needs. The end result was creation of the Vocational English-Language Training (VELT) program by the consulting firm Pacific American Institute of Corte Madera, California. Beginning-level VELT was introduced in 1981. The program focused on the English-language skills Saudis needed to perform in seven different job categories, including both critical and noncritical maintenance jobs as well as oil-plant and gas-plant operations.

Aramco Maintenance Training

n April 1980, MRPD recommended Dunwoody as the sole-source contractor to provide a complete maintenance-training program for Aramco. A former Aramco trainer, Khamis Abdalla, owner of Haitham Enterprises, became Dunwoody's partner in Saudi Arabia. The following month, the Management Committee authorized contract negotiations with Dunwoody but directed that the Training Department rather than MRPD take charge of the negotiating effort. Negotiations on a contract began in August 1980. A proposed contract called for Dunwoody to handle all operations, from curriculum writing to instruction and testing during the initial five years of the program. Dunwoody would train Aramco instructors to take over the program at the end of the five-year period. The line organizations were to continue their craft training programs until Dunwoody could be brought into the field.

A five-year start-up program employing as many as 150 Dunwoody trainers was proposed. Ali Dialdin and others felt it was an "outrageously" expensive proposal — well over $150 million — and it was rejected by management. Negotiations with Dunwoody dragged on for months without agreement. Money was the sticking point. "We negotiated with them and negotiated with them some more," Jim Nolan recalled, "but never got them to come down to anything that we would expect to pay."

Larry Tanner and some other line-organization executives had reasons, in addition to expense, for disliking the proposal. The contractor wanted to create a new maintenance-training program rather than build on what the line organizations had already started.

"I personally thought it was going to delay the development of Saudi craftsmen by around five years if we embraced the proposal," Tanner said. "They were going to start developing a program from scratch. What I wanted to do basically was start with what we already had. You know, you really get your craft-type training from doing. This would get our people to doing much earlier in the game."

Larry Tanner, 1968.

The existing maintenance training programs continued to grow and improve to the point where, by 1981, many line organization executives felt it would be preferable to handle craft training in-house rather than through a contractor. Northern Area Maintenance Training enrolled 650 Saudis by April 1981. Forty-two of its trainees had completed the first phase of the three-phase instrumentation maintenance program. Southern Area Maintenance Training had more than 1,000 trainees, with a training staff of about 70, and more than 30,000 square feet of office, classroom and shop space. The Southern Area had also developed a mainframe computer system called TRAC (Training Reporting Analysis and Certification) for keeping training records.

By all accounts there were some fierce arguments within SAMCOM and the Management Committee over the pros and cons of using Dunwoody to conduct maintenance training. The arguments in favor of in-house training were compelling: a shorter program completion time than previously planned, a training program tailored to Aramco's needs, a more cost-effective effort and, most important, a program which would have total acceptance by user organizations. Finally, in April 1981, on a recommendation of the Training Department, management decided to end negotiations with Dunwoody in favor of an in-house maintenance training program.

"No doubt, at the end of five or six years the Dunwoody program would have been superior to what we were doing," Tanner said. "But with as few skilled craftsmen as we had in the time period, we couldn't wait."

Hashim Al-Sadah during Marine Department navigational training, 1989.

Although the contract proposal had been rejected, the consultant's training concepts had not. A plan similar to Dunwoody's was adopted, using Aramco employees instead of contractors to conduct a four-stage maintenance training program. The program consisted of one year of basic ITC/ITS core courses at the AMTCs, followed by 1½ to two years of job-targeted training. The trainee then went into the field for another 1½ to two years of on-the-job training. Finally, the trainee returned to the VMTC for another 1½ to two years' advanced craft training classes. The duration of the training program varied according to job target, but averaged about five years from starting out at the AMTC to qualification for an entry-level maintenance job. Several more years of work experience and specialized training classes were required to make an employee a fully qualified critical-skills craftsman.

Abdelaziz M. Al-Hokail

"There was no question of management support for [Saudi Arab] manpower development. That was there from day one when I joined the company; it showed in the way I was brought up on the different programs, first in Engineering, then Producing, then all the different levels."

bdelaziz Al-Hokail left the small mud-brick village of Majma'ah, approximately 90 miles northwest of Riyadh, for the first time in 1959. At 16 and a half years old, he was on his way to Austin, Texas. His first dream had been to study medicine at the University of Cairo, but he was one of the top high school graduates in the Kingdom that year and had been awarded a scholarship by the Saudi government to attend the University of Texas.

It was his first time to be separated from his family, and his first time to ride in an airplane. Even the words "restaurant" and "hotel" were not part of his everyday vocabulary. His itinerary took him from Majma'ah to Riyadh by car, then by air to Jiddah, Brussels and New York. Personnel from the Saudi Embassy met him in New York, where they guided him through the bewildering maze of skyscrapers before putting him on a plane to Austin.

Even though he knew little about the United States or Texas, success crowned his efforts there. He received a bachelor of science degree in petroleum engineering at the University of Texas in 1964. He worked for about three months at the Ministry of Petroleum and Mineral Resources in Riyadh before convincing his supervisor, a deputy minister who later became minister, to let him work temporarily at Saudi Aramco — where he stayed.

His experience was similar to that of other young petroleum engineers in the company at the time: three-and-a-half years of guided development in reservoir and production engineering. Early on, he gained experience making decisions. In drilling operations, for example, a young engineer sometimes has to decide on the spot when to run casing, as the well reaches an oil zone.

When he became assistant superintendent of Producing, in Safaniya in 1971, his responsibilities included the identification of talented and motivated Saudi workers who could benefit from additional training, including out-of-Kingdom programs. He felt that many of these workers had qualities of intelligence and leadership ability, and that all they lacked was education and training to qualify them for positions of higher responsibility. By 1974 he was named manager of Southern Area Producing, and helped implement several programs to train operators.

Great increases in hiring by the company in the late 1970s forced the Training Department to expand. To coordinate this expansion and plan for the facilities that would need to be constructed, the company's Management Committee formed the Saudi Arab Manpower Committee (SAMCOM) Al-Hokail headed a task force that helped chart the course for the company's rapid Saudization of its workforce.

Al-Hokail joined the ranks of the company's executive management when he was named vice president of Employee Relations and Training in 1981. He reached corporate management level in 1982, when he was named senior vice president of Industrial Relations. Al-Hokail was named to the board of directors of Saudi Aramco in 1989 and promoted to executive vice president of Industrial Relations and Affairs in 1992. He was named executive vice president of Manufacturing Operations in 1993.

any organizations, aside from those engaged in maintenance training, started their own training programs outside the Training Department. Between 1975 and 1983, the number of formal training programs conducted by line organizations increased from 60 to 230. In just two years, 1979 to 1981, the number of people on Business Line training staffs increased from 200 to more than 1,000. These included training supervisors, coordinators, trainers, instructors, analysts, curriculum specialists, technical advisors and support personnel. At least 30 different organizations were running multiple training programs covering 140 job titles altogether for more than 4,000 trainees daily. The various organizations had allocated facilities covering about 600,000 square feet to training. Some of the facilities were larger and better equipped than those belonging to the Training Department.

Several of these training programs had outstanding records in areas such as job-qualified completions, facilities, staff, curriculum and support. For instance, during 1981 there were 29 new plants or plant additions opened in the Southern Area, an average of one every 12½ weeks. Southern Area Oil Operations (SAOO) was able to completely staff each new facility with Saudis who had qualified as operators in SAOO training programs. Materials Storehouse Operations had established a training complex in Dhahran consisting of 36 portable buildings with classrooms for up to 300 trainees, and a 20,000-square-foot warehouse for simulated on-the-job training. Forty Saudis completed the two-year Materials Storehouse training program during 1981 and moved into entry-level jobs. By 1982, the Marine Department's Training Division had produced more than 600 graduates for engineering and seaman positions. Basic and advanced courses were being taught at the newly renovated 5,000-square-foot Marine Training Center on Ras Tanura's West Pier. Terminal Oil Operation's Training Division was on track to fill more than 80 percent of its technical positions with Saudi employees within the next four years.

One of the most enduring programs was Engineering and Operation Services' Saudi Technical Development Program. It was started in 1981 to train Saudis to work as surveyors, laboratory technicians, assistant engineers, technical clerks, draftsmen and inspectors. More than 600 technicians were trained to entry- or higher-level jobs before the program was turned over to Job Skills Training 13 years later.

The expansion of line-organization training activity changed the shape of the company's training system. Until then, most training had been consigned to the Training Department. Training had been a part-time activity, with departments scheduling employees for two hours of training a day, or four hours a day for a trainee on the "fast track." Line organization training, however, was usually a full-time, eight-hour-per-day affair, designed to bring the trainee to a much higher level of competence than had been possible under the part-time system. Following this example, full-time training became more common for the Training Department in job skills and academic programs.

Line organization-dominated training did not come without penalty. The orderliness of a single training organization system was lost. The patterns of responsibility for training became unclear. Among the unanswered questions were: Who would coordinate the multitude of training programs, establish procedures, direct planning and evaluation, set priorities, fix standards,

Trainee firemen practice hose-handling skills, 1986.

qualify instructors, measure the effectiveness of training programs, monitor the utilization of training facilities, and record the progress of trainees?

Corporate Training Task Force of 1981

n late 1981, at SAMCOM's request, the Management Committee created a five-member Corporate Training Task Force to examine such questions. Les E. Goss, vice president of Community Services in Dhahran, was named to head the task force. Before the task force finished its work, Goss moved to Ras Tanura as vice president of Northern Area Oil Operations. He was replaced as the task force leader by James M. Templer, former head of Aramco Overseas Company. Other members of the task force were Vern Hebert, an industrial relations specialist; Dan Walters, superintendent of the Training Department's Curriculum and Testing Unit; and John P. "Crif" Crawford, the planning analyst whose computer programs had been used to forecast manpower needs.

Between November 1981 and March 1982, task force members interviewed 100 company managers, supervisors and training coordinators in 60 different organizations. They gathered data on each department's training workload, manpower, facilities and inventory of current programs. Using Crawford's computer forecasting program, they were able to predict the rate of increase in the Saudi work force for each department during the 1982-87 period and beyond. Some big numbers were mentioned. For example, the task force report predicted "Aramco's Saudi training and development activities will grow to almost $1 billion in annual operating expenses in 1987 and employ more than 4,000 training personnel." The report forecast that the company would hire 38,000 more Saudi employees by 1987, nearly doubling the size of the Saudi work force, and that about 70 percent of these new Saudi employees would have fewer than 12 years of education. Almost all new-hires would need extensive training before taking entry-level jobs, the report said, resulting in an "unprecedented" emphasis on training in the future.

To provide a solid foundation for the expected surge in Saudi training, the task force recommended creation of a training organization with "a corporate focal point." "A major objective of the new organization," the task force said, "should be improved service to Business Lines."

"There are a number of concerns and misunderstandings prevalent that the new training organization must address," the report said. "Many of these will be resolved by proper communications, single-organization discipline, and appropriate attention to priorities."

The recommendations were accepted by management. On May 1, 1982, the six divisions of the Training Department were reorganized into three departments and combined with the Career Development Department to form Training and Career Development (T&CD). The new organization was given responsibility for all training company-wide. T&CD absorbed the various programs and many of the personnel who were responsible for the successful line-organization job skills training.

T&CD was composed of the Central Area Training Department, headed by Ali Dialdin; the Southern Area Training Department, under Malcolm J. Werner; the Northern Area Training Department, directed by Dan Walters; and the Career Development Department, with Zaki A. Ruhaimi as director.

Robert Luttrell

he man selected to head T&CD was Robert S. Luttrell, vice president of Mechanical Services, whose organization had coordinated development of maintenance training programs. Luttrell became vice president in charge of Training and Career Development effective October 15, 1982. Bill O'Grady, director of Training for 16 years, became head of the Training Study Task Force. When O'Grady took charge of Training in 1966, the company had only 1,500 trainees, and the director of Training had a budget of several hundred thousand dollars and a General Office staff of 23. The organization Luttrell took over had 13,000 trainees, 2,160 full-time training personnel and an annual operating budget in excess of $200 million.

Robert Luttrell, in dark shirt, with trainers (from left) Khalid Abubshait, Ajan Novotny, Thamer Al-Murshed, Ali Dialdin, Omar Abdi, Elias Matouk and Steve Tolle.

Whereas O'Grady was a teacher and administrator, Luttrell was an engineer with no formal training as an educator. Luttrell joined Aramco in 1956 immediately after graduating from Virginia Polytechnic Institute with a master's degree in chemical engineering. He was hired through Aramco's College Recruitment Program, a three-year management development program very similar to the PDP for Saudi professionals. Luttrell was one of the few expatriate executives who had actually been trained by Aramco. During 26 years with the company, he had held supervisory and management jobs in districts from Safaniya to Abqaiq and had built a reputation as a superb organizer.

As a vice president, Luttrell was at the highest executive level of any Training director since Roy Lebkicher, a vice president who had a brief tenure as director of Training in the 1950s. Luttrell's training experience was in organizing and coordinating line organization craft programs. He tended to delegate responsibility for the academic side of training to others, notably Ali Dialdin.

"One of the problems we were having in Training," Luttrell recalled, "was that it wasn't structured. I thought, 'Let's put together computerized systems that would identify everybody. Let's find out how many jobs we've got to fill, how many people we've got to train. Let's tag everybody, and let's track them.'"

Significant advances toward that goal were made by the end of 1982, Luttrell's first year as head of T&CD. Electronic Data Processing (EDP) applications had been expanded to include separate computer files for personnel, facilities, vehicles, equipment and manpower budgets. Progress had been made in developing a computerized enrollment system which would facilitate future ITC/ITS enrollments.

Crif Crawford remained on Luttrell's staff. He worked on developing computer models which departments could use to forecast their Saudization and hiring needs for as long as 20 years into the future. He created two basic models: the professional model and the craft model. "Some people used them primarily for forecasting Saudization," Crawford said, "and other people used them primarily to see how many people they needed to hire to reach some goal."

More Training Facilities

The company's extensive training facilities building program had not yet caught up with the need for training space. A delay in completion of the new Dhahran North ITC/ITS complex, plus a failure to prepare 50 portable classrooms in time for the September 1981 start of classes, resulted in the deferral of training for 4,000 Saudis. They had to wait to start their training until 1982, when new facilities came on line.

The Dhahran North ITC/ITS Training Complex opened in April 1982. Set on a hillside in a 66-acre tract, and built at a cost of about $30 million, the new training center consisted of six buildings with a capacity for about 1,500 trainees. The complex included an administrative center, a teacher workshop,

The newly opened Dhahran North Industrial Training Center in 1982.

a snack bar and a library plus two ITC and three ITS buildings. There were 60 ITC classrooms and 28 shops. Each shop occupied 1,500 square feet and was equipped for training in a specific skill such as welding, electronics, instrument repair, machinery maintenance and air conditioning and refrigeration. The Dhahran craft training program would follow the same curriculum as that to be used in the Area Maintenance Training Centers then under construction. The Dhahran North ITC/ITS was located north of Al-Munirah Camp just off the al-Khobar-Abqaiq Road. It was across the road from a large camp with housing for Aramco employees, including many trainers. With the opening of Dhahran North, the ITC which the company had operated in leased space at Dammam since 1972, was closed. The Dhahran North complex was in addition to, and not a replacement for, the older ITC/ITS training buildings near the Dhahran Hospital. With completion of Dhahran North, training had about 100 ITC classrooms and 40 training shops in the Dhahran area. The company already had plans in the blueprint stage for a large-scale expansion of the Dhahran North complex.

A courtyard at the Dhahran North Industrial Training Center.

A second new ITC/ITS facility opened in October 1982 at Mubarraz in the al-Hasa area near Hofuf. The $14 million center consisted of three separate buildings with a total area of 75,000 square feet, located on a 17-acre site along al-Muhassin Road. It was part of the Mubarraz training campus, which, in the future, was to be expanded to include an Area Maintenance Training Center facility and a separate Vestibule Training Center building. The new three-building ITC/ITS complex consisted of an ITC with 33 classrooms; an ITS building with 16 shops; and an administration building with offices, teacher work rooms, a library and a general assembly hall. The Mubarraz ITC/ITS replaced an older training complex operating out of leased facilities and company-owned portables.

A unique ITC/ITS facility had been established in the former Jones-Wallace-Zahid (JWZ) construction camp at Rahima, about two miles northwest of Ras Tanura. It was the largest such facility in the company at the time. Completely renovated buildings on the 21-acre site contained 56 ITC classrooms and 34 shops. Jones Camp, as it was commonly known, had an abundance of features not found in other training centers. It was the only training center with a 1,000-person dining hall, a 500-seat theater and assembly area, a cash office, an on-site clinic, and tennis, basketball and handball courts. These features were

King Fahd ibn 'Abd al-'Aziz inaugurates Aramco's Training Center in Ras Tanura.

carried over from the days when the camp was home to several hundred construction workers, and were retained for the convenience of trainees and staff because the center was located in an undeveloped area between Government Highway and Aramco's Ras Tanura airstrip. Jones Camp opened in stages, starting with six buildings and 600 trainees in May 1979, and continuing until the renovation was completed in September 1981, when 2,400 trainees were in attendance daily.

Dan Walters was superintendent of Northern Area Training during the conversion from a construction camp to a training center. "It was a chore," he said.

191

"There was nothing out there — no roads. Overnight we went into the community maintenance business because the old Jones Camp had its own water supply, its own sewer, its own cafeteria."

Jones Camp was considered a temporary facility after the Area Maintenance Training Center concept was adopted in 1980. The site chosen for the Ras Tanura AMTC was adjacent to Jones Camp. Plans called for construction of a new ITC/ITS building connected to the AMTC. In the meantime, Jones Camp served to relieve overcrowding at the old two-story Ras Tanura ITC built near the Radhwa Gate in 1957 and the neighboring one-story Ras Tanura ITS. The ITC/ITS activities were moved into the Ras Tanura Area Maintenance Training Center in 1987.

By July 1982, the company had just over one million square feet of training space, encompassing 700 classrooms and 240 shops. Another one million square feet of training area had been approved for construction. The projects approved but not yet built included the first permanent ITC/ITS at 'Udhailiyah and a training center for Tanajib. If all went as planned, in another five years the company would have 1,028 classrooms and 466 shops in 2.2 million square feet of training space. More than 90 percent of that space would be in permanent training facilities. By contrast, in 1982, about 50 percent of Training's floor space was in portable buildings, in leased facilities or in space borrowed from other organizations.

Record Training Enrollment

ramco now operated one of the largest corporate training programs in the world, if not the largest. In 1983 about 85 percent of all Saudi employees attended training classes. Enrollment in the Training Department's academic and job skills programs reached a record high 20,500. The company's 1983 training budget was $506 million — more than 10 times the amount spent in 1975 — and a 25 percent increase in just two years. The company sponsored 1,300 Saudis for university studies, 700 Saudi graduates were enrolled in the Professional Development Program, and more than 4,000 Saudis attended classes in the company's management and professional training facilities. The full-time training staff increased from 2,630 in 1981 to 3,260 employees in 1983. In mid-1983, all female ITC teachers were let go in favor of an all-male teaching staff.

1983 was the 50th anniversary of the signing of the Concession Agreement and the launching of the oil industry in Saudi Arabia. It was a year of celebration. King Fahd ibn 'Abd al-'Aziz, who had assumed the throne on the death of King Khalid the previous June, visited Dhahran on May 16, 1983, to inaugurate the flag-adorned Exploration and Petroleum Engineering Center.

H.M. King Fahd inaugurates Aramco's new EXPEC Building, 1983.

The company newspaper, the *Arabian Sun*, devoted most of its May 25, 1983, edition to comments and reflections on the anniversary by three oil industry leaders: Ahmed Zaki Yamani, then Minister of Petroleum and Mineral Resources; John J. Kelberer, chairman of the board and chief executive officer of Aramco; and Ali I. Al-Naimi, executive vice president for Operations. On November 8, 1983, the board of directors elected Al-Naimi as the first Saudi to become president of Aramco.

In his article for the *Arabian Sun*, Al-Naimi described Aramco as "an important center of technology transfer." He wrote, "A young Saudi who joins the company and learns Aramco's various systems and procedures is assimilating

192

know-how. Technology is being transferred. This human resource then becomes a national resource. In a sense, Aramco is a national training ground."

Yamani declared that the Saudi government "did not forget the great administrative and technical heritage created by Aramco, whose chief element was the creation of a Saudi staff who are pioneers in understanding the operation and mysteries of the oil industry. "

In October 1983, during ceremonies at King 'Abd al-'Aziz University in Jiddah, an Aramco geologist, 'Abd al-Aziz 'Abd Allah Al-La'boun received the first doctoral degree in geology ever awarded in Saudi Arabia.

Aramco at 50

As part of the 50th-anniversary celebration, a man who had seen it all, Floyd Ohliger, visited Aramco. He found the changes "overwhelming." Ohliger, the oil company's resident manager in the early years, had first entered the Kingdom in 1934, wading ashore from a launch at al-Khobar. Making what was to be his last visit to Saudi Arabia at age 81, Ohliger declared that the new developments "for the country as a whole, including Aramco, in the last 10 years have been greater than in all the preceding years."

Only a year earlier, Dammam Well No. 7, where Ohliger and his contemporaries exuberantly witnessed the first commercial oil strike in Saudi Arabia, was shut down. "Lucky No. 7" had averaged 1,600 barrels of oil per day and had produced a total of 32 million barrels since March 3, 1938, when it "came in big."

Floyd Ohliger, 1983.

Another old-timer, George Rentz, could hardly recognize Dhahran in late 1983, although he had visited the community only five years earlier. Rentz, an Arabic scholar, wrote chapters on the history, geography, religion and people of Saudi Arabia for the *Aramco Handbooks* published in the 1950s. What had changed so much in Dhahran since his last visit? "The size of the buildings, the complexity of the organization and the cosmopolitan nature of the population," he replied.

Trainees visit Aramco's 1938 discovery well, Dammam No. 7, in 1976.

The cosmopolitan work force Rentz referred to included people from 57 nations. They were very different from the expatriates of Ohliger's and Rentz' time, if one can judge by the type of entertainment they booked into Aramco. Instead of the big bands and classical pianists who performed at Aramco in the 1950s and '60s, employees of the 1970s and '80s booked country-and-western

performers like Barbara Mandrell, Kenny Rogers, Billy Dee and B.J. Thomas, and rock-and-roll entertainers such as Jerry Lee Lewis and Wayne Cochran.

On January 1, 1984, Ali Dialdin was appointed to the newly created position of general manager, Training Operations. He assumed responsibility for Central Area, Northern Area and Southern Area Training Departments as well as the Program Development and Evaluation Division. At the same time, Ali H. Twairqi was named director of Southern Area Training, replacing Malcolm Werner.

The impressive new training centers opened at Abqaiq, Mubarraz and Ras Tanura in late 1983 and early 1984. They were basically the same as originally designed in 1980-81, although some frills, such as decorative water fountains, were eliminated following a slump in oil prices. The centers provided a total of 710,000 square feet of training space, with 122 classrooms and 107 shops.

193

Some Aramco projects were so large that only on-the-job training could provide the necessary experience.

Saudi machinist practices a precision cut.

Marine workers on the deck of an oil tanker.

Automotive maintenance students electronically analyze a truck engine.

An Aramco student in the language lab.

Saudi management trainees discuss problem-solving techniques.

194

Between classes at the Dhahran North Industrial Training Center.

The Mubarraz and Ras Tanura facilities were nearly identical in size. Both measured slightly more than 295,000 square feet. The Abqaiq facility was the smallest at 117,000 square feet, with 20 classrooms and 20 shops. Abqaiq was the only one of the three without a VMTC. The Mubarraz facility had a combined total of 45 classrooms and 46 shops. The Ras Tanura center included 57 classrooms and 41 shops. The largest of the Ras Tanura shops, a metal shop, covered more than 21,000 square feet.

No official name was given to these sprawling training centers. The original titles, Area Maintenance Training Centers and Vestibule Maintenance Training Centers, seemed too awkward for everyday use. Instead, they came to be called Aramco Training Centers, meaning the entire facility with space for both academic and job-skills training. The job-skills portion was known as the Job Skills Training Center. The title of Industrial Training Shop (ITS) was eliminated. But the academic-training wing of the same facility was frequently referred to by the old name — Industrial Training Center, or ITC. At the same time, it became customary to call instructors of academic subjects "teachers" and job skills instructors "trainers."

More than $3.5 million worth of tools and equipment had been ordered for training in the shops at the three new facilities. Luttrell wanted the centers stocked with equipment from the field, as well as training simulators. "We tried to make it as real-life as possible," Luttrell said. He estimated that $1 million worth of used equipment was removed from company work sites and installed in shops at the Job Skills Training Centers. "There was nothing comparable to it in other oil companies. The closest you'd get is the military, where you train people on the work site."

Technological advances were making electronics training considerably easier than it had been only a few years before. "Everyone had been worried to death that we wouldn't have enough trained people to do the electronic-technician work," Luttrell said. "But technology had changed the way you did electronic repair. Instead of having a technician diagnose a problem, the computer told you what was wrong, and you just took the old part out and put a new one in. It was a lot less complicated than doing diagnostics."

Expansion Era Ends

The era of record oil prices and record expansion ended almost as suddenly as it had begun. Consumption of petroleum products among noncommunist nations dropped by 14 percent between 1979 and 1982, at the same time that Saudi Arabia and other OPEC nations were producing oil at all-time-high levels. The price of oil on the spot market reached a high of $42 a barrel in 1981 before a worldwide oil surplus developed and prices began to plummet. Between 1981 and 1982, the Kingdom's income from oil revenues fell by an estimated $38 billion. Saudi Arabia and other OPEC nations agreed to sharp cutbacks in production and production quotas in an effort to stabilize prices. They succeeded for a time. Saudi Arabian light crude sold for $34 for a barrel through January of 1983, but thereafter prices fell precipitously, bottoming out at less than $10 a barrel in 1986. From a record-high 9.63 million barrels a day in 1980, Aramco's production slid as low as two million barrels a day at times during 1984. For the year 1985, the company averaged three million barrels a day, the lowest yearly average since 1969.

Many Saudi employees live in Doha, near Dhahran, in homes financed through Aramco's Home Ownership Program.

The fall in oil prices meant sharp reductions in the Aramco work force, particularly among expatriates. By 1985 the company payroll was about 15 percent below its 1982 level and still falling. According to one count, 19 of the company's top 80 expatriate managers, most of them Americans with long careers at Aramco, had retired or been transferred out of their jobs by the end of 1985. The Saudi component of the work force actually increased slightly, reaching a high of nearly 35,000 in 1984, before it, too, began to fall. By 1985 the company had closed all but three of the 20 recruiting offices it had run in the boom days, and veteran Saudi employees were being encouraged to take early retirement. The company began "a phased withdrawal" of families from 'Udhailiyah during 1985 in preparation for a complete shutdown of the community. Between 1982 and 1985, the number of employees at locations such as Tanajib, Khurais, Abu Ali, Shedgum and Berri was reduced by 60 percent. Only 1,600 contractor employees had company-assigned living quarters in 1985, compared to 6,400 at the end of 1984.

The 1980 forecasters had missed the mark. They had based their projections on the only data available — the momentum built up by the great expansion of the previous decade. The collapse in oil prices turned things upside down. Instead of 75,000 employees at Aramco in 1985 as had been predicted five years earlier, the total was a little more than 50,000. Instead of an enrollment of about 20,000 in the ITCs, it was 12,500. Instead of 8,000 enrolled in the ITSs, it was 3,400. Training facilities had been greatly overbuilt. T&CD took steps to mothball or reassign excess space to other departments.

But it had been a great run. Between 1970 and 1980, the company added 27 new oil fields to the 31 fields discovered up to 1970. Year-to-year increases in oil production of more than 25 percent were common during the decade. Production increased from 3.5 million barrels daily to nearly 10 million barrels a day. NGL production rose from 52,118 barrels a day in 1970 to 369,232 in 1980. In their book, *The Oil Price Revolution*, Steven A. Schneider and John Hopkins declared the Kingdom's earnings from oil sales during this period "constitute the largest nonviolent transfer of wealth in human history."

When the expansion began in 1971, the total Aramco work force was 10,707. It peaked at a record-high 61,227 in 1982. The number of Saudis on the payroll went from 8,709 in 1971 to 33,067 in 1982. Between 1972 and 1982, the company hired a total of about 38,000 Saudis to achieve a net gain in the Saudi work force of nearly 24,000.

SAMCOM's recruiting and training work and the company's aggressive Saudization policy had paid off. The percentage of Saudis in the work force went up from a low of 52 percent in 1979 when SAMCOM was formed, to 54 percent in 1982, the final year of the expansion period. This was just the opposite of the trend in the past; the Saudi component of the work force had fallen during previous expansion periods. The level of education among Saudi new-hires also improved. Starting in 1983, more than half of all new Saudis hired were high school or college graduates.

Ali Al-Naimi, looking back on the period, described SAMCOM as part of a maturing process. "From 1979 onward there was definitely a tremendous focus on training," he said. "But that doesn't mean that there was no training success before. To be able to get where we were in '79, you have to look back

196

in history and realize that people in both parts — the government part and the early company pioneers — all of them were visionaries. They believed that if you were going to have a successful enterprise where you have different nationalities, you needed to develop the human resource. SAMCOM was really nothing more than a maturing process that brought together these things. It says, 'OK, now we have the ingredients for success; let's bring them together and focus on what we want.'"

As the size of the work force continued to decline, the percentage of Saudis at all levels increased. By 1985 the ratio of Saudis in the total work force had climbed to about 65 percent. About 22,000 of the 33,000 Saudi employees were considered job-qualified at grade code eight or above. More than 9,000 other Saudis were in entry-level jobs below grade code eight.

The Aramco Board of Directors elected Ali I. Al-Naimi, seated right, as president, in 1983.

Saudis had made considerable headway into the upper ranks of the company. By 1985 there were 1,700 Saudis classified as professionals in Aramco, compared to only 197 Saudi professionals 10 years earlier. However, as impressive as the 1985 figure might sound, it amounted to Saudis occupying only 21 percent of the 8,000 professional positions in the company. Eighteen of 31 executive management positions were now held by Saudis. Ali I. Al-Naimi was president, and two of the six senior vice presidents, Abdelaziz M. Al-Hokail and Nassir M. Al-Ajmi, were Saudis. Within the next few years, Saudis would succeed to nearly all the top executive-management jobs in the company.

From Aramco to Saudi Aramco 1985-1990

"If I were pressed to state the single most important factor in the mutual respect that characterized the Saudi Aramco relationship, it would be the company's training programs."
– Robert L. Norberg

On Tuesday, November 8, 1988, the Arabian American Oil Company, better known as Aramco, ceased to be the producing agent for Saudi oil. In a meeting in Riyadh that evening the Saudi government Council of Ministers approved a charter for a new national oil company — the Saudi Arabian Oil Company.

The new charter seemed little more than a technicality since, as a practical matter, the Saudi government gained control of Aramco in 1980 when it purchased the assets of the company from American shareholders. For most people it merely meant that the company had a new name; in common parlance it was Saudi Aramco instead of Aramco. Yet, it was a notable change.

Saudi Aramco was now Saudi by law and by name as well as in practice. The company's top executives were Saudis. Seven months earlier at a board meeting in Houston, the Saudi government's Minister of Petroleum and Mineral Resources, Hisham M. Nazer, was named chairman of the Aramco Board of Directors. At the same time, Ali I. Al-Naimi, the company president, assumed the additional title of chief executive officer. Executive Vice President Nassir M. Al-Ajmi was next in the company's chain of command, followed by five senior vice presidents, all Saudis. The senior vice presidents were Nabil I. Al-Bassam, Abdallah G. Al-Ghanim, Abdel-aziz M. Al-Hokail, Sadad I. Husseini and Abdallah S. Jum'ah.

Opposite: An instructor and Saudi geologists in the field near Dhahran. Below: The new Saudi Aramco Board of Directors at their first meeting in Dhahran, March 1989.

John J. Kelberer, the company's last American board chairman and chief executive officer, stepped down in favor of the new Saudi leadership in April 1988. He remained on the board of directors and served as its vice chairman until his retirement a year later. "One of my most satisfying accomplishments as chairman," Kelberer said in a retirement statement, "was the development of Saudi nationals." During his 10 years as head of the company, the ratio of Saudis in the work force climbed from 52 percent to 73 percent, and the ratio of Saudis in supervisory positions increased from 41 percent to 77 percent.

The passage from Aramco to Saudi Aramco could hardly have been smoother or more congenial. The company newspaper, the *Arabian Sun*, announced the changeover in its November 16, 1988, edition, and printed an open letter from Ali Al-Naimi to all employees. It read, in part: "The most important asset that Saudi Aramco will have upon assuming its responsibilities will be the human resources of the current work force, its depth of experience, its technical competence and its loyalty.

"Working together, we can preserve the many Aramco accomplishments of the past, ensure the success of Saudi Aramco and continue to contribute to the future prosperity of the Kingdom. ...

"All employees may rest assured that all their current rights and benefits as Aramco employees will continue to be preserved and protected," Al-Naimi said.

For the average employee the most noticeable change, aside from the company name, was the modified company logo, an "S" inside the old Aramco "A" interwoven with the stylized letters "C" and "O." The logo was the idea of graphic designer 'Abd al-'Aziz Al-Rudwan and was approved by management in June 1989.

The Saudi Aramco Board of Directors held its inaugural meeting on March 14, 1989. For the first time since the oil company was founded more than 50 years earlier, the board of directors met without the presence of the American companies, the former partners in the Saudi concession. A long-term expatriate employee, Robert L. Norberg, assistant to the vice president of the company's Washington, D.C., office, noted

New Saudi Aramco logo (top) and the old Aramco logo.

the passing of this era in company history. In a speech at Duke University in Durham, North Carolina, Norberg said: "This month, with the first board meeting of Saudi Aramco, we witnessed the final chapter of the special relationship as we've known it for more than half a century.

"If I were pressed to state the single most important factor in the mutual respect that characterized the Saudi Aramco relationship, it would be the company's training programs. Aramco worked hard toward its own redundancy. Today, the world's largest oil company is being run by Saudis, from the board room to the drilling rig floor. Fifty-six years ago, when the hiring and training of Saudis began, the average recruit was unschooled and illiterate."

Growth of Education in the Kingdom

n the intervening 56 years a Kingdom-wide network of schools had been created and illiteracy rates greatly reduced among both men and women. For the first time, elementary schools, intermediate schools and high schools were within easy reach of nearly every home in Saudi Arabia. These schools, constructed and operated almost entirely with revenue from the sale of Aramco petroleum products, provided tangible evidence of the impact of the oil industry.

In 1933, when the concession agreement was signed, only a handful of schools existed in the Kingdom, none of them in the Eastern Province where the oil company was headquartered. King 'Abd al-'Aziz instituted an extensive school construction program in 1945 which by 1951 had raised the number of schools in the Kingdom to 226, with 29,887 students. By 1970 there were 3,107 schools of all types, with a total enrollment of 547,000 students. Then came a period of exceptional growth. On average, two new schools opened somewhere in the Kingdom each weekday for the next 20 years. By 1989 there were more than 16,000 schools in Saudi Arabia, with an enrollment of 2.8 million students. These included 86 schools, with a total enrollment of 38,600 students, constructed by the company through the Aramco-Built Government Schools

Program. More than 60 percent of Saudi students were still in elementary school. But, since 1970, the number of intermediate and high school students had risen to 737,000, a tenfold increase, and enrollment in the Kingdom's colleges and universities was 117,000, about 16 times higher than it had been in 1970. In the expansion year of 1982, the company hired more than 3,000 Saudi employees, about half of them college or high-school graduates. By 1988 the school system had evolved to the point where the company was able to restrict its hiring almost entirely to Saudis with high-school or college diplomas.

Saudi Aramco Responds to 1980 Oil Price Changes

In the mid-1980s oil prices plunged as much as 70 percent below record highs reached earlier in the decade. Saudi Aramco responded by downsizing. Total employment fell from more than 60,000 in 1982 to 43,500 in 1987. More than 14,000 of the 17,000 positions eliminated had been held by expatriates.

Training & Career Development (T&CD) experienced sharp cutbacks in staff, spending and facilities during this time. Between 1983 and 1988, ITC/ITS enrollment declined by more than 50 percent, expenditures for training dropped by one-third, the number of persons employed full time in Training fell by 45 percent, and the space devoted to training activities was reduced by half. On May 1, 1986, T&CD and Employee Relations were consolidated into one administrative area called Employee Relations and Training. Seven months later, on January 1, 1987, training operations were reduced from three to two regional departments — the Southern Area Training Department under Zaki Ruhaimi in Abqaiq, and a newly combined Northern/Central Area Training Department headquartered in Dhahran under Thamer R. Al-Murshed. The Career Development Department was kept intact.

T&CD operations were concentrated in four facilities — the Dhahran North Training Complex opened in 1982 and the huge Aramco training centers opened a year or so later at Abqaiq, at Ras Tanura and in al-Hasa near Mubarraz.

The opening of four training centers, coupled with the decline in recruitment, resulted in a surplus of training space. T&CD mothballed some excess training facilities, turned others over to different departments, and demolished still others. The Ras Tanura ITC/ITS complex built in the 1950s was torn down. The nearby Rahimah ITC, also known as the Jones Camp, was mothballed. In 1985 an Abqaiq landmark, the "Green Building," a training site for industrial workers since 1979, was turned over to Southern Area Manufacturing. The Abqaiq ITC was closed and the building utilized for storage by the Medical Department. The portable buildings that had comprised Abqaiq's Mansur training complex since 1981 were either sold at auction, mothballed or demolished. Two of three buildings in the former Dhahran ITC/ITS center were turned over to the Medical Department in 1987, and the third building (No. 552) was converted to Training offices.

The 'Udhailiyah ITC, which opened in 1984, was mothballed in 1987. Temporary buildings housing the old 'Udhailiyah ITC were sold at auction. A job skills training center in 'Udhailiyah's light industrial park was turned over for gas plant operator training. Instead of having nearly 1,500 classrooms and shops, as was called for in the business plan adopted five years earlier, Training operated only 825 classrooms and shops in 1987, of which nearly 30 percent were deemed to be excess space.

The central mosque in Dhahran, 1987.

raining's leadership changed hands in 1986. Robert Luttrell left his post as vice president of Training and Career Development (T&CD) to become a technical and financial consultant in the United States. Luttrell was succeeded by Ali M. Dialdin, who had been general manager of Training Operations. Dialdin, who retained the title general manager, became the first Saudi to head the company's training organization, still the largest industrial training organization in the world, despite recent cutbacks. At the time, T&CD employed 1,700 teachers, trainers and support personnel and ran training programs for more than 11,000 Saudi employees a year.

Ali Dialdin, fourth from right, at a day-long meeting of Aramco trainers in Dhahran.

Although his appointment as general manager was a significant moment in training history, it was not what Dialdin personally considered the highlight of his career. "I would go back to January 1, 1984, when I was named general manager of Training Operations. That was my personal milestone," he said. "That was the time major training issues first came under my control. I did not have overall training responsibility, but I did have the key area reporting to me — the Program Development and Evaluation Division. That was the policymaker. Having it report to me meant I was in charge of all proposals, new training policies, programs and testing. Guiding the future, being instrumental in changing policies and procedures, was to me very important."

When Dialdin became general manager of T&CD, the company's business plan called for a 20 percent reduction in training activity over the next five years. "We were surplusing teachers and trainers in considerable numbers," Dialdin recalled. "Every year we were down by 100 to 175 trainers and staff." Among those who took early retirement were four bilingual trainers who had helped the company weather a critical period in its history. They were Khalil Nazzal and Ghazi Nassar, both originally from Palestine; Jamil Milhem, from Lebanon; and Wadie R. Abdelmalek, from Egypt. These four were among scores of Middle Easterners hired by Aramco in the '50s and early '60s, at a time when the largely non-Arabic-speaking training staff was close to being overwhelmed by the influx of young Saudi recruits. Frank Jarvis, one of the American trainers on staff at the time, believes the arrival of bilingual instructors averted a labor crisis.

"Back then they saved Aramco's hide," Jarvis said. "I don't think many people realize how the Middle Easterners kept the company going and how they kept the company from having severe labor problems. We had no books, we had nothing to teach these young kids who kept coming to us like a flood. The Middle Easterners could teach very well the lower levels of English, arithmetic and general science," he said. "They could teach in English. More importantly, they could teach bilingually."

Downturn Reversed

he petroleum market began showing signs of renewed vigor during the late 1980s. Oil prices climbed by more than $4 a barrel during 1987, ending a five-year-long downward spiral. The company increased production in 1988 to an average of 4.92 million barrels of crude oil a day, the

highest amount in six years. At the same
time, Saudi Aramco took on some heavy
new responsibilities. At the direction of
the government, the company's exploration activities, previously limited to a
concession area in the eastern part of
the country, were expanded to cover
about two-thirds of the entire Kingdom,
an area almost as large as the combined
size of Germany, France and Spain. In
November 1988, Saudi Aramco took its
first step into the downstream oil business overseas by having its subsidiary,
Saudi Refining Inc., enter into a joint

*Signing the
Star Enterprise
agreement in
London, 1988.*

venture with Texaco to form Star Enterprise. The joint-venture partners were to
refine, distribute and market petroleum products under the Texaco trademark
in the eastern and southeastern part of the United States. That same month,
Saudi Aramco took over direct marketing of gas and crude oil in the Kingdom.
In addition, the company made plans to increase its sustained production capacity to 10 million barrels of oil per day. It would be a long-range, multi-billion-dollar project calling for reactivation of facilities mothballed earlier in the decade,
construction of new facilities, costly upgrades of existing facilities, and the addition of hundreds of new employees.

Despite these new responsibilities and expectations for future growth, total
employment remained static at about 43,000. Not surprisingly, there were questions raised about the availability of trained Saudi personnel to handle the
increased workload. Thousands of expatriate craftsmen had been let go during
the mid-1980s downsizing. In 1987 the ratio of expatriate employees in the
work force, 27 percent, was the lowest since 1975. Saudis held 130 out of the
193 department head and above positions, and three-fourths of all supervisory
jobs. But the process of Saudization in the industrial and clerical work forces had
been uneven. For example, nearly all oil and gas plant operators were Saudis.
An all-Saudi crew started up a new oil plant for the first time at Safaniya in 1983,
yet Saudis occupied as few as one-third of the skill positions in some other job
categories. Only one-fourth of Saudis in the industrial work force had reached
the journeyman level. An analysis made during 1987 of nationalities in various
job skills categories found 9,000 positions for which Saudis needed to be trained
to replace expatriates.

The number of skilled and semiskilled Saudi workers in critical crafts such
as instrument technician, electrician and machinist actually declined by three
percent in 1986, mainly due to the company's early-retirement plan. Furthermore, line organizations were unhappy with the overall quality of new employees assigned to entry-level jobs in these fields. Training, which had just assumed
responsibility for job skills training programs started by the line organizations,
studied this issue and concluded that some of those employees had neither an
aptitude for, nor an interest in, the work they were being asked to perform.

In the opinion of many veteran employees, much of the difficulty in the
Saudization of craft fields could be traced to large-scale, direct hiring during the
start-up of the maintenance training program in the early 1980s. Jim Nolan, former manager of the Maintenance Resource and Planning Department, was
involved with the maintenance training program almost from the start. He said,
"We hired many people directly onto the Aramco payroll that were not able to
make it up to the standards we wanted. We should not have put them on the
payroll. We should have put them in as apprentices and let them get through the
basics, to determine whether they should become regular employees or not."

203

is point was not lost. By the mid-1980s the idea of starting a new apprenticeship program was gaining strength. The main advantage cited by advocates of an apprenticeship program was that it would allow the company to screen out unmotivated low achievers before they became regular Aramco employees — that is, before they qualified for the same pay and benefits afforded full-fledged Aramco employees. The apprentices would receive a monthly stipend until they completed their apprenticeship or until they were dropped from the program. The company would be under no obligation to hire even those who successfully completed the program. Hiring would be strictly on the basis of the company's needs at the particular moment.

Trainee works in Petroleum Engineering's Interactive Well Selector Program, 1989.

The push for apprentices came, not from Training, but from the line organizations. "Training had been surplusing people, and the projection was for a continued downward trend," Ali Dialdin recalled. "The Operations people started raising concerns in terms of attrition, replacement of expatriates and the long lead time it took for individuals to get prepared. That's when they started really talking about going back to an apprentice program. When news of this broke out, I slowed down plans to let some people go, and held on to them."

The company's only previous experience with an apprenticeship program lasted just four years, from 1970 to 1974. That program fell victim to a high rate of attrition brought on by a sudden, rapid expansion of the Saudi infrastructure and competition for employees between Aramco and other private industries. In the end, Aramco line organizations desperate for help raided the company's own apprenticeship program, hiring Saudi apprentices as regular employees before they completed the apprenticeship program. However, the job market in 1987 was entirely different from that in the early 1970s. Record high numbers of young men were graduating from the Kingdom's high schools, but they were having trouble finding jobs. Aramco anticipated no problems in attracting a sufficient number of recruits.

An apprenticeship program was formally suggested to the Employee Relations and Training organization in early 1987. Employee Relations approval was a necessary first step to instituting such a program. 'Abd Allah Ali Al-Zayer, then in the Employee Relations Policy & Planning Department, remembered a meeting with line organization representatives: "They came to us in Employee Relations and said, 'We want qualified employees.' They said they were not getting any benefit out of employees when they had to train them. They were just wasting time with these people because they had no proper training. Teach them to speak English and how to do an entry-level job. That was it. That was what they wanted."

Al-Zayer was asked to look into the pros and cons of an apprenticeship program and submit his recommendations to the Saudi Manpower Committee (SAMCOM). He was uniquely qualified for the assignment. Al-Zayer had just moved into Employee Relations after some 36 years in Training. He began his Aramco career in 1950 as a 12-year-old teacher/trainee at the Abqaiq ITC and, through company scholarships, obtained a high school certificate and a bachelor's degree in education and English from Lyndon State College in Lyndonville, Vermont. In 1970 he became the first Saudi assistant principal of an ITC, and in 1972 he became the first Saudi principal of an ITC. Al-Zayer helped set up the

Saudi students review the use of integrated circuit boards, 1988.

original Apprenticeship Program in the early 1970s. He later served as assistant superintendent of the Industrial Training Centers and an administrator in the Career Development Department.

His inquiry into the feasibility of a new apprenticeship program took about six months. Al-Zayer worked with Zaki Ruhaimi in Career Development and "a lot of people from line organizations and Training." The first goal was to determine the needs of the line organizations. "We wanted to base our program on their needs. The goal was to meet their requirements, satisfy their needs," he said.

"A lot of research had to be done in order to see what was needed to have a really successful program. The company was going to invest a lot of money in it, so it was worth spending time on. We had to be sure of it. At least I had to be sure of it before I went to SAMCOM and said 'this is what I think about an apprenticeship program.'"

In the course of this research Al-Zayer and his coworkers found large imbalances between the number of Saudis in various Job Skills Training programs and the number of employees needed to completely fill those positions with Saudis. For instance, there were fewer Saudis being trained as advanced technicians than were needed for the Saudization of that field. The advanced-technician category included such diverse job titles as materials supply controlman, assistant engineer and petroleum technician. The researchers found fewer trainees than needed for advanced craft positions, including such positions as machinist technician and senior electrical technician. They also found a shortage in the number of Saudis being trained for clerical work.

Conversely, a few job skills programs had substantially more Saudis in training than the number of men needed to fill available positions. The largest such discrepancy was in the skilled-craft category, which included Saudis in training for jobs such as welders or instrument repairmen. The researchers also found more trainees than needed for Saudization of the skilled-operator and semi-skilled operator categories.

On July 7, 1987, Al-Zayer presented his findings to SAMCOM. He proposed that the company launch an apprenticeship program "in order to increase Saudization of jobs in grade codes three to 10, particularly those of a higher technical level in crafts, operators and clerical." He recommended a two-year program limited to high school graduates who had maintained a grade average of 70 or better and whose scores on the company's mental abilities and aptitude tests "indicate a high potential of success in the apprentice's assigned career track."

Hamad Al-Juraifani, left, and Fahd Garawi, right, congratulate Muhammad Abu Sharifah, the first Saudi refinery supervisor, who retired in 1986.

"The proposed program," Al-Zayer cautioned, "will not produce fully job-qualified employees, but would provide an opportunity to screen out apprentices for program failure, lack of interest or any other justifiable causes during the period of higher than normal attrition.

"In addition to the real needs of the company," he added, "there is an increased social obligation for the company to offer training and/or employment to Saudi youth in an economy that has a restricted number of job openings elsewhere."

As to an imbalance between the number of jobs and the number of Saudi trainees on certain training tracks, he said: "It can be concluded that many current trainees have been misplaced in their career path relative to Aramco's needs." Some trainees might transfer from oversubscribed training tracks to those tracks with too few trainees relative to available jobs, but, even if they do, Al-Zayer wrote, "it is still clear additional trainees are needed to accomplish the stated goal of Saudization in the technician, advanced craft and clerical job skills category."

The entire SAMCOM presentation, Al-Zayer remembered, took only about half an hour, and SAMCOM "accepted it right away." The proposal went next to the Management Committee, which approved it with equal speed, and no questions asked, he said.

It seemed an ideal time to begin the new Apprenticeship Program. Growth was at a manageable pace. Aramco recruited only 373 Saudis in 1987, compared to more than 3,000 Saudis hired during some years in the 1970s. There was a surplus of Saudi high school graduates in the job market, enabling the company to limit recruiting for the Apprenticeship Program to Saudis with a high school diploma, an unprecedented high academic standard for the industrial work force. In the old 1970s program the company had to accept apprentices with as little as a sixth-grade education. Because the new generation of apprentices would start from a higher academic level, they could complete the program in less time. Therefore, the new program needed to be only two years long, compared to the 1970s version, which began as a five-year program and was later reduced to three years.

New Job Descriptions

 ome important innovations from the Training Department, although not directly related to the Apprenticeship Program, provided support for it. One of these was the introduction of new job descriptions for industrial and clerical workers in grade codes three to 11, the very type of jobs for which apprentices would be trained. The job descriptions listed minimum academic requirements and years of experience required at each grade code level for 1,100 job titles covering about 27,000 Saudi workers. It was the first update of job descriptions in 20 years, and it was completed just weeks before management approved the new program.

The new descriptions "certainly helped" in formulating the Apprenticeship Program, Al-Zayer said, "because it gave us information on developing manpower. We were not talking loose, or talking about something we didn't know about. We were talking about job requirements that were in front of us."

Ahmad Al-Dossary prepares a statistical analysis, 1990.

The job descriptions were the outgrowth of painstaking work by Virginia Charlton and Issam Abu-Zaid in the Training Department, and by Eugene W. Bain, then corporate job advisor for the Organization & Industrial Engineering Department (O&IE). About 2½ years earlier, Charlton and Abu-Zaid began a needs analysis survey of some line organization jobs in order to determine the amount of English, math and science required for specific positions. Their work led to some significant curriculum changes, including the creation of upper-level English and math classes that were more closely work-related.

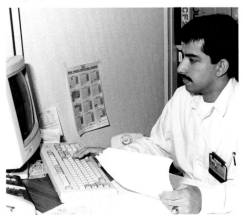

The original Charlton/Abu-Zaid needs analysis covered about 110 nonsupervisory jobs and four supervisory positions. When Bain saw the results of their work, he started thinking about a much larger survey.

**'Abd Allah Ali
Al-Zayer**

*"I was ambitious
and wanted to
advance as much
as I could. A lot
of my friends
would have a
nice social life
in the evenings,
but I would study
until 11 p.m."*

hen 'Abd Allah Ali Al-Zayer joined Aramco in 1950, he was just 12 years old. The youngster from a village in the Qatif Oasis on the Gulf Coast went to work as an office boy in Abqaiq in the building that housed the Industrial Training Center. There he began studying English. "It was very strange, like a different world," he said.

Al-Zayer was the son of a farmer. His previous education had been in the mosque school in his home village of Um al-Hamam. "We had religious men at the mosques, and they taught the Quran, some arithmetic, reading and writing," he said. Poor transportation at that time ruled out getting a more formal education in nearby towns. "It was difficult for somebody living in one of the Qatif villages to go to any school in Dammam or al-Khobar or even Qatif, which was about six miles from my village," he said.

Although the minimum age to join the company was 18, "We didn't have any birth certificates, so if you were eager to start work you would just say you were 18," Al-Zayer said. "My parents left the decision about joining Aramco up to me."

Al-Zayer transferred to Community Services in Dhahran in 1951 to be closer to his family. He studied for six hours after work each evening at the Aramco school in nearby al-Munirah and then joined Training to become a student and a trainee/teacher at the ITC. He earned a high school certificate in Beirut in 1959; a bachelor of arts in English and education at Lyndon State College in Lyndonville, Vermont, in 1964; and a fifth-year diploma in administration from Boston and Vermont universities — all under company sponsorship. "I was ambitious and wanted to advance as much as I could," he said.

Al-Zayer returned to Dhahran as an advanced ITC teacher. He was the first Saudi principal of an ITC before progressing to assistant superintendent of the ITC and Industrial Training Shops. He transferred in 1978 to the Career Development Department, first as superintendent, then as administrator.

Al-Zayer also helped pioneer the initial Apprenticeship Program in 1970. That program stopped after four years. A new Apprenticeship Program began in 1989 and served as the basic platform for bringing young Saudis into the work force. "One thing I'm really proud of was getting this program started up again," he said.

In Career Development, Al-Zayer helped oversee a major expansion in the number of employees studying abroad to some 500 per year in the early 1980s, compared to only about 50 per year before 1970. "I'm proud to have been involved in the expansion years with the training and advancement of Saudis," he said.

Al-Zayer moved to Employee Relations in 1987 as administrator of the College Relations and Professional Recruitment Division. He took early retirement in 1989 and offered this advice to young Saudis in Saudi Aramco: "Set goals and work hard to achieve those goals."

An apprentice welder and his instructor.

"I looked at the information that the training organization had gathered, and I looked at our jobs and was able to see there were certain jobs clustered around the same academic requirements. I was able to use that information, which was similar to our existing minimum job requirements, and say 'Let's try to standardize this across the board.'"

By "across the board" Bain meant almost all positions occupied by some 27,000 persons in the industrial work force. He took it as a personal, part-time project. Bain had no computer, so he made lists of jobs on large sheets of paper. Eventually, he was able to consolidate most of the 1,100 industrial positions into 18 job families. Employees in the same job families performed similar tasks using similar skills. Transferred into a computer program, the job descriptions became known as the *Full Job Academic and Experience Matrix*, better known as simply "the matrix." It listed the minimum academic requirements and years of experience at each grade-code level for each job family. In Bain's words, "It said if you want to be at this level, you're going to have to have this much English, this much math and this many years with the company.

"Inconsistencies had worked their way into the job descriptions over the years," Bain said. "For instance, it was not uncommon to find that a salary code 10 job required fewer years of experience and less education than a salary code nine job, and that didn't make sense."

The new job descriptions raised the question of what to do with Saudi employees who did not meet one or more of the matrix's requirements for the jobs they currently held. For example, what about the Saudi foreman whose English is not at the level required in the job description? The foreman might be doing an excellent job otherwise, but had to rely on a clerk to write letters for him in English. Such issues were resolved by "grandfathering" Saudi employees into the jobs that they held in July 1987, when the matrix was adopted. However, before advancing to the next highest job level, employees had to make up any of the matrix requirements they lacked.

"There was a lot of resistance to the matrix because it meant some adjustment in salary structure," Dialdin said. "The majority of managers wanted jobs with high qualifications because higher qualifications meant higher salaries, so, in many cases, we were overteaching."

New Tests Introduced

The introduction of a new series of aptitude and mental ability tests, known as the General Aptitude Test Battery (GATB), gave the apprenticeship program a further advantage. In 1983 GATB replaced the old Job Aptitude Test Battery (JAT), which had been the standard aptitude and ability testing battery for recruits since the 1960s. By 1987, upgrades and refinements of the Arabic-language version of GATB greatly reduced instances of new employees being assigned to jobs for which they had little or no aptitude.

Dialdin estimated GATB sliced a full year off the time required to complete the Apprenticeship Program. It did so by eliminating the need to expose an apprentice to a variety of introductory job skills courses before deciding which type of job the apprentice was best suited for. Matching aptitudes and jobs was not enough for Dialdin. He also wanted to know what kind of work the trainee would find interesting. Consequently, in 1990, Training began administering the Vocational Interest Test (VIT) to apprentices. VIT was an interest inventory that compared the test-taker's likes and dislikes with those of successful people in more than 1,000 different types of jobs.

Trainers were supposed to go strictly by GATB and VIT test results in targeting apprentices for a job, precisely to avoid disheartening results such as those experienced on one occasion by Abqaiq Training. Dale Saner recalled: "We had a class of five instrument technicians. Four out of the five failed the Phase Two training program. We had a heck of a time trying to find out what the problem was. We finally went back and looked at the GATB, and of the four who failed, three should never have been instrument technicians, according to the GATB. They weren't even close. The other one was marginal, like the lowest-ranked thing he was qualified for. We practically wiped out a whole group of trainees because they somehow got into the instrument training program even though the GATB said they should not be there."

Quality Assurance for Training

I n 1983 Training created a Quality Assurance Unit. Its mission was to determine how effective Training programs were in preparing Saudis for their targeted jobs. The unit evaluated ongoing programs and conducted follow-up reviews with graduates and their supervisors. Quality Assurance was the only Training unit reporting directly to the general manager of Training.

Bill Valbracht

The original two-man unit was composed of Bill Valbracht, an engineering psychologist with experience in developing military training programs, and Charles Clock, a former school psychologist from Hartford, Connecticut. Valbracht and Clock spent most of 1983 and part of 1984 setting up the new GATB program. Next they looked into Job Skills Training, a program recently transferred to Training from the line organizations. They discovered Job Skills trainees faced a jumble of tests, usually written by the men who trained them. The trainers may have been good technically, but they could be very poor test writers, often writing English as a second language. "Sometimes you couldn't even fathom what the question was, much less try to answer it," Valbracht said. The first and, in Valbracht's opinion, the most important thing Quality Assurance did for Job Skills Training was to put a credible testing system in place, helping to standardize Job Skills Training throughout the company. "When people graduated, we knew they had the knowledge and skills that we put them into training for," Valbracht said. "Those two, the GATB and Job Skills testing, are probably my finest memories of Training," he said. "They had a lasting impact."

Richard Arons

In July 1988, a new one-man Quality Assurance Unit, Corporate Quality Assurance (CQA), was created. On request of a line organization, it examined and rated the organization's own training efforts. This job was filled first by Bill Larson, followed by Frank Christopher and Guil Mullen.

Innovations like new job descriptions, improved aptitude and interest testing, and standardized Job Skills Training, directly or indirectly strengthened the Apprenticeship Program as well as all other training efforts.

Setting up the Apprenticeship Program

B efore the Apprenticeship Program could begin, ground rules and procedures needed to be established and a curriculum approved. The job was given to a specially created Training Department task force, the Curriculum Committee for the Apprenticeship Program, headed by H. Richard Arons, superintendent of Job Skills Training in Ras Tanura. Ahmad Ajarimah, supervisor of the Academic Curriculum Unit, was a full-time member of the committee. Steve Tolle from the Ras Tanura Job Skills Curriculum Unit was an active member. Dale Saner of Job Skills Training in Dhahran also participated.

The task force was formed in the autumn of 1987. In January 1988, SAM-COM received and approved a list of task force recommendations. Some of the recommendations generated controversy within the training community, particularly the proposal for "intensive," seven-hour-long English-language classes during the first half of the first academic year. Intensive, day-long training in a single academic subject was something entirely new to Aramco training. Trainees on the regular track were normally scheduled to attend ITC classes for only two hours a day; those on an accelerated track spent a total of four hours in various classes. At the two-hour-per-day rate, it took as long as four years to cover the English material needed to qualify for an entry-level job. On the intensive training schedule the same material would be covered in just one academic year.

On Training's recommendation, applicants for the Apprenticeship Program had to be high school graduates with at least a 70 grade point average. In addition, to qualify for the program, candidates had to test at the equivalent of ITC English level three or higher and score at least 55 on the GATB screening test. Candidates for the crafts, operators and technician tracks needed to place at the equivalent of level four on the company's ITC math test. A level-three math minimum was established for clerical apprentices.

Apprentices were to be paid SR2,500 (about $667) a month to start, and given increases for completion of each stage of the program up to a top salary of SR4,000 ($1,066) a month. The company would provide housing for apprentices living outside a 50-kilometer (31-mile) radius of the training facility. Apprentices would have 28 days of vacation a year, bus transportation to and from the training site, and free medical care, excluding dental and optical. They would not be eligible for a housing allowance or other benefits given regular Saudi employees. It was estimated that the company would have considerable savings from reduced salaries and benefits by hiring these individuals as apprentices rather than as regular employees.

The seven-hour-long intensive English-language schedule was approved despite stiff opposition. "There was a lot of discussion and a lot of resistance to it," Arons recalled. Many traditionally trained ITC instructors were opposed, fearing the students would become bored spending such long hours on a single subject. But to other trainers, intensive language training was a proven commodity, not a new concept. The U.S. military had used intensive language programs for years; so had the U.S. Peace Corps. Full-day English-language programs were used at the state-owned National Iranian Oil Company and by Saudis in training at the government air base in Dhahran.

"We weren't recommending something that hadn't been used before," Arons said. "It was a proven thing, but it was not proven within Aramco."

The idea of an intensive English training program for Aramco apprentices originated with Thamer R. Al-Murshed, director of the Central/Northern Area Training Department. "I thought high school graduates really don't need six or seven years of English," Murshed said. "I know a lot of people, Saudis and others, who went with a high school diploma to the United States or Europe and within one year learned the language to the point where they were able to pursue academic studies in that language.

"So, perhaps, all we needed to have was an intensive program to put pressure on these apprentices. Let them sink or swim. This is how the idea was born. I shared this idea with my management and staff. They said, 'OK, come up with a proposal.' Steve Tolle and Ahmad

Thamer Al-Murshed, right, congratulates graduate Jameel Al-Dossary during a 1988 training awards ceremony.

210

Ajarimah did the legwork. Steve Tolle and all of them were extremely helpful. I just came up with the idea."

The "sink or swim" attitude was reflected in testing procedures. Testing was tied to objectives. "The apprentices had to meet a certain level of proficiency or they didn't pass," Arons said. "If that meant the whole group failed, the whole group failed. It wasn't like norm-referenced training where you might pass 90 percent of the class. They either met the standard or they didn't."

Arons gave Ali Dialdin high marks for standing behind the task force's recommendations. "I admire him for having the nerve to take a system that was so different from what we had in the past and implementing it. If it had failed, it would have been an expensive failure. But it didn't fail. It did very well."

An Aramco production crew shoots a scene for a training film in the Media Productions studio, Dhahran, 1988.

Following another task force recommendation, the old semester system was abandoned and the academic year was divided into four 54-day "phases," or quarters, as they were often called. Under the intensive training timetable, there would be 432 hours of training time for each 54-day phase. That equaled the amount of time available in an entire academic year under the old two-hour-per-day training schedule. The opening two phases of the Apprenticeship Program were devoted almost exclusively to English-language training. The next two phases were divided about equally between English and other subjects. In the first year of their apprenticeship, trainees received 1,728 periods of instruction, including 1,080 periods of English. Apprentices in craft and technical fields received 432 periods of basic job skills training, 162 periods of industrial mathematics and 54 periods of remediation/acceleration (REMAC). REMAC included individualized instruction for low achievers, and enrichment material for high achievers. Second-year apprenticeship training patterns varied according to which of more than 80 target jobs the apprentice was aiming for. Apprentice craftsmen spent the entire second year in job-specific training under line organization supervision, but technician and operator trainees took English classes during their second year along with job-specific training. Most of the technician and operator apprentices enrolled in English level 6B and 6A, where they learned to read, write and speak about topics such as industrial safety, using hand tools and operating machinery, along with some English-language instruction in the use of computers. Several line organizations, the Marine Department, for example, created specialized English programs for apprentices.

The task force also laid out a clerical training pattern consisting of eight 54-day phases over the two-year period. Apprentices on the clerical track followed the same intensive English-language training pattern as the industrial apprentices did during the opening two 54-day phases. After that, clerical apprentices began a regime of academic and business courses not required of trainees for the industrial work force. These included courses in advanced reading and writing, business skills, typing and computer operation.

Apprenticeship Program Approved

n January 1988, Ajarimah presented the task force's report on the Apprenticeship Program to SAMCOM and it was approved. The next step was to sell the program to line organizations. Ajarimah toured company facilities during the next several months, explaining the new program to virtually anyone who was interested. He appeared before audiences as large as 150 and as small as three or four. He spoke to vice presidents, group leaders, supervisors, trainers, and training coordinators, representatives from the whole spectrum of Aramco professions and trades. He visited Dhahran, Ras Tanura, Abqaiq, al-Hasa and Yanbu', giving his slide presentation to more than one audience in several of those locales.

"There was a great level of interest," he said. "We told them we were going to reduce costs by $11 million, optimize training time and terminate the failures. Those were the benefits of the pilot Apprenticeship Program, as it was then called."

One of the strongest selling points for the program was the promise that the time it took graduates of the program to become job-qualified would be reduced by about a year. The reasoning behind this expectation was simple: graduates of the program would be hired as regular employees and enter established Job Skills Training programs. They would be better educated and better qualified than any previous trainees; therefore, they could cover Job Skills Training material at a faster pace than previous trainees. After entering Job Skills Training, it was calculated, employees should qualify for entry-level jobs as a clerk in a total of 2½ years, as a gas and oil plant operator in four years, and as a craftsman in about five years.

Screening for the first group of apprentices began in May 1988. From that time on, all Saudi candidates for jobs in the industrial work force were supposed to begin as apprentices. They would have to earn a spot on the regular industrial payroll by successfully completing the Apprenticeship Program. Projections called for 644 apprentices during the first year of the program and 700 to 800 per year after that. The initial screening was a disappointment. Of 806 applicants tested for an August 1988 start-up date, only 145 qualified for the program. In a second batch of 406 candidates tested for a November 1988 start-up, only 80 qualified for the program. In total, 1,020 candidates were tested during 1988, and only 225, or 22 percent, met company standards. At that rate, Training would have to test 3,000 candidates a year in order to find the required 700 to 800 apprentices.

Clearly, adjustments were needed. SAMCOM agreed to a recommendation to lower the English placement criteria from level three to level two for all candidates starting in 1989. Based on previous test results, this would allow about 50 percent of those who passed the GATB test to join the program. In order to bring candidates up to the English three level, however, the apprenticeship had to be lengthened by three months, to a total of 27 months.

An instructor explains complex refinery piping to Saudi trainees.

Saudi Aramco took on many more apprentices than the 700 to 800 per year originally anticipated. The program began with 363 apprentices in 1988, increased to 628 in 1989, and climbed to 1,655 in 1990. During the initial two years of the program, 7,700 candidates applied for enrollment and 6,418 applicants were tested. Only 2,200 of those tested made it into the Apprentice Program. Nearly half the tested applicants failed to meet the English placement requirement. Many of those who qualified for the program dropped out. Up to 50 percent of the candidates quit the program in the first four months, but only 17 percent of those who stayed with the program longer than four months were lost through attrition.

The reason for a high rate of attrition at the start of the program, Dialdin said, was pressure. "There was pressure on the apprentices all day. Saudis who had been attending classes in high school for maybe five hours a day had never experienced anything like it. The individuals who were not willing to come to work, or to go to school, or were not able to be disciplined, disappeared. Either we fired them, or they dropped out — because they could not take the pressure."

Intensive English for Apprentices

The first group of apprentices attended classes in the Dhahran ITC at the request of Nimr Atiyeh, then assistant superintendent of that ITC. Thamer Al-Murshed asked Atiyeh's opinion about intensive English-language training sessions for apprentices. Atiyeh said he told Al-Murshed: "It is against educational practice. However, it is an idea that is worth trying. I am willing to try it. If you want it to succeed, start it at my ITC."

Salahel-Din El-Mahdi, who had been teaching English in the ITCs for about 10 years, called the first group of apprentices definitely the best students he had yet encountered. "They were really motivated. There were some really fast ones. Things that were scheduled for half an hour they finished in five minutes."

The intensive, seven-hour-long classes were for students only. Instructors were supposed to follow the normal teaching schedule — five hours teaching in the classroom and three hours in preparation for future classes — but staff shortages often required them to put in longer hours. Such was the case with El-Mahdi, who taught English to the same group of apprentices for seven periods each day.

"The schedule was very demanding, very hectic," he said. "Teaching seven hours left you only one hour to prepare the next day's lesson. We usually stayed an hour or so extra." He spent the seven hours standing rather than sitting in front of his class. "You had to stand up, not because you wanted to teach properly, but because you had to keep an eye on the students. They were young and a little bit frisky. We didn't want any horseplay."

El-Mahdi remembers "many problems with absenteeism" and dropouts during those years. The sons of Saudi Aramco employees tended to have a better attendance record and lower dropout rate than others, he said. "Apparently the fathers who were company employees kept pressing their sons to continue, telling them the many advantages of working for Saudi Aramco, getting a house and things like that."

The first 91 apprentices graduated in June 1990 and became Saudi Aramco employees after a 90-day probation period. The graduates included 47 Saudis apprenticed as clerks, 16 oil/gas plant operators, 15 craftsmen and 13 technicians.

The company accepted 1,446 new apprentices in 1991 and 2,071 more in 1992. By that time there were more than 3,000 apprentices in the two-year program. It was too much to handle, even in the large new Aramco Training Centers. At Ras Tanura, part of the former Rahimah ITC (the old Jones Camp) was reopened to handle the flood of apprentices. The 'Udhailiyah ITC, opened in 1984 and mothballed in 1987, was reactivated for apprentices. Similarly, a portion of the Abqaiq ITC that had been closed for several years was opened again for apprentice training.

A job skills instructor diagrams an electrical generator.

Engineer Salman Al-Agnam examines a plant flow graphic, 1989.

The academic record of those who stayed in the Apprenticeship Program until they graduated was outstanding. Ninety-eight percent of them passed, and 76 percent earned As or Bs. They were exceptionally well received on the job. Steve Tolle remembers reports from the field declaring that graduates of the program were better qualified than some 10-12 year veterans.

"The gas and oil plant operator program got especially dramatic results," he said. "Craft training was good, but there was not such a dramatic improvement."

By late 1992 it appeared there were more apprentices being trained than there would be jobs available to them when they graduated. Consequently, on January 31, 1993, the company stopped accepting new candidates for the Apprenticeship Program. A total of 7,025 people had entered the program since it started on August 20, 1988. Of that total, about 64 percent, 4,515 people, graduated and entered the work force by year-end 1995. About one-third of those who enrolled, 2,388 apprentices, had been lost due to attrition or failure.

The program sold itself to the line organizations on the basis of quality, Dialdin said. "I was under a lot of pressure. Executive management was concerned about too much attrition. I told them if you want quality, this is how we're going to get it. There is no other way. Some of it was a hard sell in the beginning. We just learned to live with it.

"I'm proud of the accomplishments in the Apprenticeship Program. I'd say nearly all of the recipients of apprentice graduates spoke very highly of the results."

Saudi Clerical Training

n the early 1980s, responding to pressure from upper management, Training began a campaign for the Saudization of the clerical work force. This was the last of the company's large job families to be targeted for Saudization. At the time, most of the company's 6,000 clerks were expatriates, primarily Asian nationals. There were only a few training programs related to clerical work. The Training Department offered a course called Office Practice, taught at English level five and designed to acquaint office workers with business machines and office protocol. The department also had courses in supplementary clerical subjects such as typing, business math and basic accounting. Several other organizations, among them Community Services, Finance and Materials Supply, ran short-term, on-the-job training programs for clerical workers, but there was no structured, company-wide system to prepare Saudis for clerical positions.

In 1983, Elizabeth Babb of Training's Academic Curriculum Unit polled nine departments to determine the skills most frequently required of clerks. English-language typing, using a telephone and operating a photocopier were nearly unanimous mentions. Filing documents and completing forms came next on the needs list. Based in part on these findings, the Office Practice course was amended and lengthened from two to four semesters. It was piloted at the Industrial Training Centers in Ras Tanura and Dhahran in the fall semester of 1984. But the course remained general in nature, designed to prepare trainees for all types of office jobs, not just clerical work.

Ali H. Twairqi

"Aramco provided the environment for success in training — the setup, the atmosphere to succeed was there, in the form of good teachers, good facilities, good supplies, good laboratories and a well-run organization."

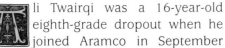li Twairqi was a 16-year-old eighth-grade dropout when he joined Aramco in September 1966. He entered the Company's Technical and Supervisory Training Program as an administrative and supervisory trainee. While he was happy to be an Aramcon, he didn't look forward to the training part of the job.

"You go around strutting like a peacock because you joined Aramco," he said. "The family was proud because they would get free medical treatment and be able to use company facilities. These materialistic things were the focus of somebody coming to the company. Training was not on the list. For somebody who dropped out of school, the last thing you wanted to think about was going back to school."

He was sent first to Abqaiq for a regimen of four hours of training and four hours of work. His first job was as a ticket taker at the Abqaiq Recreation Theater. Twairqi's assignment was later changed to the Abqaiq Library.

Twairqi was targeted to fill an administrative job, such as clerk, but the door was open for further advancement—based mainly on continuing training and education. He was assigned to live in the Mansur Camp section of Abqaiq, where you carried your own mattress and bedding to your assigned, non-air-conditioned room. The students sometimes slept outside at night to keep cool, and did their own cooking and housecleaning.

To his surprise, school turned out to be exciting. The company's training system was based on American learning theory, which moved each student forward as he completed topics and subjects. Unlike in the Saudi system, students repeated only those portions of the curriculum that they had not yet mastered. He was impressed to find that all the books and supplies he needed, were available on the first day of class, and with the professionalism of the instructors. The laboratories were stocked with every ingredient and tool required to perform experiments. Students were even assigned their own microscopes.

In 1971 he joined the Advanced Training Program in Dhahran on a full-time basis and was selected in 1972 for out-of-Kingdom training. He went to the Penn Center Academy in Philadelphia, where he earned a high school diploma, and then went on to graduate with a bachelor of science degree in business administration from Sam Houston State University in Huntsville, Texas, in 1978.

Back in Saudi Arabia, after less than one year as a staff advisor in Training, he was assigned to Aramco Services Company in Houston to oversee students sent to the U.S. under the company's College Degree Program training.

Twairqi completed the Management Development Seminar (U.S.A.) in 1982 and the Executive Program at the Darden School of Business at the University of Virginia in 1988. Over the years he held a number of management positions, including manager or director of Training, Community Services, Personnel, Employment and Employee Relations Policy & Planning. He was named director of Career Development in 1995. In 1996 he became general manager of Training and Career Development for Saudi Aramco, the first graduate of Company training programs to hold that position.

bout this time a difference of opinion developed over the best way to teach clerical subjects — whether clerical skills were best learned during on-the-job training, as were other industrial jobs, or whether it should be taught in the classroom as an academic subject, like typing. Ali Dialdin, who had recently been named general manager of Training & Career Development, decided that clerical training was a job-skills function. With that settled, the development of a clerical training program proceeded at a rapid pace.

In the three months from April to July 1985, Alf McLean, a vocational analyst who earlier developed a training plan for Office Services, and Jim Hanson, a contractor with previous experience as an office manager, created a 180-hour course designed especially for training Saudi clerks. They called it Basic Business Skills, and it was conducted in as much of a hands-on setting as possible. In the classroom each trainee had his own office area equipped with a standard-sized desk, pencil sharpener, filing cabinet, adding machine, typewriter and a telephone through which the trainee could talk to his instructor and no one else. Two classrooms of this type were set up at the Dhahran North Training Center. The new Basic Business Skills course was piloted there on July 27, 1985.

Alf McLean

A major problem had to be overcome before classes could begin. Training had no instructors with the knowledge and experience to teach a clerical course. What's more, the instructors assigned to the Basic Business Skills program were teachers of academic subjects — "chalk-and-talk" instructors, McLean called them — unaccustomed to the hands-on type of instruction required in job skills training. "It was a problem," McLean said, "that gave me nightmares." He solved it in an unconventional way. Video monitors were installed in the classrooms, and each lesson was preceded by a videotaped demonstration of the proper procedures and techniques for various clerical tasks. Then, instead of telling trainees what to do, instructors helped them to perform the day's tasks.

"It was a very successful method," Dale Saner, then head of the Job Skills Curriculum Unit, said. "We were able to take teachers who had no subject knowledge and have them use a video, so we had standardized clerical instruction throughout Aramco — quite a unique program, I think."

Dialdin agreed, although it was an expensive program. "We could not take more than eight in a classroom, compared with 15 to 16 in an academic setting," he said. "That limited our ability to take trainees into clerical training. It was much more expensive, but to me it was worthwhile because it was so supportive to the trainee himself."

Saudi trainees at an ITC computer center, 1988.

The first half of the 180-hour Basic Business Skills course was presented in the fall quarter of 1985. Meanwhile, McLean and Hanson rushed to complete

the second half in time for the start of the January 1986 quarter. The final module contained Training's first computer course, a 20-hour PC introductory course. Programs and classrooms similar to those at Dhahran North were opened in Ras Tanura and Abqaiq over the next two years. They formed the nucleus of the first company-wide clerical training program for Saudis.

Still, it wasn't long before it was realized that the Basic Business Skills course by itself was not sufficient preparation for any but the most fundamental clerical chores. Such tasks as budgeting, operating plan preparation, and mastery of advanced computer techniques were too much to cover in a 180-hour course. In early 1987 a task force composed of members from Management Services and Employee Relations analyzed the training needs of the clerical work force in response to a directive issued by Aramco President Ali Al-Naimi. A task force survey found Saudis now occupied about 52 percent of the 6,200 clerical positions in the company, a substantial increase from the early 1980s. Judged by the company's job descriptions, however, the survey found only 18 percent of the Saudi clerks could be considered job-qualified. Many of those designated as clerks were not actually performing clerical work. Instead, they were doing the type of work more often associated with administrative aides. The task force's report concluded, "Each one of the 3,588 present [Saudi clerks] incumbents needs varying degrees of training to become fully qualified and competent."

That same year, 1987, SAMCOM approved a five-phase Corporate Clerical Training Program based on the same format used in the company-wide Job Skills Training programs. Phase One provided basic English, level 4B, and basic math and typing, each at level 3B. The Basic Business Skills course became Phase Two of the program. Phase Three provided on-the-job training using Aramco Job Training Standards (AJTSs) for job certification. Phases Four and Five were for advanced clerical training. By 1988 there were more than 500 participants in the new Clerical Training Program. Aramco Job Training Standards had been developed for 13 high-priority clerical jobs employing more than 1,100 Saudis. McLean continued to produce and refine clerical training programs along with Mesrob K. Tashdjian, who for more than 10 years, starting in 1986, was the Job Skills Curriculum Unit's coordinator for clerical training.

College Degree Program for Non-Employees (CDPNE)

In the spring of 1986, at about the same time as a new Apprenticeship Program was being considered, SAMCOM received a proposal to initiate a College Degree Program for Non-Employees (CDPNE). The CDPNE and the Apprenticeship Program had one important feature in common — participants would have to earn their way onto the regular Aramco payroll by completing a rigorous training program. As with the Apprenticeship Program, the main argument in favor of CDPNE was that it would free the company from any obligation to employ program failures and dropouts. It would be a major departure from the company's existing college scholarship programs. Participants in those programs became regular Aramco employees, with all the privileges and job security that that entailed, before they went off to college. CDPNE participants, on the other hand, would not qualify for regular employee status until they had received a four-year degree in a company-approved field of study.

Luay Ismail was the first graduate in the College Degree Program for Non-Employees.

For years Aramco had been presenting college scholarships to promising young Saudis, patiently grooming them for positions in the professional work force. The grade code 11 to 14 professional employees included engineers, geologists, computer experts and other college-trained professionals who, collectively, possessed the knowledge needed to build, operate and maintain an oil company. In 1986, only 27 percent of the company's professionals were Saudis, a very low ratio compared to an average of 60 percent Saudi in other segments of the work force.

One of the most distressing aspects of the College Degree Program (CDP) was the rate of attrition. A rapid buildup of CDP enrollment in the early 1980s had been followed by a similar increase in the number of CDP dropouts and failures. By the

217

Omar Abdulhamid was the first Advanced Degree Program participant to receive a Ph.D. from M.I.T.

mid-1980s, as many as half of all CDP participants had left the program without obtaining a college degree. A study by the Career Development Department found the highest attrition rate among freshmen and sophomores majoring in technical and scientific fields. Unmarried students were more likely to drop out than married students. There was little difference in attrition between those enrolled at colleges and universities in Saudi Arabia and those attending schools overseas.

Attrition was tied to concern over an increasingly burdensome cost of the college scholarship program. When the first Aramco-funded college scholarships were issued in 1951, the cost of tuition plus room and board at colleges and universities in Lebanon and Syria was $1,800 a year. In the next 20 years the company awarded fewer than 500 college scholarships to employees, but with the coming of the 1970s expansion era, Aramco's scholarship program grew steadily larger and more expensive. The number of employees on scholarship increased from 90 in 1970 to 283 in 1975 and to 320 in 1979. The company introduced several new types of college scholarship programs in the late 1970s, grouped together under a new administrative title, the College Degree Program. They covered all scholarship programs leading to a bachelor's or higher degree. With the additional programs and aggressive recruiting of candidates, the number of Saudi employees on scholarship more than tripled in the next four years, rising from 465 in 1980 to a high of 1,452 in 1983. By 1983 the cost of maintaining an employee on scholarship ranged from $30,000 to $50,000 a year, depending on salary, marital status and whether the student attended school in-Kingdom or overseas. The price was 15 to 30 times more than it had been when the program began. Considering the high attrition rate, it was a very risky investment.

The sharp decline in oil prices in the mid-1980s forced Aramco to institute a series of stringent cost-cutting measures, and scholarship programs were not spared. By 1987 the number of employees on scholarship dropped to 734, about half the record-high 1983 level. Between 1983 and 1985 the list of colleges and universities in the United States approved for CDP participants was cut from 127 to 33. All schools on the shortened list had strong engineering and technical programs. CDP administrators were directed to strive for a mix of 90 percent technical and science majors to 10 percent majors in nontechnical subjects such as business administration and accounting. That narrowed the field of scholarship candidates largely to those with good grades in math and science and an aptitude for technical subjects.

In 1985 the company reduced the amount of compensation it paid to CDP participants by about one-third. Employee Relations Policy & Planning (ERP&P) compared the amount scholarship students received from the Saudi government and private industries in the Kingdom with Aramco's compensation levels and concluded the company's payments were unnecessarily high. The salary and living allowance paid by Aramco to CDP participants, above and beyond the cost of tuition and registration fees, ranged from about $36,700 (SR137,583) a year for married students at schools overseas to $24,700 (SR92,622) a year for unmarried students at colleges and universities in-Kingdom. ERP&P reported that scholarship participants returning to Aramco after graduation from colleges overseas often "experience a reduction in total compensation when they are promoted to a grade code 11 entry-level job." In accord with ERP&P recommendations, payments were reduced to $26,460 (SR99,226) a year for a married scholarship students overseas and to $14,470 (SR54,265) for unmarried students studying at in-Kingdom colleges and universities. These changes brought Aramco's rates in line with those of other employers in Saudi Arabia.

Ali Dialdin

"O'Grady told me, 'We might have you with Training for a year and then send you back to Industrial Engineering.'"

Ali Dialdin did not go back to Industrial Engineering, as Training Director Bill O'Grady suggested. Instead, he remained with the Training Organization and became general manager of Training & Career Development in October 1984. Dialdin was the first Saudi Arab put in charge of Aramco Training, the largest industrial training organization in the world.

At the time, Training employed more than 1,700 people, managed nearly one million square feet of training space and was responsible for 20,000 Saudis being trained in company classrooms or attending colleges and universities on an Aramco scholarship. As always, T&CD's main job was to satisfy the human resources needs of the company and train Saudis to replace expatriates at all levels in the work force. Dialdin fulfilled that role well. During his years as head of T&CD, the ratio of Saudis in the work force climbed from 58 to more than 80 percent.

Training was not the field Dialdin envisioned for himself when he joined the company. As a young man he had been a primary-school teacher and principal in his hometown of Medina, but when the government awarded him a college scholarship, he set out to become a geologist. Dialdin attended San Diego State University, earned a degree in geology and joined Aramco's Industrial Engineering Department at Ras Tanura in April 1968. Eleven months later his supervisor, Bill Griffin, was transferred to the Training organization and asked Dialdin to join him a few weeks later. All training activities had just been consolidated under William O'Grady, the director of Training.

"I'll never forget," Dialdin said, "O'Grady told me I was coming at a time when employment was down and Training was going to have a very small number of people. He said, 'We might

have you for a year and then send you back to Industrial Engineering.'"

Instead, Dialdin remained with Training for 30 years. His department's Job Skills Training Centers were the first such centers in the Middle East to be accredited by the internationally recognized Accrediting Council for Continuing Education and Training. He also organized and implemented the Apprenticeship Program in 1988 and the testing, placement and establishment of training programs for more than 10,000 former Samarec employees during the integration of Samarec and Saudi Aramco in 1993.

Dialdin stepped down as general manager of T&CD in September 1996, but remained with the company as general manager of the Human Resources Task Force. Later that same year, he became the first Saudi Arab chairman of the prestigious International Federation of Training and Development Organizations (IFTDO). Fourteen years earlier he had been the first Saudi Arab member of IFTDO.

"No company so far as I have seen has, overall, as complete a training system as Saudi Aramco does. In most parts of the world they rely on people coming out of industrial trade schools. What you see is patches of training. They concentrate, for example, on management training or they do some on-the-job training. They do training in pieces. You don't see an overall system. Nobody in the world, no single company like Saudi Aramco has 700 people going to school at one time in various parts of the world. No company compares in vocational training. No company trains maintenance technicians in the refinery, for example. They hire them."

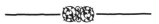

CDPNE Approved

ustafa al-Khan Buahmad, director of the Career Development Department, went before SAMCOM on May 31, 1986, to propose creation of the College Degree Program for Non-Employees. He was facing a cost-conscious group of Aramco executives. Buahmad's presentation stressed the savings that could be expected from CDPNE. In the words of his report, the two primary benefits of CDPNE would be:

"1. The Company will no longer be obligated to offer jobs to college dropouts or failures. Although difficult to quantify, this feature of the program will realize considerable savings, since, based on experience, the rate of CDP attrition through failure and dropouts has been 40 percent to 50 percent.

"2. The proposed program will realize further reductions in the costs of Aramco's CDP. It is estimated that $26,000 per degree will be saved, $67,000 per an OOK (Out-of-Kingdom) married participant's degree. An estimated amount of $1.4 million can be saved during a 5½ year program period based on a class of 60 with an assumed 30 percent attrition rate."

CDPNE participants were to receive a monthly stipend, or living allowance. Buahmad recommended a stipend of $666 a month (SR2,500). This was about half the average salary for regular CDP enrollees. Buahmad also recommended an allowance for married CDPNE participants of $320 (SR1,200) a month, whether the student attended a school in-Kingdom or overseas. This was the same amount as had been paid to married CDP students attending schools in Saudi Arabia, but the married allowance for CDP participants at schools overseas had been twice as much, $640 a month.

Mustafa al-Khan Buahmad

SAMCOM members were already familiar with the CDPNE concept, and they quickly endorsed the proposal. Buahmad, who happened to be secretary of SAMCOM at the time, had kept the committee briefed at each step in the process of putting together a program proposal. "This was not the first time a college degree program for nonemployees had been discussed," he said, "but it was the first time an in-depth study had been made on the idea and a formal proposal made." Management committee approval followed SAMCOM's endorsement, and CDPNE became effective on January 1, 1987, about seven months ahead of the Apprenticeship Program.

Saudi honor students, Saad Al-Sharani and Salem Shehry, on the campus of the University of New Orleans.

Approval of CDPNE meant the College Fast Track program created in 1979 would be phased out. All new college scholarship participants, other than regular employees coming out of the ITCs, would be nonemployees. Minimum qualifications for admittance into CDPNE were similar to those for the College Fast Track program. Candidates had to be graduates from the high school science track with an 80 or better grade average. Those already in college would need a 2.5 or better grade average to qualify for a company scholarship.

A degree program for nonemployees needed some regulations that had not been necessary when all scholarship participants were full Aramco employees. For instance, CDPNE participants taking a job at Aramco after graduation from college would be required to remain with the company for a period of time equal to the time they spent in the scholarship program. Students who dropped out of the program without company consent would have to repay the company for the costs they incurred. Those who flunked out of college or were dropped from the program by the company would not have to repay program costs. Anyone who successfully completed CDPNE and was not offered a job by Aramco was relieved of any obligation to repay the company.

220

Although it was to be phased out, the company's College Degree Program for regular employees had to be judged a success. In the previous seven years alone, CDP had produced more than 960 college graduates for the company's professional ranks, an average of more than 135 per year. Few, if any, other business organizations in the world could say as much.

College Preparatory Program (CPP)

All college scholarship candidates were required to successfully complete the company's year-long College Preparatory Program (CPP). This program began operations in 1983 at the old Dhahran ITC. The company had been sending scholarship recipients to specialized schools in the United States for up to a year prior to their college entrance examinations, primarily to improve the candidate's English-language skills. Training's Program Development & Evaluation Division created CPP as an in-Kingdom replacement for preparatory courses overseas. In its first year, CPP enrollment was limited to 36 College Fast Track students, all recent graduates of Saudi high schools. But the next year, 1984, management decided that all scholarship recipients, other than those already attending a university, should complete CPP before applying to a school using English as the language of instruction. That included King Fahd University of Petroleum and Minerals (KFUPM) next door to Aramco's Dhahran headquarters. Management's decision resulted in a fourfold increase in CPP enrollment, to 120 students, during 1984. The in-house program was relatively inexpensive. CPP cost the company about $8,500 per student a year, compared to about $11,000 per year for a similar program at KFUPM and more than $20,000 per student a year at specialized schools in the United States. CPP also served to filter out sizable numbers of low achievers. Between 1983 and 1987 a total of 283 students enrolled in CPP, out of which 104 failed to complete the program. CPP moved in 1986 to permanent headquarters at the Dhahran North Training Center.

Sami Al-Ajmi (left) and Saud Al-Zahrani, at graduation exercises in Maryland, studied computer engineering technology.

The original CPP curriculum provided 1,440 hours of instruction in six phases. Of that total, 970 hours, or about two-thirds of the time, was devoted to English-language instruction. A team of U.S. educators from the National Association for Foreign Student Affairs (NAFSA) reviewed the CPP in 1988 and judged the program's graduates "very well prepared for admission to a U.S. university."

In 1992 the program was expanded to 1,560 periods, and in 1994 it was increased to 1,800 hours of instruction over five phases of 45 teaching days each. In the 1994 version, as in the original program, about two-thirds of the time, a total of 1,215 hours, was given to English instruction. Most of the remaining time was divided equally between mathematics — algebra, trigonometry and calculus — and science, including chemistry and physics. Classes in computer skills, report writing and oral presentations, plus orientation to overseas universities, were also included.

During the 1987-88 academic year, the first full year of CDPNE, the company recruited 109 high school graduates for the program. Forty-two of them were terminated or withdrew before completing the College Preparatory Program. Six of the remaining 67 recruits failed the entrance examination to KFUPM in 1988 and were terminated. Only 61 of the original 109 actually entered a university, 56 of them at KFUPM and five at universities overseas.

At first the number of students enrolled in the old College Degree Program dwarfed the size of CDPNE's enrollment. In the 1987-88 academic year there were 826 Saudis attending colleges and universities through the CDP, compared to 61 participants in the nonemployee program. Upper management and Training, nevertheless, considered CDPNE to be the program of the future. During 1988 SAMCOM directed that the ITC college-track program be phased out. The committee gave some 440 ITC college-track participants until 1990 to complete their requirements for a scholarship. No new enrollments would be accepted after that. The College Fast Track program had already been closed to new participants. That left CDPNE as the only college scholarship program that would accept new candidates. After 1990 no more college scholarships were awarded to regular Saudi employees, except in unusual cases and only after thorough review by the College Degree Program Committee. By 1995 there were about 650 CDPNE participants, compared to fewer than 100 students in the fast-fading CDP.

Aramco-sponsored university graduates gather in Dhahran, 1988.

Developing the Saudi Professional Work Force

audizing the professional work force had been, and would continue to be, an agonizingly slow process, as Ali Al-Naimi, then the company president, explained to an interviewer from *Focus* magazine in 1987.

"Aramco needs about 6,500 professionals, not including teachers and medical doctors," Al-Naimi told the interviewer. "On the payroll today, there are probably 2,200 to 2,300 Saudi professionals. So the gap between the need for professionals and the availability of qualified Saudis is about 4,500 positions, again, excluding physicians and teachers.

"For an employee to get on first base as a professional, he needs a university degree and at least three years' experience. That takes about eight years, total. The maximum number of qualified Saudi employees we are able to get from our own development programs and from universities is between 200 and 250 recruits per year, and that's a high estimate.

"This means," Al-Naimi concluded, "for the next 20 years Aramco will probably continue to require expatriate professionals" to fill professional positions for which Saudis are not yet available.

His assessment may have been somewhat pessimistic. In August 1987, when Al-Naimi was interviewed, Saudis held about 35 percent of all professional positions in the company. Eight years later, at end of 1995, the ratio of Saudis in professional jobs had risen to 56 percent.

Training Department Reorganized ... Again

n July 1, 1989, the Training Department was once again reorganized, the sixth general reorganization since 1952, when Training became a full department. The geographical orientation established in 1982, with training directed locally in each of the three districts, was dropped. Instead, leadership was centralized in Dhahran and the departments organized along functional lines. A Job Skills Training Department was created with Zaki A. Ruhaimi as director. The Training Organization was subdivided into six divisions, each headed by a superintendent. The superintendent of the Program Development & Evaluation Division and superintendents of the two academic training divisions reported

Zaki A. Ruhaimi

directly to the general manager, Ali Dialdin. The superintendents reporting to Dialdin were: Zubair A. Al-Qadi, Program Development & Evaluation; Suleiman A. Al-Rasheed, Northern/Central Area Academic Training; and Mohammad M. Al-Hajri, Southern Area Academic Training. The Job Skills Training superintendents reporting to Ruhaimi were: Said I. Al-Ghamdi in Dhahran, Saad A. Ghurmalla in Ras Tanura, and H. Richard Arons in Abqaiq.

In 1990, Ahmad A. Ajarimah was named acting superintendent of Northern/Central Academic Training. Abdel-Monim H. (Naeem) Al-Saif became superintendent of Job Skills Training in Dhahran, and Said Al-Ghamdi took the place of Arons as superintendent of Southern Area Job Skills Training in Abqaiq.

"The reorganization in 1989 really paid off," Dialdin said. "We were suffering from the problem of one manager who wants to do it one way and the other manager who wants to do it a different way. I was always trying to mediate. We could not function that way. It had to be centralized. Job Skills Training had to be the same all over the company. Now we had a single Job Skills manager. When I gave directions, I gave directions to one person rather than to three individuals."

For Dialdin the reorganization was one change in a series of changes or "focus points," which, in the course of a decade, completely transformed the company's training system. The process had been triggered by the report on training issued in the autumn of 1978 by Hugh Goerner, then company president, and George Larsen, former vice president of Exploration and Development. The Goerner-Larsen report found wide variations in the quality and quantity of training within the company. The company was changing and growing rapidly. Training could not keep up with the needs of line organizations, especially in the area of job skills. "The report was a major eye-opener," Dialdin said. "Corporate management realized the shortcomings of the fragmented training that was going on in the company." The result was the formation in 1979 of the Saudi Arab Manpower Committee.

"SAMCOM was really the major thrust, a milestone in giving direction to Training," Dialdin said. "That was the key focus point. The era of the '70s was an era of tumult. We didn't have the programs and we didn't have the facilities. Training was deadlocked. We just did the ITC part, the academics, and let the line organizations do the rest.

"SAMCOM focused our attention on how training should be done. The first step was to take individuals from whatever level they were at and train them to be competent on the job. That's where the job matrix came in. We began focusing on qualifications of individuals and certification to international standards as described in the job matrix.

"Another major change came when we began to utilize the General Aptitude Test Battery (GATB) and the Vocational Interest Test (VIT). With the application of these tests, we were able to predict where we were going to put a person. We zeroed in on his interests and put him in a job category that suited him best."

The development of Saudi schools was an enabling factor in the transformation, Dialdin said. "Over a 10-year span our training programs evolved from very basic to advanced because we were able to depend on the new recruit's background. They came from a high school system that is very strong in science and math."

The development of more and better Saudi schools allowed Saudi Aramco to become a much more selective employer. Job candidates without prior work experience

Saudi trainees at Dhahran's Management Training Center.

had to prove themselves by completing either the Apprentice Program or the College Degree Program for Non-Employees. That requirement grew out of lessons learned in the '70s.

"Then we had employees with limited abilities and little or no interest in school. So they dropped out and their training stopped, which limited their job capabilities," Dialdin said.

A New Training Philosophy

By 1988 the idealistic goal of training each Saudi to his highest potential, a byword in Aramco Training for 40 years, had been set aside. Recruits taken into the Apprenticeship Program were told they would not be candidates for a college scholarship. They would be trained for an entry-level job, and if they performed satisfactorily, they would be trained for a job at the next highest level, and so on.

"They used to call it 'training to full potential,'" Dialdin said. "The word 'potential' was used very loosely. People under this heading just went from one

A management training class in presentation skills.

training program to another — clerical to mechanical to electrical. So we got away from this 'highest potential' scheme. It was an unattainable target. You have to have an initial target for an individual. You have to have more focus. You can't always keep changing the target.

"Now we train to the job requirement. An employee's job title is cross-checked against the job requirements, and the required training is provided for him, no more, no less. That has served us real well in terms of streamlining our training operation. We were able to control the cost of training. When we send someone to training, there is an objective, there is a job in mind for him. And that, of course, is the basic philosophy of our industrial training."

Almost everyone attending classes on company time — apprentices, regular employees, college scholarship candidates — takes training for a full day. By contrast, for years, training had been a part-time activity — two to four hours of ITC classroom or shop time a day.

Lower-Level ITC Phased Out

In May 1989, SAMCOM endorsed a proposal by Training to phase out ITC level one through five classes for all regular employees within five years' time. By then all regular employees would have completed these lower-level classes. Thereafter, the only trainees requiring lower-level ITC classes would be apprentices. The phase-out was to start with level one classes in 1990 and end with level five classes in 1995.

Line organizations were encouraged to give their employees a chance to complete the courses they required. The response nearly overloaded Training's facilities. ITC enrollment reached 7,500 in 1988, about 2,000 more than Training had expected. The next year, ITC enrollment exceeded predictions by 1,500 trainees. In January 1989, to accommodate the surge in enrollment, Training initiated an intensive eight-hour-per-day English-math program for regular employees. In the next two years a total of about 3,500 regular employees were brought up to standard in math and 1,200 achieved minimum requirements in English. The pass-fail ratios for regular employees on an intensive training schedule were exceptionally good. About 85 percent of the English students and 95 percent of the math students passed their courses.

224

audi Aramco entered the 1990s with a payroll consisting of 43,688 employees from 50 nations, including 32,106 Saudis. Between 1980 and 1990 the ratio of Saudis in the work force had increased from 54 percent to 73 percent, the highest Saudi ratio in 20 years. Saudis held nearly all of the company's top management positions, 76 percent of the supervisory positions, 77 percent of the industrial jobs and about 60 percent of the professional, grade code 11 or above, jobs.

The Training Organization had plans to increase staff size during 1990 to accommodate the rising numbers of apprentices. But on August 2, 1990, all plans disintegrated. On that day, Iraq invaded Kuwait, precipitating the Gulf Crisis. The Gulf Crisis impacted Training mostly in the area of personnel. In common with other organizations, Training had problems recruiting and retaining employees during this period. Training ended the year with 877 employees, about 120 fewer people than planned at the beginning of the year. At the same time, the company accepted 1,655 more apprentices, about 400 more than had been expected. The result was a serious shortage of instructors. "If the availability of teachers remains limited," Training's 1990 accountability report cautioned, "accommodation of the projected yearly input of 1,500 apprentices will become questionable."

The Gulf Crisis ended in early 1991. Training immediately began enlarging its staff, hiring mostly contractors rather than regular employees. By the end of 1992, nearly 200 additional instructors had been employed, bringing the total of teachers and trainers to about 580. The teaching staff included professionals from Saudi Arabia, the United States, Australia, Britain, Egypt, the Philippines, Jordan, Ireland, Palestine and the Sudan, but it was not a large enough staff to keep up the expanding workload. The company added 2,000 new apprentices in 1992, bringing the total number of apprentices in training to 4,289. To handle these numbers, the allowable class size had to be increased from 17 to 20 or more students. At the same time, teachers and trainers worked longer hours. They averaged about 6.5 hours a day in the classroom compared with the normal schedule of five hours in the classroom and three hours in preparation for classroom work. Training expected 1,600 more apprentices to be hired in 1993, but, as often happens, things did not work out as expected.

Floating oil booms protected coastal installations during the Gulf Crisis, 1991.

Saudi Aramco at 60
1990-1996

"Saudi Aramco's business has become much more complex. It has become fast-paced, computerized, digitized. The linkage now is with a worldwide network. We want to continue to be the intermediary between what is happening outside of Saudi Aramco and what the company requires."
– Ali Twairqi, General Manager, T&CD

The year 1993 marked the 60th anniversary of the Concession Agreement, signed on May 29, 1933. In those six decades, that agreement had led to the discovery and production of nearly 60 billion barrels of oil and the creation of Saudi Aramco, one of the largest oil producing and exporting companies in history. By 1993 Saudi Aramco managed 60 oil and gas fields, including the world's largest onshore field, Ghawar, and the world's largest offshore field, Safaniya. The company was caretaker of the largest oil reserves in the world, an estimated 258 billion barrels — one-quarter of the world's known total. No longer limited to exploration in the eastern part of the Kingdom, the company had discovered major oil and gas deposits in central and southern Saudi Arabia and on the Red Sea coastal plain in the western part of the Kingdom. Saudi Aramco entered the anniversary year with 46,000 employees from 50 countries. About 75 percent of the employees were Saudis, the highest ratio in 17 years.

The 60th anniversary celebration took on added luster after Saudi Arabia was recognized as the number one oil producing nation in the world. Saudi Arabia had long been the leading exporter of crude oil. Now, according to oil industry reports, the Kingdom was also the world's leading producer of crude oil. For years, both the Soviet Union and the United States ranked ahead of Saudi Arabia in production, but the breakup of the Soviet Union, combined with increasing production in Saudi Arabia and declining output in the United States, propelled Saudi Arabia into first place. The lead widened during 1993, when, for the first time, the Kingdom's output of crude oil exceeded the combined total output of all the states of the former Soviet Union. International oil-industry publications estimated Saudi Arabia's crude oil production during both 1992 and 1993 at an average of more than eight million barrels per day. About 97 percent of that total came from Saudi Aramco-managed fields.

Opposite: Central Dhahran. Below: Employees convert seismic data into a computer image of underground structures.

227

n array of world-class-sized industrial works stood as active memorials to Saudi Aramco and its predecessors, Casoc and Aramco. The list of projects built and operated by these firms in Saudi Arabia included the largest sea-water treatment plant in the world, the largest offshore gas-oil separator plant in the world, the largest-diameter oil storage tank in the world, and more than 7,000

Marjan GOSP 2 lies offshore in the northern Arabian Gulf.

miles of pipeline crisscrossing the Arabian Peninsula — more than enough pipe to link Dhahran with Saudi Aramco's Stateside affiliate, the Aramco Services Company, in Houston, Texas. The pipeline network included the longest and most advanced computer-monitored natural gas liquid (NGL) pipeline ever built — the 725-mile-long East-West Oil Pipeline, running across the Arabian Peninsula from Shedgum in the Eastern Province to Yanbu' on the Red Sea. The pipeline could deliver up to 290,000 barrels of ethane and NGL daily to Yanbu'. It paralleled a 750-mile-long crude-oil pipeline system capable of delivering five million barrels of oil daily to Yanbu'. Abqaiq, near the giant Ghawar oil field, was the largest crude oil processing center in the world. Ras Tanura, the oil refining and shipping terminal on the Arabian Gulf, was one of the largest oil ports in the world. The crude oil tank farm at Ras Tanura stored more than 30 million barrels of crude oil products, nearly five times the average daily oil production of the entire United States during 1993. The Master Gas System was the giant of them all. Aramco, on behalf of the government, designed and built the system in the 1970s to provide energy for industrial use all across the Kingdom.

The company's industrial training program probably belonged on the "world's largest" list. No one could say with certainty that it was the world's largest, but Ali M. Dialdin, the company's general manager of Training and Career Development, had attended training conferences and studied industrial training programs all around the world, and he was convinced.

Control room at the 'Uthmaniyah Gas Plant, 1996.

"Nobody does training like we do," he said. "When I talk about taking in three thousand or more apprentices a year, it floors them. Nobody takes young kids and trains them the way we do. Others get them from colleges, vocational schools, and so forth. No one has 30,000 to 35,000 training class periods a day. Only large universities do that. And nobody has more than 950 teachers and trainers and nearly 800 support personnel in training. In terms of size, we are the world's largest industrial training program. So far I haven't heard any challenges to that statement."

Aramco Training Developed Aramco Executives

he Training program had another distinction that was, most likely, unique in the world. What other industrial training program could say that the company's top two executives got all their formal education, from primary school ABCs to college degrees, through the company's industrial training and scholarship programs? Such was the case for both Ali I. Al-Naimi, president and chief executive officer of Saudi Aramco, and Nassir Mohammed Al-Ajmi, the executive vice president. They, and many other Saudis in the

Nassir Al-Ajmi, left, presents a leadership award to Abdullah Al-Helal, 1991.

executive, managerial and professional ranks, were products of Aramco training.

The generation to which Al-Naimi and Al-Ajmi belonged straddled two very different eras in their nation's history. Between childhood and middle age, they witnessed and contributed to the transformation of Saudi Arabia from an impoverished desert Kingdom to a modern economic powerhouse.

Al-Ajmi, who retired from Saudi Aramco on May 1, 1993, was born in a desert encampment 50 to 60 miles southwest of Hofuf. When he was hired by Aramco in 1950, at age 15, Al-Ajmi was illiterate, as were almost all Saudi new employees at that time. His first job was cleaning the oil and mud off used vehicle parts in the engine rebuilding shop of the Dhahran Transportation Department. Al-Ajmi received his basic education in the company's Industrial Training Centers, earned his high school diploma while in Beirut on a company scholarship, and received a bachelor's degree in business management from Milton College in Milton, Wisconsin, on another company scholarship.

Ali Al-Naimi's story was much the same. He, too, was born into a nomadic tribe and had no formal schooling before coming to Aramco. Al-Naimi entered the company's Jabal School in 1947, and through company training and scholarship programs, earned a master's degree in geology from Stanford University.

After his retirement, Al-Ajmi authored a book, *Legacy of a Lifetime*, in which he paid tribute to the company's training program and its trainers: "If the oil sector has been the economic lifeline of Saudi Arabia," he wrote, "Saudi Aramco's brightest success story has been in the area of human resources development. The structured training and development programs in place today were developed and maintained over the years by Aramco. Every Saudi Aramco executive, from the president down to operating management levels, is the product of Aramco's training programs.

"Through the constructive application of these training programs, Saudi Aramco today has the depth of skills, discipline and the best technical and managerial competence in the country. The credit goes to many managers who dedicated their professional careers to the training and development of Saudis throughout Aramco's history. They may have remained nameless, but their legacies and human endeavor live forever."

Time Takes a Toll

audi Aramco, at age 60, was no longer a young company. As if to accentuate the passage of time, this anniversary year brought news of the death of three men prominent at the beginning of the oil industry in Saudi Arabia: Floyd Ohliger, "Soak" Hoover and William Mulligan. Ohliger was the company's top man in Saudi Arabia in March 1938 when the discovery well, Dammam No. 7, came in, proving that there was oil in commercial quantities under Saudi Arabia's sands. Ohliger died in Pineville, Pennsylvania, on January 24, 1993, at the age of 91. A few months later, J.W. "Soak" Hoover, one of the first geologists in Saudi Arabia, died in Houston, Texas, at age 89. In 1934 Hoover selected a rock

Aramco pioneer "Soak" Hoover and family visit the Oil Exhibit in Dhahran, 1978.

Bill Mulligan in 1978 retirement photo.

formation called the Dammam Dome as the site for Well No. 1, the first test bore by the company in Saudi Arabia. Another early employee, William E. Mulligan, had been a popular speaker, storyteller, and author of many highly readable magazine and newspaper articles on the life and times of Aramco. Without him, much of the history of the oil industry in Saudi Arabia would have been lost. Mulligan joined Aramco in 1947 and stayed for 32 years, working in the company's Government Relations and Government Affairs offices. He died at Manchester, New Hampshire, in December 1992.

Aside from marking the 60th anniversary, 1993 looked like it would be an ordinary year, at least by Saudi Aramco standards. The company planned a modest increase of about 500 employees during the year. Most new employees were expected to be graduates of Training & Career Development's Apprenticeship Program. In its first four years the Apprenticeship Program took in more than 7,000 candidates, making it one of the largest programs of its type anywhere. Management considered the possibility of handling as many as 12,000 apprentices by the end of 1997; however, other events intervened.

Samarec Merges with Saudi Aramco

n June 14, 1993, the Saudi government Council of Ministers decided to merge government-owned Samarec (Saudi Arabian Marketing & Refining Company) into Saudi Aramco. Samarec, headquartered in Jiddah on the Red Sea coast, more than 700 miles west of Saudi Aramco's headquarters, was responsible for domestic refining, international product marketing and the distribution of petroleum products throughout the Kingdom. The government's Information Minister, Ali Al-Sha'ir, announced the council's decision. "All oil refineries and distribution facilities will be brought under the Saudi Aramco company," he said. "By this decision, Saudi Aramco will become a complete oil company, incorporating the different phases of production, refining, transportation and marketing in a single national firm." Although there had previously been talk of consolidating the Kingdom's petroleum activities, the merger announcement caught Saudi Aramco by surprise. During a speaking engagement a few days later, Saudi Aramco's president and CEO, Ali I. Al-Naimi, confided that he first learned of the council's action while listening to a television newscast. The merger was sanctioned in a Royal Decree issued by King Fahd ibn 'Abd al-'Aziz on July 1, 1993.

The integration, as Saudi Aramco chose to call it, posed an unprecedented challenge to the company. Samarec facilities included refineries at Jiddah, Riyadh and Yanbu' plus a share of three other refineries in which Samarec was a joint-venture partner: at Jubail with Shell, at Yanbu' with Mobil, and at Rabigh with Greece's Petrola. Saudi Aramco also took over a Kingdom-wide network of 34 bulk plants and airport refueling units. In addition, the company acquired Samarec's international marketing system with offices in London, Tokyo, Singapore and New York. Overnight Saudi Aramco expanded from a mostly Eastern Province operation, with all its major facilities no more than a few hours' driving time apart, to a company with facilities and operations all over the Kingdom. Saudi Aramco's international activities also expanded because of the merger.

Evaluating and Training Thousands of Samarec Employees

long with a string of facilities, Saudi Aramco was responsible for some 11,500 Samarec employees — about 10,000 regular Saudi employees and 1,600 contract employees. The company was obliged to offer all Saudi employees of Samarec a job at the same pay scale, although not necessarily in the same field, as they had with Samarec. Saudi Aramco's work force

The former Samarec refinery in Riyadh.

was going to be increased by as much as 25 percent, depending on how many Samarec employees elected to join the company. The responsibility for testing and training each of the new employees to see where they would fit into the Saudi Aramco system fell on the shoulders of Training & Career Development.

The pool of potential new employees included 370 executives in positions from manager to senior vice president, 1,366 supervisory personnel, about 3,500 refinery workers, and some 4,800 employees of Samarec distribution facilities. Instead of 47,000 employees as originally expected, Saudi Aramco's Operating Plan for the coming year was revised to allow for a manpower level of 58,900, the highest number since the peak of the expansion period a decade earlier.

It took Ali Al-Naimi less than 24 hours after the merger announcement to create the Saudi Aramco Integration Task Force and dispatch it to Jiddah to work full time on the transfer of Samarec operations to Saudi Aramco. The Task Force was composed of 11 Saudi Aramco executives plus their selected staff members. It was led by Abdelaziz Al-Hokail, executive vice president for Industrial Relations and Affairs. The Task Force was headquartered in Samarec-owned offices in the Musaidiyyah neighborhood and in offices leased by Samarec at the Jamjoom Building, a glass-walled office tower near the waterfront north of Jiddah's city center. The Training Organization was represented by Zubair A. Al-Qadi, superintendent of the the Program Development and Evaluation Division in Dhahran, and Dale Saner, superintendent of Southern Area Job Skills Training in Abqaiq. They were part of the Industrial Relations section of the Integration Task Force reporting to Ahmad Al-Nassar, general manager of Community Services. Saner and Al-Qadi arrived in Jiddah in late June 1993 and settled in a small office on an upper floor of the Jamjoom Building.

The Task Force's first priority was to complete the "due diligence" phase of the merger. Representatives of each Saudi Aramco organization were responsible for inventorying the property as well as all the duties and obligations being transferred to Saudi Aramco from their counterpart organization in Samarec.

"We were going through all their training operations to identify what they did, the people they had, what kind of training they had, what contracts and obligations they had outstanding," Saner said. "We had a mobile team — people in Jiddah, Riyadh, Dammam and Yanbu'. They had to go out and do physical inventories of all facilities, all capital assets, down to video recorders, televisions and such things." Howel Darney and Brian Shepley supervised Training's due-diligence team, which eventually produced an inventory list of more than 3,500 separate items.

Zubair Al-Qadi

Samarec's Training and Manpower Development Department (T&MD), the counterpart of Saudi Aramco's Training Organization, employed 346 persons, including 240 regular Saudi employees and 105 contractors. T&MD was headed by a vice president. Directly under him were three directors: a director of Manpower Development, a director of Training Support and a director of Training Centers. The T&MD staff included 12 managers and about 145 instructors and assistant instructors. Samarec ran a three-year-long apprenticeship program for about 190 trainees, plus developmental programs for about 100 recent graduates of colleges

Former Samarec administration building, Jiddah, 1996.

and technical schools. In addition, Samarec offered short-term, one- to five-week courses in management development; supervisory development; computers; and fire, security and safety standards. The Saudi Aramco review team found that Samarec training had about as many employees as it did trainees.

Samarec operated full-time training centers at Dammam, Jiddah, Riyadh and Yanbu', plus small, part-time training facilities at bulk plants and airplane refueling units around the Kingdom. The four main training centers had a total of 135 classrooms and 27 job skills training shops with an estimated capacity of 1,755 trainees. The original training buildings at Dammam, Jiddah and Riyadh were identical two-story, steel-frame buildings constructed in the early 1980s. The Dammam facility was virtually unchanged since its construction, but the facilities at Jiddah and Riyadh had been considerably expanded and enlarged. The Jiddah Training Center consisted of the original classroom building plus three newer buildings, a small clinic and a warehouse. At Riyadh, an annex containing offices and classrooms had been built onto the original training building, and job skills classrooms had been added to the interior of the original center. The Yanbu' training facilities, owned by the Yanbu' refinery, included three small shops and a medium-sized shop considered suitable for entry-level job skills training.

In early July 1993, Saner made a "get acquainted" tour of Samarec's Jiddah Training Center. He conveyed his impressions of the center to the Integration Task Force in succinct terms:

"Excellent facilities."

"Adequate training equipment in some areas for Saudi Aramco Phase II."

"Excess staff for trainee workload."

"*Very small* trainee enrollment versus facility capacity."

"Training supervisory staff wear Samarec *shirts and ties!*"

"Trainees wear uniforms as do operator and maintenance personnel in the plants."

While the due-diligence team did its work, Saner requested and got approval for a team of Saudi Aramco teachers and trainers to review Samarec training programs firsthand. The team was composed of Bill Granstedt, principal of the Dhahran North ITC; Malik M. Kalimullah, supervisor of Southern Area Instrument Training in Abqaiq; and 'Abd Allah M. Al-Yami, supervisor of Northern Area Mechanical and Metals Training in Ras Tanura. During the week of July 14-21, 1993, they looked at facilities and monitored classes in Dammam, Jiddah and Riyadh, spending one day at Dammam and two days at the other locations. The following week they spent two days at the Yanbu' training center.

No one expected Samarec and Saudi Aramco to be completely compatible, and they were not. Some differences were rather easily overcome. For example, Samarec used the metric system of measurement. Saudi Aramco used the American system of feet and inches. The problem was solved by distributing conversion tables that listed the American system's equivalent of a metric measure. The two companies also used different electrical systems. Samarec operated on 220-volt electrical power, while Saudi Aramco ran on 110 volts. The remedy was to install transformers at hundreds of electrical outlets to convert 220-volt power to 110 volts. Other differences could not be handled with such dispatch. For instance, some Samarec shops were inadequate for Saudi Aramco's Phase II job skills training, and all of them were too small for advanced, Phase IV, job skills training. Management decided that Samarec personnel would have to be transferred to Yanbu' or to the Eastern Province for the duration of their job skills training, which could take a year or more. In addition, it was found that expensive alterations were needed to bring Samarec training facilities in line with Saudi Aramco safety standards.

Samarec Training Practices

It was no great surprise, therefore, when the Granstedt/Kalimullah/Al-Yami team also found differences in training practices between Samarec and Saudi Aramco. The most glaring difference between the two organizations was in the level of English-language usage. The majority of Samarec's training superintendents, supervisors and trainers had little English capability.

At Riyadh, Granstedt observed the daily 50-minute English-language classes for Samarec apprentices who were already halfway through the three-year apprenticeship program. "I was rather disappointed by the fact that they had spent a year and a half getting some English-language training, but they could essentially not communicate in English," he said. "The English program was really more technical English than it was communicative English. It was more like studying English out of a physics book. They could identify things. They could follow the instruction as long as they had a picture in front of them to help fit together concepts. But in terms of asking questions, they had very faulty grammar and really very minimal communication skills, so we saw that program as being quite deficient."

Saudi Aramco always insisted that its employees learn basic English because English was the medium of technological transfer in the oil industry. A large part of Saudi Aramco's training time was devoted to English-language instruction, and most classes were given in English. At Samarec almost all instruction was in Arabic, with English names used for equipment and machinery parts. Al-Yami discovered that Samarec job-skills training was conducted in Arabic even when the instructor was a British contractor speaking to the class in English. "What happened," Al-Yami said, "was they brought in some bilingual Egyptians. The Egyptians translated into Arabic what the teacher was telling the class in English. This happened nearly everywhere" in Samarec job skills training.

Employees calibrate a training simulator that models the control system for gas-turbine engines.

By mid-summer, Saudi Aramco's Integration Task Force had grown to more than 100 people. Training's full-time contingent had expanded from two to eight. Ibrahim Taha, supervisor of Operator Training in Southern Area Job Skills; Naeem Al-Saif, supervisor of Job Skills Training in Dhahran; and Brian Shepley of Facilities Planning joined Saner and Steve Tolle, supervisor of Ras Tanura's Curriculum Group, as full-time members of the team in July. In early August, Ron Visconti, T&CD Policy and Planning Coordinator; Salih Al-Ghamdi, assistant principal, Dhahran College Preparatory Center; and Muhammad Mater of the Contracts Group moved into offices at the Jamjoom Building. Others who were with the team for shorter periods of time included Mohammad Mahjoub, supervisor of the Academic Curriculum Unit, and Khalid Shehabi, head of the Contracts Group.

In late July the Saudi Aramco Integration Task Force ordered T&CD to implement academic placement testing and general aptitude testing for all Samarec employees on an "expedited basis." It was a tall order. Ali M. Dialdin, general manager of Training & Career Development, called it the Training Organization's "biggest undertaking ever."

"These [Samarec] men had all been initially trained to one standard, and now, as employees of a new company, they would have to be exposed to another," Dialdin said.

"We had to discover the vocational training and development needs of these new employees, and we had only a short time in which to work, since most of these men had ongoing responsibilities in up-and-running facilities."

Fouad Dagher, supervisor of the Program Development and Evaluation Division's Testing and Evaluation Unit, arrived in Jiddah in early August 1993 with a

Fouad Dagher

staff of eight to oversee the testing of Samarec employees. Tests were to be administered to all of Samarec's Saudi personnel below the rank of manager — a level comparable to a Saudi Aramco superintendent. The field included about 8,200 industrial employees, 190 apprentices and 1,620 college-degreed professionals. Apprentices and Samarec employees with a high school education or less would be given the same series of tests taken by all nonprofessional Saudi Aramco employees. This series was composed of the General Aptitude Test Battery (GATB), designed to indicate a person's mental abilities; the Vocational Interest Test (VIT), to pinpoint particular interests; and the company's English and math placement tests. Samarec employees with a college degree would take only one test, the English-language placement test for PELP (Professional English Language Program) participants.

The integration of some 10,000 Samarec employees awaited the results of these tests. "They were counting on us," Dagher said, "because they could not start training without placements" to indicate where Samarec people would fit into the Saudi Aramco system. Each of the 8,000-plus Samarec men who did not have a college degree spent two days taking tests. The GATB and the VIT were given the first day, the English and math placement tests on the second. Large groups of 180 to 200 people were tested at the same time and in the same place. At Jiddah, the training center's auditorium became a testing room. In Riyadh, test-takers sat at desks in the long hallway outside the training center's classrooms. College-degreed employees were tested separately in smaller groups. It took a total of about 9½ hours to complete the testing of nonprofessionals, and about one hour, 40 minutes to do the placement testing of college graduates.

Most tests were scored on site by test administrators. The results were sent to Dhahran for entry into T&CD's Testing and Evaluation Database, a tedious job requiring entries in 25 to 35 separate data fields for each individual. Every record contained the name, badge number and years of schooling of the individual, plus his test results. The GATB alone produced 12 test scores, each one of which had to be entered in a separate data field. Remarkably, test results were usually available to the appropriate Training Center and to the employee's organization within 24 hours after testing was completed. "People worked far into the night to give trainers something to work with," Dagher said.

Minimum passing test scores were not established, but Saudi Aramco required most of its employees to complete level four in English and math before they began training for a specific job. Graduates of English level four were expected to have a vocabulary of about 3,300 technical and nontechnical terms, be able to read and write simple messages, understand instructions and hold conversations in English on familiar topics. Math level-four graduates would be able to add, subtract, multiply and divide whole numbers; work with decimals,

Dale Saner

fractions and mixed numbers; find percentages and square roots; read graphs and solve basic geometry problems. Test scores for college graduates were at four placement levels — A, B, C and D — with level C being the level all Saudi Aramco college graduates were expected to reach. Level D was equivalent to advanced English required for some job classifications.

By August 25, three weeks after Dagher's group arrived in Jiddah, the first 1,000 Samarec employees had been tested. The results were surprising. According to Saudi Aramco's standards, more than 90 percent of the nonprofessionals tested required more English-language training, and 85 percent needed further instruction in math. Fifty-five percent placed at English level one, the beginner's level. Among Samarec's college-degreed personnel, 74 percent would need more English-language training; almost one-third tested at the beginner's level in English.

Saner, in his weekly report to the full Integration Task Force, declared: "Based on English placement results to date and on discussions with refinery and distribution managers, there is a very significant training workload ahead for both academic and job-skills programs." He estimated it would take three to five years to raise the average nonprofessional Samarec industrial worker to the English and math levels required for his job in Saudi Aramco.

Saudi Aramco Replaces Samarec Training

Even before the first test results were in, Saudi Aramco's executive management decided to replace Samarec training programs and the majority of Samarec training personnel with Saudi Aramco training programs and personnel. The decision was based on the recommendations of Ali Dialdin and Ali H. Saleh, vice president of Employee Relations and Training. They agreed with the T&CD work group's finding that "Saudi Aramco training programs were of superior quality" to those of Samarec. Nearly all the 150 contractors working for Samarec's training organization were to be let go. Regular employees of the training organization who were not considered qualified trainers by Saudi Aramco standards would be assigned another type of work. Saudi Aramco teachers were to move into Samarec's training facilities and begin classes by September 4, the start-up date for the new semester at training centers in the Eastern Province.

To implement these orders, a new training department, the Central Region/Western Region Training Department (CR/WR), was created. Zubair Al-Qadi was named director of the new department. CR/WR covered an enormous area, spanning the Kingdom from northern border to southern border and from the Red Sea east to Riyadh. The new department had four divisions: the Jiddah, Riyadh and Yanbu' training divisions, plus a Career Development Division headquartered at Jiddah. (Samarec's former Dammam Training Center came under the supervision of Central Area Training in Dhahran.) Ahmad Ajarimah, who had been superintendent of the Academic Training Division in Dhahran, became the first superintendent of the Jiddah Training Division. F.S. Al-Aboudi was named superintendent of the Riyadh Division, with Bill Granstedt as assistant superintendent. Saner served as superintendent of the Yanbu' division for the first few months, followed in December 1993 by Charlie Head, who had been assistant superintendent of Job Skills Training in Jiddah. Head's division not only inherited Samarec's training operations but also assumed responsibility for training activities and buildings formerly operated in Yanbu' by Saudi Aramco's Southern Area Manufacturing Division.

Al-Qadi remembers the department's early days as busy and somewhat disorganized. "We were coming to an area where we did not have a lot of facilities. We did not have housing, we did not have proper vehicles, we did not have

Trainees in an ITC chemistry lab.

a system set up, so it was chaotic for a period of time.

"Almost three-fourths of the Hyatt Regency Hotel was occupied by Saudi Aramco staff. We'd be working together throughout the day, and in the evening we'd go back to the hotel and talk business. In fact, we were in conference rooms sometimes up to 1 a.m. discussing business. It was really an incredible time. We had nothing but business going on 24 hours a day."

Saudi Aramco's Apprenticeship Program was suspended in June of 1993 for the duration of the Samarec training effort. A list was made of 47 Saudi Aramco ITC teachers who were to be transferred from the Eastern Province to the new CR/WR Department, 34 to the Jiddah Training Division and 13 to Riyadh. The list included 31 English teachers, among them four senior teachers of English and six PELP instructors. Twenty-four of the English teachers were assigned to Jiddah and seven to Riyadh. In addition, six math teachers were transferred to the new department, four being sent to Jiddah and two assigned to Riyadh. Yanbu' retained the same teaching staff as it had under Southern Area Manufacturing.

Ajarimah arrived in Jiddah from Dhahran on August 13, shortly after executive management announced its decision to apply the Saudi Aramco training system. As superintendent of Jiddah Training, his job was to organize the staff, establish training programs and get classrooms ready for the new semester, scheduled to begin in less than three weeks.

"We worked day and night," Ajarimah said. "I would go to the training center before the morning prayer [at dawn]. The security guards had to open the gates for me. We'd work until six or so in the evening, then we'd go back to the hotel and work until midnight.

The first class for Samarec employees was held at the Jiddah Training Center on August 22, just nine days after Ajarimah and his team began work. It was a Professional English Language Program (PELP) class enrolling 60 college-degreed professionals for intensive English-language training. To set up a program in so short a time was, Ali Dialdin said, a "remarkable achievement."

There was still a race to prepare for a much larger group of trainees who would be reporting for classes at the start of the new semester on September 4. They would not be young apprentices or new-hires, as trainees typically were in the Eastern Province. Instead, the majority of those targeted for training were veteran Samarec employees, some of whom resisted being assigned to training.

The Central Region/Western Region Training Department officially came into existence September 1, 1993, and classes began as scheduled at Jiddah, Riyadh and Yanbu' on September 4. The training centers used the intensive, all-day training format that had proved so successful in Saudi Aramco's Apprenticeship Program. Trainees studied a single subject all day. No part-time academic training was offered. All training was devoted to English-language instruction, mostly at the beginning level. Enrollment on the first day of classes for the new department totaled 391 trainees — 191 at Jiddah, 139 at Riyadh and 61 at Yanbu'. Classes began in Dammam on September 15 with 120 trainees.

By year's end, total enrollment in the CR/WR department had climbed to 1,565, including 245 college graduates, 185 apprentices and 1,135 nondegreed employees. Any concerns about the abilities or desire of Samarec employees to

Abdallah
S. Jum'ah

"You earn a lot of goodwill if you know something about the history of the place, the politics of the place."

bdallah S. Jum'ah was the first president and chief executive officer of Saudi Aramco who had not been an engineer or geologist. His major field of study at the American University of Beirut was political science. It was the type of background needed to lead the company into a new era of growth.

When Jum'ah became president in 1995, Saudi Aramco had evolved from a production company with few business interests outside Saudi Arabia into an international, integrated oil company in partnership with firms in Asia, Europe and North America.

Jum'ah moved into the company's top position from the post of senior vice president of International Operations, an office he had occupied since that business line was created in 1991. He brought with him ideas on new directions for the company's training programs.

"We are interacting with Koreans, Japanese, Greeks, Chinese, South Americans, people of many lands," he said. "Training should involve language capabilities, understanding of other peoples, appreciation of other cultures.

"We are looking for people with an international perspective, people who are business oriented. Understanding international issues is very, very important today. You earn a lot of goodwill if you know something about the history of the place, the politics of the place. You can make smarter business decisions if you understand what's happening in the country."

Jum'ah, the son of a pearldiver, grew up in al-Khobar, in the shadow of Aramco's Dhahran headquarters. His family was among the first 100 settlers in al-Khobar, then a cluster of homes on a pristine beach along the Arabian Gulf. After graduating from Dammam High School, Jum'ah went to the American University in Cairo (AUC) under one of the 60 college scholarships annually funded by Aramco. The company financed the scholarships, but, at that time, the AUC and the American University of Beirut (AUB) selected the recipients. Jum'ah spent his freshman and sophomore years at AUC, where he completed an extensive English-language program.

"I know when I was interviewed in high school and asked what I wanted to do, I said I wanted to go into medicine. But things didn't work that way. I think getting into Cairo during the political turmoil of the 1960s made political science an interesting field."

Political problems prevented him from getting a visa to Egypt for his junior year at AUC, so he transferred to the American University of Beirut. When Jum'ah returned to al-Khobar after graduation from AUB in 1968, he had a choice either to attend the University of Petroleum and Minerals or go to work for Aramco. He settled on Aramco.

From then on it was an upward climb. Along the way he completed the company's Management Training courses and the Program for Management Development at Harvard University. Jum'ah's march up the job ladder included positions as general supervisor of the Current Affairs Division for Government Affairs in 1970; an assignment as manager of Power Systems Operations for Saudi Consolidated Electric Company (SCECO) in 1979; vice president for Power Systems and managing director of SCECO; then senior vice president for Saudi Aramco Industrial Relations; senior vice president of the new International Operations business line in 1991; and, finally, Saudi Aramco president and chief executive officer in August 1995.

accept training were quickly dispelled. In general, they were outstanding students. Eighty-nine percent got passing grades, and 65 percent earned As or Bs during the first semester.

A Job Well Done

In the 1993 year-end Accountability Report, Ali Dialdin praised those who participated in the merger effort. "Judged by the most stringent standards of excellence," he wrote, "these accomplishments during the period from August to December 1993 were outstanding."

Many out-of-the-ordinary deeds were performed during that time. Most of them involved men who worked much longer hours than normal. Ahmad Ajarimah not only gave of his time but put his money on the line as well. It happened one weekend when there was an error in reservations for Saudi Aramco flights from Dhahran to Jiddah. Trainers who had come east for the weekend were told no seats were available on Saudi Aramco flights that would get them back to Jiddah in time for classes on Saturday. So Ajarimah used his personal credit card to purchase tickets on a Saudia Arabian Airlines flight from Dhahran to Jiddah for 32 teachers and trainers at a cost of about SR8,000 ($2,130). His actions were greatly appreciated by his superiors, since, without the teachers, 400 to 500 students would have been dismissed from class, creating mass confusion. He was, of course, reimbursed by the company in a timely manner.

Using his own money to buy tickets and worrying later about being reimbursed was not "something I would have done in ordinary times," Ajarimah said. "But when you are interested in getting a mission accomplished, then you have to think of all conceivable alternatives before you say it can't be done."

Ahmad Ajarimah

Company-wide, 7,700 employees enrolled in training programs during 1993, an increase of only 350 over the previous year. The shutdown of the Apprenticeship Program largely offset the enrollment increases brought on by the Samarec merger. Training administered a total of nearly 32,000 placement and aptitude tests and printed and processed more than 186,000 tests during the year. The department also distributed 152,000 textbooks to nine training centers, a record for one year. Twenty-four Samarec trainers had been retained — 13 job skills trainers and 11 from the academic staff.

Samarec was declared fully integrated with Saudi Aramco on January 1, 1994. The test scores and academic records of former Samarec employees were entered into Saudi Aramco's computer files. The names of nearly 10,000 employees were moved from Samarec to Saudi Aramco identification numbers. The two companies merged into one seamless entity.

The integration, followed in 1994 by the addition of facilities at the Rabigh Refinery, north of Jiddah, more than doubled the number of Training locations. In two years, Training had expanded from five main locations, all in the Eastern Province, to 10 main locations and seven training satellites scattered throughout the Kingdom. Training acquired about 200 more classrooms and shops, bringing the organization's total to nearly 730 classrooms and shops.

In the rush to staff these new facilities, Training had turned to contractors instead of directly hiring employees. The contractors were expatriates recruited by independent employment agencies and assigned to positions with Saudi Aramco. The company saved the time and effort required to locate, recruit, process and bring its own employees to Saudi Arabia. Training took in 150 contractors in 1993, compared to only 30 the previous year. Between 1992 and 1996, the ratio of contractors to regular employees in the Training Organization increased from three percent to 30 percent.

The merger effort left those who had been involved in it with a feeling of a job well done. Bill Valbracht, superintendent of the Training Support and Quality Assurance Division, described the general reaction to the Samarec challenge:

"It was really a case of employees responding in a very positive way. Management told them, 'We need to get set up for training. Make it happen.' And people did all kinds of wonderful things — arranging to have things airlifted over to them in Jiddah, trucks loaded with furniture and other supplies going to schools there. It was just a great success story. I think it was one of the finer moments for T&CD."

First Attitude Survey Conducted

Ali Dialdin felt the time was right to go ahead with a project he had been thinking about for some time: the first-ever system-wide survey designed to measure the attitude of Training Organization employees toward their jobs and the department. In mid-1994 Dialdin proposed such a survey. The idea was approved at a meeting of Training Department directors chaired by Ali H. Twairqi, then acting general manager of T&CD. Twairqi cautioned the directors that the survey results could not be shrugged off, that management must be prepared to respond to issues raised by the survey. Valbracht, an industrial psychologist, oversaw the design of the survey document. The Organization and Industrial Engineering Department distributed it and tallied the results.

The T&CD survey with 91 questions was distributed to both regular department employees and contractors in the late summer of 1994. The results were announced in December 1994. Of the 1,400 employees surveyed, 599, or 43 percent, responded. The responses indicated several problems, chief among them lack of upward communication, overly rigid adherence to rules, refusal to acknowledge problems and resistance to change.

"These were troubling results," Valbracht said. "They fell mostly in the area of communication. We were really poor in that area. Along with communication comes participation. If you can't communicate, you can't participate. The Samarec integration showed us what real employee participation could do — how good people could feel about being a Saudi Aramco employee."

Training's top management discussed the survey results with employees during unprecedented face-to-face communication meetings in the local communities. Those representing management at the first of these meetings on January 2, 1995, in the Ras Tanura Training Center were: M. Yusof Rafie, vice president, Employee Relations & Training; Ali Twairqi, acting general manager; Steve Hardy, director of the Job Skills Training Department; and Saeed M. Al-Ghamdi, director of the Academic Training Department. They answered questions about a variety of issues, including eight-hour teaching schedules, overtime, salary, vacations and rotational assignments. Rafie assured the teachers and trainers of management's willingness to consider all their concerns.

The survey revealed that some supervisors had withheld information they were supposed to share with employees. Ali Twairqi advised all supervisors in written directives that they must give employees access to the Saudi Aramco General Instructions, the Training Operations Instruction Manual, and (for Saudi Aramco regular employees only) the Industrial Relations Manual. "We were shocked to find this information being kept with the supervisors and not shared," Valbracht said. The directives also reiterated that supervisors must meet annually with each employee as part of the employee's performance appraisal process. Employees had to be informed of their performance category and given a copy of the appraisal form once they signed it.

M. Yusof Rafie

239

Total Quality Management

n the aftermath of the survey, Dialdin, newly returned from a leave taken for health reasons, instituted a new system for dealing with work-related problems. The system was based on a seven-step process created by Organizational Dynamics Inc., a leading U.S. consultant in Total Quality Management (TQM). TQM was an outgrowth of the system W. Edwards Deming and Joseph M. Juran introduced to Japan in the late 1940s with spectacular results. It sought to involve employees at all levels in making improvements and solving problems. Management Training conducted its first Quality Improvement Team Facilitator workshop at Ras Tanura in March 1995. The 24 participants, representing five Training departments (Academic Training, Career Development, Job Skills Training, Program Development & Evaluation, and Training Support & Quality Assurance), learned a seven-step process for group problem solving. They, in turn, were supposed to help their local Quality Improvement Teams handle issues such as those raised by the survey.

Ahmad Ajarimah, superintendent of Academic Training in the Central Area, considered the formation of Quality Improvement Teams to be a by-product of the Samarec integration effort. As the first head of Academic Training in Jiddah, Ajarimah was on the front lines of the integration process before he moved back to Dhahran in 1994.

"After the integration, I saw more effort on the part of first-line supervisors to show a sense of caring for their employees, more than I saw in the first 17 years of my service with the company. There was for many years supervision modeled after a generation of supervisors and managers who were not very considerate and caring. I think this was a significant change in organizational climate.

"There was also more awareness of providing service to the user and taking the user into account. You are providing a service to a client, almost," he said. As examples of what he meant by more service-oriented training, Ajarimah cited a six-week English course specially designed for grade code five and six security guards, and English courses created for bulk-plant operators at former Samarec facilities.

"Samarec was the springboard for these changes," he said. "It took us out of our normal comfort zone. It worked as a catalyst, I think."

Job Skills Training Centers Accredited

n the midst of the integration effort, but unrelated to it, T&CD's five Job Skills Training Centers in the Eastern Province received a singular honor. In September 1993 the Job Skills Centers became the first industrial training centers in the Middle East to be accredited by the Accrediting Council for Continuing Education and Training (ACCET), a U.S.-based organization recognized by the U.S. Department of Labor. After reviewing the report of the team

The 1990 class of Ras Tanura Job Skills Training graduates.

**Ali I.
Al-Naimi**

*"The biggest
drive was the
desire for
education
instilled in me
by my older
brother."*

li I. Al-Naimi's career could have come out of the pages of an inspirational novel. He came to Aramco as an illiterate Bedouin boy and advanced to president and chief executive officer of the company. Then he left Saudi Aramco to become Minister of Petroleum and Mineral Resources for Saudi Arabia, the largest oil producing country in the world.

How did Ali Al-Naimi achieve such success? "I think I know," Al-Naimi told an interviewer. "I believe the reason one gets where one finally winds up is outlook, how he goes about doing things. Maybe in my case it was luck, maybe just timing, but the one thing I believe is to really do your best and not worry about your next promotion."

Al-Naimi's older brother, 'Abd Allah, took Ali to the Jabal School for the first time when Ali was only nine years old. "My brother convinced me to commit my life to getting an education."

So insistent was 'Abd Allah about his younger brother's education that once, when Ali got in trouble and was told to leave class, 'Abd Allah had him sit outside the window to listen to the lecture. 'Abd Allah died at age 17. Ali Al-Naimi took over his job as an office boy and continued to attend the Jabal School, just as 'Abd Allah would have wanted. Fifty years later he could still name some of his instructors, people like Jerry Dunbar, Vince James, Bud Krueger, 'Abd al-Hafiz Nawwab, Powell Ownby, Marvin Shippey and Helen Stanwood.

In 1947 Al-Naimi was hired as a junior clerk in Aramco's Personnel Department, where he quickly learned typing and shorthand. "I had many opportunities in those days to make more money," Al-Naimi said. "I insisted on foregoing them for the sake of education. My biggest drive was the desire for education instilled in me by my older brother. The turning point in my career was 1953. I was sent on a summer program to the American University of Beirut. I got exposed to math, physics and chemistry. I found it to be very, very fascinating. I discovered an ability that I didn't know I had."

Shortly after returning from AUB, Al-Naimi visited the assistant general manager of Exploration. "I told him I wanted to become a geologist. He said, 'Why do you want to be a geologist?' I smiled and said, 'Because I want to be president of the company.' I said it facetiously, but there was truth in it. I knew that all the presidents before that time were geologists. Actually, I realized later that I liked geology because I like being outdoors, and I still do."

For the next 10 years the company kept Al-Naimi in school virtually full time. He earned a high-school diploma from International College in Beirut in 1957 and attended AUB for the following two years. In 1959 he was awarded a company scholarship and was sent to the United States, where he earned a bachelor's degree in geology at Lehigh University in 1962 and a master's degree in geology from Stanford University in 1963.

Thereafter, Al-Naimi climbed steadily, rising in just 10 years from foreman in 1968 to senior vice president.

His training had not stopped. Al-Naimi completed the Executive Program in Business Administration at Columbia University in 1974, and the Advanced Management Program at Harvard University in 1979.

Al-Naimi was elected to the Aramco Board of Directors in July 1980 and was named executive vice president in 1982. On January 1, 1984, he became the first Saudi president of Aramco and was made president and chief executive officer of Saudi Aramco in 1988. In August 1995, King Fahd appointed Al-Naimi as Minister of Petroleum and Mineral Resources for Saudi Arabia.

"I am a product of Aramco training," Al-Naimi said in a 1986 speech to the American Businessmen's Association. He could have paid no higher tribute to the company's Training Organization.

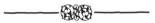

that had evaluated the company's job skills training, ACCET president Roger Williams wrote: "The high number of standards rated superior by the on-site evaluation team was, in my experience, as unprecedented as it is indicative of the high goals to which Saudi Aramco aspires." The original ACCET accreditation was for three years. In 1996, the ACCET renewed its accreditation for five years, the maximum allowable under council rules.

Ali Twairqi becomes General Manager of T&CD

On September 15, 1996, due to health reasons, Ali Dialdin, general manager of Training & Career Development Department, moved to an advisory capacity as head of a Human Resources Task Force. He had presided over the Training operation for 12 years and had headed T&CD in the previous 10 years. He had become a well-known and internationally respected figure in the profession of human resource development.

Ali Twairqi, director of Training's Career Development Department, succeeded Ali Dialdin as general manager of Training & Career Development

Ali Twairqi

(T&CD). He took over one of the largest and most diverse training organizations in the world. T&CD had a staff of 900 teachers and trainers and 600 support personnel. The organization operated 26 different types of training programs, with a total enrollment of 9,400 employees. The two largest programs — job-related academic training and job skills training — enrolled more than 8,000 Saudis, all of them attending classes for at least four hours a day.

In addition, the Apprenticeship Program, suspended during the Samarec integration, resumed in the fall of 1996 with about 1,000 candidates. T&CD was also responsible for 1,000 Saudis who were either attending college on company scholarships or being readied for college in Training's year-long College Preparatory Program. An additional 850 recent college graduates were getting work experience through T&CD's three-year-long Professional Development Program. Training was also responsible for arranging short-term assignments at schools both in Saudi Arabia and out-of-Kingdom. Saudi Aramco trainees went to virtually every country in the English-speaking world, including the United States, Canada, Great Britain, New Zealand and Australia.

T&CD operated training facilities with 720 classrooms and shops occupying a total of about 1.5 million square feet. These included 10 full-time and seven part-time training centers spaced throughout Saudi Arabia, an area about equal in size to the United States east of the Mississippi River. Six full-time training centers were located in the Eastern Province of Saudi Arabia, one at Riyadh in the central part of the Kingdom, and centers at Jiddah, Yanbu' and Rabigh on the Red Sea coast in western Saudi Arabia. T&CD also operated 10 libraries, known as Training Resource Centers, with a total of about 60,000 books plus current magazines and newspapers. The library catalogue included more than 3,000 training books and manuals published by Saudi Aramco, most of them created since 1980.

Ali Twairqi had a unique perspective on training. He was the first head of the department to have experienced the company's industrial training system from the trainee's point of view. He joined Aramco in 1966 as an eighth-grade dropout. As a new Aramco recruit, he was required to work half a day and attend training classes for half a day. His first job was collecting tickets for the movie at the Abqaiq theater and sweeping up after the movie ended. He found Aramco's training classes challenging, and he responded to the challenge. Twairqi progressed through the various industrial training levels, earned one of the company's coveted out-of-King-dom scholarships and, following graduation from Penn Central High School in Philadelphia, received a degree in business administration at Sam Houston State University (Texas) in 1978.

"When I joined the company, Saudi schools did not offer enough programs so that students could be assimilated easily into the company. It was a very small world, a very condensed environment. It had nothing in it related to an oil company. So the company took responsibility for bridging the gap between what the schools offered and what the company required," Twairqi said.

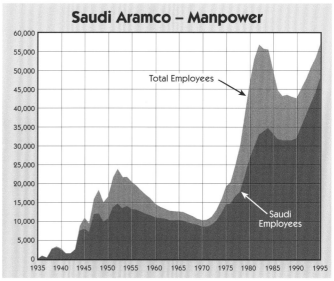

"Today the high school graduate is much better prepared in both math and science. He has a lot of general knowledge of world issues. Invariably, he has traveled. At the same time, Saudi Aramco's business has become much more complex. It has become fast-paced, computerized, digitized. The linkage now is with a worldwide network. We want to continue to be the intermediary between what is happening outside of Saudi Aramco and what the company requires. That's a very fluid, very dynamic process that we have to continually monitor and maintain."

Saudis Reach Record Employment Levels at Saudi Aramco

By 1996 the Saudization of the company's work force had reached one of its highest levels in company history. About 80 percent of the nearly 58,000 Saudi Aramco employees were Saudis. It was the highest ratio of Saudis to expatriates in 26 years, and the third highest since World War II. About 87 percent of the industrial work force was Saudi, compared to just over 50 percent in 1980. Saudis comprised 75 percent of the clerical work force. Saudis had made especially notable progress in the grade codes 11 to 14 portion of the work force — the college-trained professionals and highly skilled industrial workers. By 1996 this key segment of the work force was more than 59 percent Saudi, compared to only seven percent Saudi in 1980. Eighty percent of the company's department heads and 70 percent of the supervisors were Saudis. Saudis held 28 of the 32 top offices in the company, compared to 18 of the 31 top positions in 1984. The president and chief executive officer of Saudi Aramco, the three executive vice presidents, all five senior vice presidents and each of the company's 15 vice presidents were Saudis.

Harry Snyder's prediction, which had seemed so farfetched 50 years before, had come true. The largest oil company in the world was now managed and operated by Saudis. Training had made it possible. The great by-product of oil proved to be knowledge, a never-ending resource.

Saudi Aramco Training Locations

1994 represents the greatest geographic extent of training activities to date.

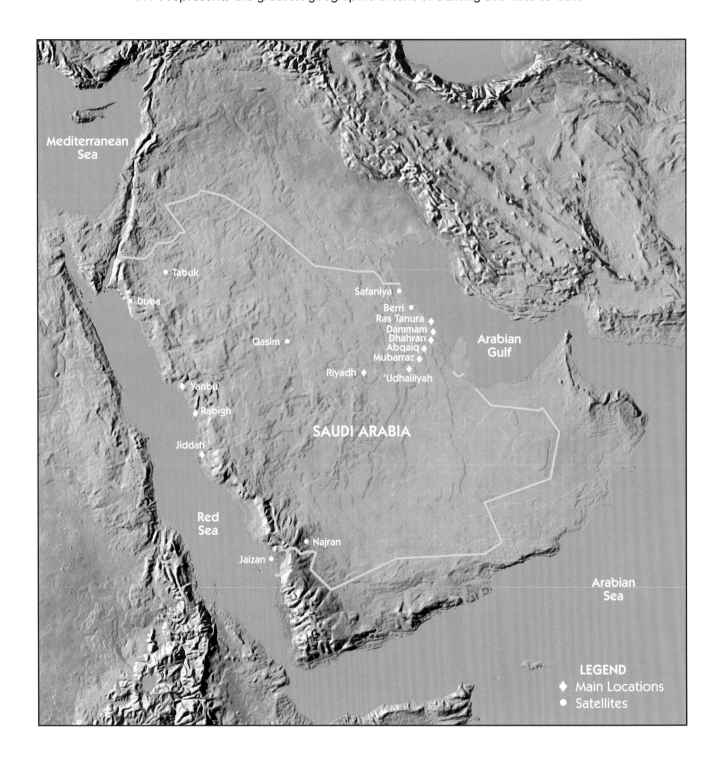

Mediterranean
Sea

Tabuk

Duba

Safaniya

Berri
Ras Tanura
Dammam
Dhahran
Abqaiq
Mubarraz

Qasim

Arabian
Gulf

Riyadh   'Udhailiyah

Yanbu'

Rabigh

SAUDI ARABIA

Jiddah

Red
Sea

Najran

Jaizan

Arabian
Sea

LEGEND
◆ Main Locations
● Satellites

Appendices

Appendix A
The History of Computers at Saudi Aramco

"We couldn't get them off [the computers]. They came during their lunch hour, and many even asked for permission to use the CAI lab after school hours."
– Bill Granstedt

The use of computers at Saudi Aramco mirrored the development of computer technology in the rest of the world. It began in the 1940s with the introduction of electrical tabulating machines. For some 40 years thereafter, computer use was limited to a relatively small group of company specialists. That changed with the introduction of desktop computers to Aramco in the early 1980s. In the next 15 years the number of computer stations in the company grew from a few hundred to 30,000, including 2,000 within the Training Organization.

Operators of the first electrical tabulating machines used a keyboard similar to a typewriter keyboard to punch holes in rectangular pieces of cardboard. A computer read the holes as letters, numbers or special characters. The Keypunch Segment of the Data Processing Operations Division produced punch cards for programs that executed tasks ranging from printing paychecks to tracking the cost and quantities of materials used by Aramco.

By 1968 the Data Processing Division was producing hundreds of thousands of cards a month. At the time, most keypunch operators were Saudis enrolled in third- and fourth-level English at the company's Industrial Training Centers.

The first mainframe computers purchased by Aramco for installation in Saudi Arabia arrived in Dhahran in the early 1950s. These bulky machines had as little as 40K of memory. The IBM 4,000-unit and 16,000-unit Model 1401 computers, acquired in the late 1950s, were used in processing the company payrolls, financial and cost-accounting systems, personnel statistics, and material supply records. Highly technical applications such as rendering of seismic data and reservoir simulation were done on computers at Aramco facilities in the United States, Great Britain and the Netherlands, or contracted out to specialized companies.

Early Use of Computers by Training

The computerization of training records began in the late 1950s. Instructors copied attendance records from index cards onto forms called "data entry rosters" and mailed a carbon copy of the rosters to Dhahran each month for entry in a computer file. At the end of the semester, final

Students tour Aramco's Data Processing Center, Dhahran, 1961.

grades were recorded on a similar form and submitted to Dhahran for computer entry. The resulting printouts, containing attendance records, test scores and final grades were returned to the Training Centers and stored in record books.

In September 1962, 'Abd al-Rahman Al-Barrak became the first Saudi employee to qualify as a computer operator. He had started working at Aramco in 1952 as a payroll clerk and transferred to Computer Processing in 1961. He qualified as an operator of both the 4000-unit and 16,000-unit IBM Model 1401 computers.

A new computer system capable of processing digitally recorded seismic data was flown to Dhahran aboard a chartered DC-6 in early 1966. The Texas Instruments Automatic Computer and Digital-Analog Readout System weighed nine tons and required a staff of 11 men to operate. The equipment was installed in a specially designed, air-conditioned and humidity-controlled 2,500-square-foot section of the Administration Building. To accommodate its many electrical cables, the floor of the computer room was raised eight inches and the equipment set on removable floor panels.

First Training Database Developed

The first computer system devoted entirely to Training was introduced in 1975. The Industrial Training (IT) system was created, operated and maintained by the Electronic Data Processing Department (EDPD). It managed data on former and current enrollees in the company's Industrial Training Centers and Industrial Training Shops, as well as those enrolled in driver training courses, management training courses and later, the Professional English Language Program. The IT system recorded academic course completions and test scores plus on-the-job training program completion. Data was keyed into the system by EDPD operators from information mailed to the department by Training Center registrars. EDPD mailed the printout back to the registrar.

Training Record Handling in the 1980s

By 1980 Aramco operated a centralized mainframe data center with 300 remote terminals, not yet enough to have much of an impact in a company of 50,000 employees. By one estimate, more than 70 percent of the company's information gathering and processing was still untouched by computer technology. This was certainly the case in the Training Organization. In the early 1980s, teachers and trainers were recording enrollments, attendance, test scores and course completions by hand on 5- by 7-inch index cards, much as they had been doing since the first company schools opened some 40 years earlier. With ITC/ITS enrollments totaling nearly 15,000, record keeping was a daunting chore.

Alice Sealy, then in academic training at Abqaiq, remembered: "We'd come in every semester, mainly during the summer [when school was not in session], and do mounds and mounds of work getting enrollment cards ready. We'd use a couple of classrooms, and we'd take stacks of cards, books and records, and we'd go in and do all this checking and cross-checking to make sure a trainee was eligible for the courses his department wanted him to take."

Once the enrollment requests were cleared, a second batch of 5- by 7-inch cards was created for use in recording attendance. Teachers needed an attendance card for every student in each class. Since many students enrolled in more than one class and had several different teachers, multiple attendance cards often had to be prepared for the same student. There were two types of attendance cards: white cards for regular trainees and salmon-colored (pale orange) cards for shift workers who were paid overtime to attend classes after their regular working hours.

"We'd bring in all hands who were back from vacation to work on the attendance cards," Sealy said. "We would spend weeks ahead of a semester start-up getting these cards ready for the day classes began."

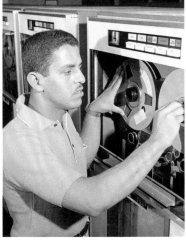

'Abd al-Rahman Al-Barrak was the first Saudi computer operator, 1962.

Microfilming relieved a paper-storage problem, 1970.

After the semester began, a third Training record was started, this time using large sheets of white paper with brown lettering called Data Trac sheets. The student's identification number, status and other data were recorded by hand and the sheets delivered to keypunch operators for entry into the IT computer system. At the end of the semester, the student's grades and training status were recorded on the Data Trac sheet and sent to Dhahran for entry into the IT system. A carbon copy of the Data Trac sheet was retained for Training's files. The information entered into the IT system was printed out on wide green-and-white-striped sheets with perforated edges known as "green sheets." These sheets were stored in ledger books in the Training Center library and referenced to determine the eligibility of students for classes in the next semester.

"We dealt with a lot of paperwork and cards during the semester," Sealy said. For each student in every semester there were at least two sets of index cards, two Data Trac sheets, and a "green sheet" from the IT system. A minimum of five different record sheets per student multiplied by some 15,000 students equals at least 75,000 new record sheets created each semester and stored for possible later reference.

Other departments had similar paper problems. Materials Supply, for example, utilized order books that took up as much space as several office desks. The situation improved in the 1970s when the company began transferring paper records to microfilm. By 1981 there were 42 information retrieval systems containing more than 16 million pages of microfilmed information — almost 300 pages per employee.

The sheer weight of the paperwork was not the only problem with Training's index card system. For instance, Sealy recalled, "If you reached the end of the semester with more names or fewer names than you started with, you had to find out why. When a trainee was transferred during the semester, he would come to the office and take his enrollment card and attendance card with him to his new location. We had a lot of problems tracking down where these trainees ended up."

TRAC: New Training Database

In 1980 the EDPD's Computer Applications Division developed a new trainee tracking system for Southern Area Maintenance employees. The system was called TRAC, an acronym for Training Reporting, Analysis and Certification. TRAC recorded the progress of trainees for some 400 to 500 jobs within the Southern Area. It generated eligibility lists for pay raises, class enrollments, graduations, ITC/ITS grades, absences and on-the-job training performance. In 1982 the Training Organization took over TRAC and expanded it to cover some 5,000 trainees in several different organizations, among them Southern Area Producing, Finance, Pipelines and the Shedgum Gas Plant.

TRAC provided an introduction to computers for Dale Saner, then newly arrived at Aramco from the National Iranian Oil Company and working for the manager of Maintenance in Southern Area Oil Operations. Saner was asked to write the specifications for job descriptions in an expanded TRAC system. The Abqaiq Computer Applications Group was assigned to do the programming. Saner generated piles of paper, trying to devise a weighted formula to determine when, according to the company's policies, an employee would become eligible for a step up the job progress ladder.

"I had all these matrixes that I was doing by hand when I met a computer programmer named Michael Joye," Saner said. "He [Joye] had just bought a TRS 80 [Tandy Radio Shack 80] computer, and he had the original spreadsheet — VisiCalc, I think it was called. He showed me how to set up a spreadsheet. I thought that was the most amazing thing in the world. I'd change a number and just click, and the computer would go through and calculate 25 different versions of the formula. Spreadsheets are what sold computers to me. Other than that, computers were just word processors like electric typewriters."

One result of this exercise was the development of a mathematical formula that generated a number known as the "Y" value. It was a distinguishing mark of the TRAC system. The "Y" value was supposed to pinpoint employees who were ready for promotions and merit increases and to flag those trainees who were not progressing satisfactorily.

TRAC and the IT system had serious drawbacks. In both systems, data provided by Training had to be given to operators for computer entry, increasing the chance for error. TRAC records extended back only 12 months, and TRAC could not access Saudi Development planning files. Although the IT system was more comprehensive, it was seldom up-to-date, and it did not capture most on-the-job training course completions and work history. The IT file was also difficult to modify to meet changing requirements. For such reasons, both TRAC and IT were considered interim systems, of use only until a more complete Business Line tracking system could be developed.

EXPEC Computer Center Opens

A summer student runs a 16-color graphic program, 1984.

In late 1981 the company's overseas computer operations were consolidated in a newly built computer center in Dhahran. The center was housed in a three-story, 74,000-square-foot annex to the Exploration and Petroleum Engineering Center (EXPEC) then under construction in Dhahran. Ground had been broken for the EXPEC building and computer annex in June 1980. The computer center opened in December 1981, several months ahead of the EXPEC building. The new center housed some of the

most sophisticated electronic equipment available, including advanced reservoir simulators, precision geological map plotters and three of the largest computers then made — the IBM 3033s. The opening of the EXPEC Computer Center brought to Dhahran more than 250 computer operators, programmers and technicians, by far the largest contingent of computer experts assembled by one company in Saudi Arabia.

The company's first in-Kingdom computer training program began that same year. The six-month-long course, called Computer Fundamentals, was designed for recent college graduates who would be working as computer analysts and programmers. The course was taught by Computer Services, a branch of the company's Computer Communications & Office Services Organization (CC&OS).

Computer Advisory Committee (CAC) Established

The major impetus for introducing computers into the workaday world of Aramco came in early 1983 when the company established a Computer Advisory Committee (CAC) under the chairmanship of Ali Al-Naimi, then executive vice president of Operations. CAC's mission, to quote Aramco's 1983 Annual Report, was "to take the power of the computer resource outward to Aramco's business units and provide organizations and individuals with the means to exploit and control this productive tool." By year's end, the company's mainframe computer system had been expanded to a total of 1,370 terminals, more than four times the number of terminals available just three years earlier. The enlarged mainframe system was capable of responding to anywhere from 250 to 270 users at a time. But CAC's attention did not focus as much on the mainframe system as it did on the advent of a new type of computer.

The Advent of the Personal Computer

Computers designed for individual users at a single station first became available in the late 1970s. These self-contained machines, originally called microcomputers, became known as personal computers (PCs) or desktop computers. By whatever name, they revolutionized the field. Desktop computers had many advantages over computers tied to the mainframe. For instance, at Aramco, operators on the mainframe needed a security clearance because they had access to confidential information stored in the company's central computer. Users of desktop computers had access only to information entered into their own machines, not to data in the mainframe system, so there was no need to run a security check on every employee or trainee who used a microcomputer. In addition, a much greater variety of software was available for desktop computers than for mainframe computers, and desktop software was much cheaper than its mainframe equivalent. Installing new software for a desktop machine might cost several hundred dollars, while the cost of installing a new program on a mainframe computer could run into hundreds of thousands of dollars. The large mainframe computers were, of course, many times more expensive to purchase and maintain than desktop computers.

The first desktop computers purchased by Aramco were astonishingly expensive compared to what they would cost only a few months later. One desktop configuration, acquired by the Job Skills Training Department in Dhahran in 1982, was comprised of:

- An Apple II Computer
- A black-and-white monitor
- 2 Apple 5¼" disk drives
- 1 Vista Dual 8" floppy drive (500 KB each)
- 64K of memory (K = 1024)
- 1 Daisywheel printer.

The cost of this package, purchased without software through a local vendor, was about $20,000 (SR75,000).

The high cost of microcomputers discouraged Training's management from purchasing very many of the new machines at first, but that didn't necessarily keep them from multiplying in the workplace. Some employees brought their own computers to work. Others ordered computer parts mailed from the States at company expense and assembled the parts into computers on their arrival in Saudi

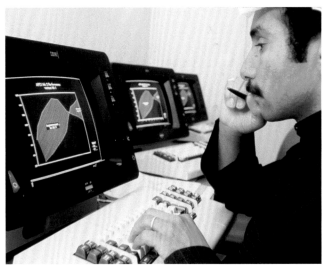

Trainee studies a computer-generated performance curve.

Arabia. In one case, computer parts were delivered under the name of "light-weight, temperature-sensitive equipment." Such chicanery ceased soon after IBM personal computers and IBM "clones" were introduced, driving prices sharply down.

Professional Employee Records Computerized

In 1980 a Career Development Department computer programmer, John Boughner, and a Saudi Development advisor, Larry Cantwell, plus a cadre of data entry clerks, began to compile the records of professional employees in a mainframe computer file. They created the Career Development Grade Codes 11-14 Review System and loaded it with information on each individual's educational background, management and professional education courses, prior work experience (both Aramco and non-Aramco), departmental appraisals, career path, promotability forecast and immediate and potential job targets. By 1982, when the Career Development Department merged with the Training Department, the GC11-14 system contained the records of several thousand employees. "They told us who was promotable and how soon, and what should be the job and job targets to get the individual up to speed," Boughner said. Using information from the system, Career Development tracked the progress of Saudi professionals and regularly contacted department training coordinators to ask if an individual was following his prescribed development plan, and if not, why not.

Computers Utilized in Training

During 1983, desktop computers began to appear in significant numbers within the Training & Career Development Organization. That was also the year computers were first introduced into T&CD classrooms, the year the first dedicated word processors were delivered to curriculum writers, and the year in which four new computerized trainee-tracking systems were initiated, reducing the paperwork involved in keeping training records.

In early 1983, Training established Computer Assisted Instruction (CAI) centers at Dhahran and Ras Tanura for participants in the Professional English Language Program (PELP). Eight Apple II-Plus computers were set up in the Professional English Language Center at the Dhahran ITC (Building 552). Ten Apple IIs were installed in a PELP classroom at the Ras Tanura ITC. PELP participants were given two hour-long introductory lessons on operating computers and using CAI programs. Then they were allowed to run CAI programs for one hour per week during study hall periods. These programs were intended mainly for remedial or acceleration purposes, but some instructors felt threatened by the new technology. T&CD's announcement at the opening of the CAI centers stressed that these programs were meant to supplement classroom instruction, not replace teachers.

Computer-Assisted-Instruction Lab at the 'Udhailiyah ITC, 1985.

The first CAI programs were limited to instruction in English reading and grammar. They used tutorials, drills and question-and-answer sessions to test comprehension. Questions were answered by pressing the letter key corresponding to one of the multiple-choice answers on the computer screen or by typing out simple one-word answers. Shelly Liberto, a teacher of English as a Second Language, and Phillip Markley, a contractor who had organized a CAI laboratory at Ohio State University, established pilot CAI programs at Dhahran and Ras Tanura starting in January 1983. Bob Lowry and Vincent Antonelli catalogued the CAI software, reviewed the software, placed it in the curriculum and maintained the Dhahran laboratory. Bill Granstedt, who set up the Ras Tanura CAI lab with Matt Wagner, was surprised at the way trainees took to computers. "We couldn't get them off [the computers]," he said. "They came during their lunch hour, and many even asked for permission to use the CAI lab after school hours." The response was so positive that by September of that year, Training had plans to make CAI available to all English, math and science students within five years.

New Tracking Systems Developed

During 1983, T&CD inaugurated four new, independently developed, trainee-tracking systems to go along with the older IT and TRAC systems. One of the new systems, the ITC/ITS Current Enrollment System (CES), validated enrollment requests, produced enrollment reports, generated attendance reports, and updated course completion data without the need to use Data Trac sheets. A second new system, called the Computerized Saudi Development Training Plans system, automatically created developmental plans for individual Saudi employees based on their ITC/ITS records. This system allowed Training to do away with the multiple-copy "green sheets," which were difficult to correct and used inordinate amounts of paper.

A third tracking system, the Manpower Forecasting Model, was upgraded from the mainframe to personal computers and expanded using Excel and Lotus 1-2-3. This system enabled line organizations to predict how many new employees they would require and what skills they would need for as long as 20 years into the future. In addition it gave managers an easy way to estimate the average years of experience in the future work force, a crucial consideration in planning for Saudization. The system used a manpower forecasting system first developed in 1978 by planning specialist John P. "Crif" Crawford and maintained by him since then. By using personal computers, Crawford was able to split the tracking system into two models, a professional model for the grade code 11-plus employees and a craft model covering the industrial work force. The versatile desktops also allowed users to illustrate their data with diagrams and graphs, something not easily accomplished on the main-frame system.

The fourth computerized Training system started that year, the College Student Tracking computer system, recorded the progress of employees in out-of-Kingdom (OOK) Training Programs in the United States. The data was compiled by Aramco Services Company in Houston and sent to the Career Development Department in Dhahran on magnetic computer tapes. The information was then shared with the Business Lines sponsoring OOK trainees. The system tracked more than 1,000 Aramco scholarship students at more than 100 colleges and universities in the U.S.

In 1983, Curriculum Development became one of the first organizations to receive company-ordered desktop computers. Eight Wang dedicated word processors were installed in the department during 1983, with the prediction that they would nearly triple the department's production capacity. In the beginning, data entry operators ran the Wangs. Curriculum writers continued using typewriters and turned their finished work over to the operators for computer entry. By the end of the year, there were about 75 computer stations within Training, including those in CAI computer labs, but it was only a trickle compared to the perceived need. A report to SAMCOM titled "Progress Report: Training and Career Development Computerization Activities" estimated T&CD would need an additional 1,500 computer units by 1988 at a total cost ranging between $3.75 and $7.5 million, depending on how much and how fast computer prices fell.

Training Develops Its Own Computer Expertise

As often as not, Training discovered the computer expertise it needed within Aramco rather than having to bring in experts from overseas. Chellakon Selvaraj, a typist and stenographer from Madras, India, was one such discovery. He joined Aramco as a clerk for the Materials Supply organization in 1977. After six months as a clerk, Selvaraj said, "I got bored. It was a routine job. Anybody could do it. I wanted something challenging. Materials Supply had a lot of computers, so I asked to learn computers." He began doing data entry and soon started writing simple computer programs. "Since I was so interested, so curious, they started teaching me on the job," he said. Soon he was taking computer courses offered by the Electronic Data Processing Department (EDP). "I took lots of in-house courses." In a year he became an assistant programmer, and in three years a full programmer working alongside expatriate consultants earning generous salaries.

In 1982 Training was looking for a computer programmer. Indira Goradia, a computer systems analyst in EDP, recommended Selvaraj. Hashim Budayr, a veteran trainer, was in charge of a project to develop a new trainee-tracking system. He decided to hire Selvaraj.

"Budayr told me we [Training] have a big project," Selvaraj recalled. "He said, 'We have ITCs and hundreds of departments all over the Eastern Province but no way to communicate with them about enrollments, absences, etc. We should find a way to communicate.' But he did not go into a lot of detail, so I went to the major ITCs at Abqaiq, Dhahran and Ras Tanura and talked to senior teachers about what they needed." Selvaraj returned to Training's offices in Dhahran and resumed work, using Materials Supply's computer system as a model. "When I took the job I didn't sleep. I used to work

Chellakon Selvaraj

up to 11 o'clock at night. I'd go about four o'clock in the evening to get something to eat. I'd get an idea and come back to the Tower Building to work. It was a hard job putting Training's rules into computer language."

Within five months, Selvaraj created a prototype of what became known as the Current Enrollment System (CES), the most comprehensive training tracking system yet developed for the ITCs and ITSs. The program was refined and introduced gradually into the training centers during 1983. At the end of its first full year, CES held the records of nearly 19,000 trainees, and the system could be accessed through some 600 mainframe terminals.

Corporate Training Management System (CTMS)

Almost immediately, Hashim Budayr formed a group to work on an even more advanced tracking system. This effort culminated in the Corporate Training Management System (CTMS), which replaced the functions of both the CES and the IT computer systems in 1987. Unlike the development of earlier tracking systems, the design of CTMS was not a hurried, one-man show. At least a dozen people worked on the project over a period of about three years. Budayr traveled to the United States looking, unsuccessfully, as it turned out, for an off-the-shelf program to replace CES. Then two senior program analysts, Selby Robeson and Tony Van Uden, were assigned to write the specifications for the new system. They worked on the project for nearly two years before the specifications were ready to be turned over to a team of computer programmers. Full-time

An instructor, right, watches a trainee respond to a simulated malfunction at a plant in Safaniya, 1985.

programmers working to develop CTMS included Terry Kanakis, Peter Hitchin, Mike Wallace, Jim Hale, Brent Holmes and Tom Adam. Others who worked on the project part time, often on an overtime basis, included Dave Ryder, Doug Jamieson, Ted Adamopoulos and Rutaro Mutoro. The programmers worked for another seven to nine months before CTMS was ready for testing. Testing required another eight months. CTMS was finally introduced company-wide in January 1987. It was the largest and most complex system developed for Training. In creating the system, all sorts of Training rules and regulations had to be codified in computer language. Many alternative scenarios had to be considered as well. For example, what if a trainee took a course and dropped out? When would he be eligible to take the course again? What if he flunked the course instead of dropping out? What would a failing grade or a dropped course do to his cumulative grade average?

CTMS validated enrollment requests, produced enrollment reports, generated attendance records, and updated course completion data, all functions similar to those performed by earlier tracking systems. CTMS also issued daily attendance reports; recorded interim test scores, final grades and promotions; and provided a suggested schedule for future training courses. It stored course completions data for employees in grade codes three to 14, out-of-Kingdom as well as in-Kingdom courses. It covered all Saudi Aramco employees, expatriates and Saudis alike, and was accessible through the mainframe computer system to all organizations. Progress reports, attendance records, interim and final grades, all formerly mailed to sponsoring Business Lines by the training centers, were now available on-line.

At its introduction, the CTMS database was said to hold 835,000 training records covering all active Saudi employees who had at least one course completion in the previous seven years. Budayr declared that development of CTMS meant that "the days when supervisors and registrars spent hours checking eligibility, preparing schedules for both teachers and trainees, recording absences and reporting grades are gone forever."

Ali Dialdin, general manager of T&CD, cited "innumerable" uses of the new system for the Training Organization. "We in Training can now plot daily classroom usage, teacher assignments and course locations and times, and view by the hour, if desired, the numbers of trainees in our centers, as well as pull up individual information on a single trainee."

Jim Hanson was another fortunate "find" for the Training Organization. A former linguistics instructor at Adrian College in Adrian, Michigan, he was contracted by Training in 1984 to design an instructional course for teaching Saudi craftsmen to become Job Skills trainers. While he was working on that project, Training learned that Hanson also had experience running a business office (he once headed a Human Services office for three counties in southeastern Michigan, and he had accumulated some 30 hours of credit for college-level business courses). So Hanson was contracted to develop the curriculum for the first Basic Business Skills (BBS) course for Saudi clerical candidates. He was given just four months to complete the 190-hour BBS curriculum. In the summer of 1985, while he was rushing to finish the BBS work, Training discovered that Hanson was also knowledgeable about computers.

"As early as 1972 I started trying to program some stuff for my teaching," Hanson said. "In the late '70s I got into doing stuff in FORTRAN [an acronym for Formula Translator, the first high-level computer language]. I believe it was Alf McLean who came to me. Dale Saner got involved, too. They said, 'Now we're starting to use computers. Can you generate a computer curriculum for the BBS as well?' The answer was yes."

Everything connected with the BBS course had been a rush job. Hanson completed the first four units of the course in mid-August 1985, about two weeks before classes started. Then, from August to December, he hurried to complete the second half of the course, including Training's first computer module.

That first effort was a 20-hour PC Introductory Course launched in January 1986 as the eighth and final module in the BBS course. Trainees learned to turn on the

computer and were given appropriate diskettes and a specific task to enter and edit data, save the program, and print the results in Basic, Multi-Mate or Lotus 1-2-3.

Carol Friday

"When we started, there were almost no computers in the Training area," Hanson recalled, so among the things I had to do was to order computers, buy them, walk them through the whole process and get them up and ready to go in about two months." He arranged through Aramco Services Company in Houston to purchase 30 to 40 IBM computers, enough so that there was a computer at each desk in the two Dhahran North clerical training classrooms. "I can't remember the total cost, but I remember it was well over $100,000, because we had to get special permission to spend that amount. By that time, PCs had come down to about $3,000 per unit."

Ahmad Ajarimah, head of the Academic Curriculum Unit, asked a computer novice, Carol Friday, to write the PC Introductory Course textbook using everyday English instead of computer jargon.

"What I basically did was rewrite what Jim had written into user-friendly text and give examples," Friday said. She wrote the course book on the mainframe computer from copy produced by Hanson on his own Compaq portable, or "luggable" computer, as he called it.

For the 1987-88 academic year the PC Introductory Course was retitled Basic Business Skills Applications. Hanson lengthened the course to 85 hours. The additional hours allowed for more computer practice time and also made room for a final project involving complex and very large spreadsheets. Hanson was assisted on this project by two self-taught computer programmers, Fawazz I. Ayoub, an Aramco English teacher from Jordan, and another Training employee, Don Prades.

Computer Training for Apprentices

In 1988 the company began a two-year Apprenticeship Program for clerical trainees.

The company hoped to train some 2,500 Saudi apprentices to replace expatriate clerks. The Basic Business Skills Applications course provided only entry-level computer training for clerical apprentices. For the 1988-89 academic year, Hanson designed three new computer courses that became prerequisites for advancement by Saudi clerical trainees. The courses were titled PC 1 (Advanced MultiMate), PC 2 (Advanced Lotus 1-2-3) and PC 3 (Advanced dBase III Plus). Trainees who completed these courses were expected to be familiar with more than 150 commands, features, functions and utilities common to MultiMate; know how to produce spreadsheets, databases and graphics; and be able to collect, store and manage information in a database.

The essentials of a long-range computer training curriculum were now in place, although additions and changes would be made over the next few years. In 1989, for example, PowerBase replaced dBase III Plus, and in 1990, the three courses designed by Hanson were lengthened to 180 hours each, or two periods per day for a 90-day semester. By 1987, Computer Assisted Instruction was available at all ITCs and the program offered a catalogue of more than 600 software programs, about 80 percent of them written in-house.

Hanson, on assignment at Management Training during the late 1980s, participated in another first: the first use of laser printers at Aramco. Laser printers produced camera-ready copy, eliminating the labor-intensive typesetting and paste-up chores previously required to prepare copy for printing.

In 1989, Training inaugurated for participants in the Professional English Language Program (PELP), a 30-hour computer course called Introduction to Personal Computing Using AppleWorks. Carol Friday and Max Larsen designed and developed the program. AppleWorks was one of the least expensive, easiest to use, and most common

Fawwaz I. Ayoub

computer programs of its time. The course gave participants an introduction to using the database, the word processor and spreadsheets. The following year, 1990, a much longer AppleWorks program was introduced for College Degree Program (CDP) candidates. The new AppleWorks course occupied 97 class periods over the final two phases of the CDP program. CDP candidates were also given a 30-hour course in BASIC (Beginner's All-purpose Symbolic Instruction Code), a programming language popular then for developing desktop computer programs.

Macintosh Computers Gain Popularity

Macintosh computers came into vogue at Aramco in the late 1980s at the expense of IBM-type PCs and operating systems. Hanson was skeptical of the Macintosh at first, but "once we hooked it up to a laser printer, it became very clear where things were going. The print quality, the output and what you were able to do in the way of graphics at that point was just light years away from what IBM could touch." In addition, Macintosh computers were easier to operate than other microcomputers of that time. Macintosh users could use a "mouse" to point and select the file or application they wanted, rather than typing instructions on the keyboard as required in the IBM DOS system. The "Mac" became the computer of choice for many Aramco managers, but several years passed before Training had the commercial hardware and software in place to teach popular Macintosh software programs to clerks and clerical apprentices. Meanwhile, many managers were using the Macintosh computer system while the clerks who were supposed to assist them used the IBM system, which was incompatible with Macintosh.

PC Training Expanded

In a four- to six-month period during 1992, three curriculum writers, Fawwaz I. Ayoub, Carol Hallett and Mona El-Ramli, designed, developed and implemented Training's first instructional programs for Microsoft Word for Windows 3.0, Microsoft Excel for Windows, and dBase

IV. These three new programs, titled Computer I, II and III, were introduced into the ITCs and at the Special Training Center (STC) in time for the August 1993 semester. They replaced the MultiMate, Lotus 1-2-3, and PowerBase for IBM computer course implemented in 1988. Trainees in the course used Dell 486 P/20 microcomputers.

Abdallah Al-Jishi

The new computer courses were required for apprentices and trainees undergoing clerical training, and open to anyone else whose job required basic computer skills. The courses were two periods a day for regular employees. Apprentices in the Computer I course had an intensive schedule of three periods a day for 54 days, a total of 162 classroom periods. At the conclusion of these courses, trainees were supposed to be able to use 150 commands, utilities, features and functions common to Microsoft Word for Windows. In addition, they were taught to create spreadsheets and graphs, design mailing labels and create complex database reports using dBase IV. That same year, 1993, Mona El-Ramli designed a 40-hour introductory course for the College Preparatory Program (CPP) on word processing for Macintosh computers. An introduction to Excel 4.0 was added later, increasing the CPP course to 80 hours.

All the major programs and systems discussed so far were created by self-educated computer programmers drawn from the ranks of trainers. Chellakon Selvaraj came to Aramco as a clerk; Jim Hanson was a curriculum writer; Carol Hallett joined the company as a secretary; Carol Friday was an English teacher; Fawwaz I. Ayoub taught English and math; and Abdallah Al-Jishi, the only Saudi in the group, had been an assistant ITC principal. They learned computers in their spare time because, as Al-Jishi put it, "We like high-tech things." Training did have one university-trained computer expert, Tom Jasper. After service in the U.S. Air Force, where he tracked satellites, and 2½ years

working in electronics for the Saudi Air Force in Riyadh, Jasper returned to school and took a degree in computer science at the University of North Florida in Jacksonville.

His first major task after joining Aramco in 1982 was to convert the TRAC system from a program limited to Southern Area Maintenance into a tracking system usable by other organizations. "The interesting part about this is that program support came from Ras Tanura," Jasper said. Lynne Reinsch, a computer programmer in Ras Tanura, and Jackie Driver, a former secretary who was training to be a computer programmer, interviewed users and outlined the system's specifications. Jasper, in Abqaiq, about 80 miles south of Ras Tanura, developed the program with the assistance of Saleem Uddin Khan, a mathematician with a computer background. When the project team needed to get together, they met in Dhahran, roughly midway between Abqaiq and Ras Tanura. "It was a funny setup, but it was the way we had to operate back then," Jasper said. "We didn't have enough manpower or knowledge or anything all in one place, so we spread it out all over the Eastern Province."

After he moved to Dhahran in the mid-1980s, Tom Jasper created Individual Development Plans (IDPs) for several different organizations. These powerful management tools had a significant impact on Saudization of the industrial work force. IDPs laid out the specific academic courses and specific job skills courses Saudi trainees needed to complete in order to advance up the job ladder. For the first time, a manager had access to up-to-date

Operators at a distribution control center in Shedgum Gas Plant.

information on the status of training for his entire Saudi work force and for each individual in the work force, but that was not all. For instance, a manager, looking at computer projections on the experience level of his work force a few years hence, could decide to moderate the Saudization process so he wouldn't have too many inexperienced workmen all at once.

Two data entry clerks, Christine Young and Janet Wade, taught training coordinators in the field to use the IDP systems and monitored the coordinators' reports. Jasper designed and implemented IDPs for the Southern Area Maintenance, Community Maintenance, and Pipelines, among others. These plans were local versions of the company-wide program he worked on next.

Company-Wide Computer System Developed

Several people had envisioned a company-wide computer system to track employees and apprentices as they progressed along a prearranged career path. Christine Young voiced the idea in the early 1980s, but perhaps only a computer professional like Tom Jasper could see the full range of possibilities. Such a system could record training, licensing and certification activities required for job qualification in the industrial work force. It could predict enrollments. It could tell line organizations just how many training steps an employee needed to complete before he was considered job-qualified. It could be used to forecast how much Saudi manpower would be available to each line organization, in what fields, at what skill levels, in the decade ahead. It could assist line organizations in fine-tuning their training and recruitment goals. It could make the independently operated line organization tracking systems obsolete. Those systems used information collected from mainframe files and reentered into the organization's own computer file, an inefficient duplication of

Rady Al-Ghareeb

effort, prone to error, often incomplete, and usually out-of-date.

Dale Montieth and Rady Al-Ghareeb, both in Dhahran Job Skills Training, may have been the first to take the idea of a comprehensive computer tracking and forecasting system and develop it into a practical proposal. Montieth had become familiar with large personnel tracking systems while working in communications with the U.S. Air Force. In August of 1987 he wrote a short memo to Dale Saner saying he could design a system to track the retraining of employees, information not then available. In early 1988, at Montieth's suggestion and with his guidance, Al-Ghareeb, a program analyst and an accomplished computer user, began writing down ideas for such a system. His report of 50-plus pages was the first step toward what would become known as the Corporate Individual Development Tracking System (CIDTS).

About that same time, Zubair Al-Qadi, another self-taught computer enthusiast, developed a sophisticated Macintosh-based tracking system for the Computer Communications and Office Services (CC&OS) Department. Al-Qadi's system used a large spreadsheet to lay out individual five-year training plans. It assumed that every trainee should have a unique, customized training plan. It was a powerful management tool, but it was a stand-alone system. All data had to be entered manually. Eventually the system became too large for the Macintosh to handle effectively. Therefore, CC&OS requested permission to develop a system that would provide the same tracking functions on the mainframe. This request created the opening Tom Jasper used to present a detailed outline for a company-wide tracking system. Ali Dialdin, general manager of Training & Career Development, gave the go-ahead for the design and development of the system that became known as CIDTS.

Originally, CIDTS was expected to be an add-on to CTMS, but during the design phase, the company acquired a new standard COBOL code generator and development tool called TELON. Programs created by TELON were unable to communicate with existing CTMS on-line transactions. At this point, CIDTS separated from

CTMS and became a full system in its own right.

It took two years to develop CIDTS. Jasper led the project development team. Team members included Saner and

Ghannam Al-Dossary, left, and Sa'ud Al-Zahrani examine a computer system board.

Saleem Uddin Khan. The chief programmer was Douglas Jamieson, an Irish contractor who had studied philosophy, tried teaching and then entered a government retraining program to become a computer programmer. CIDTS was an especially complex system in terms of handling and manipulating data. Saner thought Jamieson and the two other members of the programming team, George Demakakos and Hadi Al-Ali, did "an incredible job" of pulling the system together. Hussain Al-Mahr tested the system before it went into production. The CIDTS implementation team included Rady Al-Ghareeb, Paul Muczynski, Michael J. Bubonovich and Christine Young.

Training launched a campaign to win line organization acceptance of the new system. "We traveled north and south talking to all kinds of people," Saner said. "A lot of people had done their own training computer programs. Some were on the mainframe; some had PC systems. We actually found people whose job was to get a printout from CTMS, go over it, and sit down at a PC and re-key in all the training history for all the Saudis in their department, 200 or more people sometimes. That happened in several places. Training data was so disparate."

The central idea of CIDTS was that Saudi industrial and administrative trainees, grade codes three to 14 nonprofessionals, should be measured against a fixed standard. This approach contrasted to the idea that each trainee should have his own unique training pattern. CIDTS declared that all trainees in the system had to meet the requirements of the company's job matrix and successfully complete all the tasks listed in the Aramco Job Training Standards (AJTSs) in order to be considered job-qualified. The program allowed some room for customized training schedules, but, in general, everyone was expected to meet the same minimum standards.

CIDTS was implemented in 1991. Industrial Services became the first large organization to commit to the system. Computer Communications & Office Systems also enrolled in CIDTS during 1991. By year's end, the system was being used to track trainees in the Apprenticeship Program and the GOSP Outside Operator Program. Several other large organizations were on the CIDTS waiting list. By March 1992, nine departments were using the system.

Richard J. Klun, in Training at the Ras Tanura refinery, found that "CIDTS did away with a lot of politicking. It told you exactly who the best guy was for promotion by objectives, not by personalities or friendships. CIDTS goes in and pulls out all the information on tasks and ITC completions. Then we add in all the time-keeping records, and it gives a report on everybody in the organization and tells you exactly where everyone stands."

Training also operated the Professional Development Tracking System (PDTS), which was developed in 1984 as a stand-alone system geared to developing Saudis for promotion. This system kept training records on regular professional employees

Saudi Aramco was one of the first companies to utilize a Cray II super-computer.

Mohsen Al-Utaibi wears special glasses to view an aerial map in three dimensions.

and individualized CD-ROM instruction in addition to self-study video and laser disc courses. Any employee could book up to two hours of training time during or immediately after normal working hours. The program was considered the first step toward the ultimate computer-based training program, one that would be available, whenever needed, through desktop computers in offices throughout the company.

In the first 15 years after the EXPEC computer center opened in June 1980, its computing capacity increased 300 times and its on-line storage capacity grew by a factor of 10. In that same time period, the Training Organization developed an extensive computer network, one of the largest, if not the largest, in the company. CIDTS serviced more than 100 departments and tracked 28,000 Saudis, about 60 percent of the entire industrial work force. PDTS tracked 9,000 Saudi professional and highly skilled industrial craftsmen. MicroComputer Training geared up to handle 3,000 trainees a year. CTMS, with 1,800 users, was the second largest system in the company, behind only Materials Supply's PIDTS. In 1996, people attending the annual In-Service Training Day program found a choice of nearly 100 different exhibits, presentations and workshops conducted by Training personnel on subjects such as interactive computer programs, graphing tool calculators and computerized process control systems. Saudi Aramco Training was keeping pace with the fast-moving field of information technology.

with a college degree, as well as company-sponsored nonemployee college students, and skilled employees who were in grade codes 11 to 14.

In 1995 the Training Organization took over company-wide MicroComputer Training, a program initiated by the Computer Communications & Office Services Systems Department in 1992. The program operated out of classrooms in Abqaiq, Dhahran and Ras Tanura. It was open to all employees who needed to upgrade their desktop computer skills. Paul Kirwin, yet another of the self-taught computer experts, headed the MicroComputer program.

In early 1997, Training & Career Development established Learning Resource Centers in the Tower Building at Dhahran and at the Management Training Center at Ras Tanura. The learning centers contained some of the latest interactive

Appendix B
English Instruction in the Saudi Aramco Training Curriculum

"All of a sudden I realized what we were trying to build on in the classroom. The dialogue began to make sense to me! It was real! I could tell what the teacher had told us was true. This was how they [Americans] talked."
– Muhanna Al-Muhanna, age 16

At Aramco and later at Saudi Aramco the development of Saudi employees was based on a foundation of good English language skills. Employees used English in almost all written and oral communication within the company. It was the primary language used in the industrial training classrooms and during on-the-job training. No employee could hope to complete the company's training programs and move up the job ladder without a working knowledge of English.

The great majority of Saudis hired during the first 45 years of the company's existence could not read, speak, write or understand English. The high school graduates who entered the work force in large numbers starting in the 1980s had received some English-language training at Saudi government schools before joining the company. A few could carry on a limited conversation in English. Most knew some grammar and had a vocabulary of up to several hundred words, but they had difficulty in using what they knew. Independent reading or writing was out of the question for nearly all Saudi recruits.

English-language training had come a long way since the early, informal, on-the-job instruction, but it remained a long, slow process. By 1980 the curriculum contained seven levels of English classes — four in the basic program and three in the advanced program. A trainee normally took three years to complete the basic English program, although selected trainees could complete it in 1½ years on an accelerated schedule. The Advanced English program, levels five through seven, took another two to three years to complete. The Elementary English program comprised 1,440 hours of instruction. To complete all seven levels of English required up to 2,520 hours and five to six years of instruction.

The Industrial Training Centers (ITCs) devoted far more money and manpower to English-language instruction than to any other training program. In 1980, for example, an estimated 82 percent of the ITCs' resources, some $12 million, went into the English program. Out of 480 instructors in the ITCs that year, 420 were teachers of English.

An English-language teacher with Jabal School students, 1950.

Saudi Aramco's Changing Needs for English-Language Instruction

Even this generous allocation of resources could not guarantee success. By the late 1970s, training had experienced nearly a decade of increasingly large enrollments. Facilities were overcrowded, teachers were in short supply, and trainees were, on the average, younger and had less schooling than recruits of a decade earlier. At the same time, Training was under increasing pressure to develop trainees faster and to a higher degree than ever before.

All Saudi employees, except those few who did exceptionally well on the English placement test, had to complete the four-year Elementary English program. In 1978 only six percent of new employees scored high enough to bypass the program. The other 94 percent started Elementary English anywhere from the program's lowest level, 1-A, to its highest level, 4-B. Elementary English ran on two tracks: a regular track and an accelerated track. A trainee on the regular track would be in English class for two hours a day for up to 36 months, a total of eight semesters, to complete the basic program. Trainees who demonstrated a special aptitude for learning English went on an accelerated track. They spent four hours a day in English class and completed the Elementary English program in 18 months, half the time of their regular-track counterparts. The same curriculum was used for all trainees, regardless of track.

Saudis targeted for supervisory positions requiring extra language skills or for professional jobs calling for out-of-Kingdom training expanded their English skills in intermediate and advanced classes, levels 5-A through 7-B.

Elementary English was designed to give trainees all the written and spoken English language skills they needed for entry-level technical and clerical positions. On average, 65 percent of the Elementary English graduates were assigned to Operations, 30 percent to Maintenance and five percent to other departments.

But in the 1970s these line organizations began complaining, with increasing frequency and frustration, that Elementary English was not doing the job, that, in fact, most graduates did not have the communication skills necessary to handle entry-level positions. They found trainees especially lacking in job-related vocabulary. Instructors in other ITC disciplines also faulted the Elementary English program. Math teachers, for example, estimated they spent up to 40 percent of their time in first-year courses teaching English, because their students did not have the vocabulary necessary to start math training. Students were not introduced to words such as "addition," "subtraction" and "division" until late in their first year of English.

"We were catching a lot of criticism at this time," Dan Walters, then superintendent of Planning and Programs for Training, remembered. But no one came up with a better training method.

"Everyone knew of some training success story elsewhere in the world. In Iran they were supposed to have some huge program that made them proficient in English in nine months, but when we investigated, we found out they had a different input of trainees. Most of their trainees were either high school graduates or college graduates, and they weren't totally proficient in English at the end of nine months, either. But these stories put pressure on us to make English training more job related and do it faster.

"One part of their training in Iran, we found, was so narrowly focused we referred to it as caterpillar training, after the Caterpillar Corporation," makers of heavy-duty earth-moving equipment.

English teacher Tom Curtis in a classroom with a traditional seating arrangement, 1957.

"It was really just survival vocabulary" for those who worked with the firm's equipment.

"At Aramco the input was different," Walters said. "When we brought guys into the English program it wasn't just to get them job-proficient. It was to prepare them, if they were able, to continue on and go to college."

Since no one came up with a better method, the English-language curriculum continued as it had been, although, as the history of training at Aramco suggests, change was always close at hand.

A Variety of English-Training Methods

Over the years, Training had experimented with a number of different English-language teaching systems and combinations of systems. One of these, first developed in the 19th century and known as the Direct Method, dominated English teaching at Aramco following World War II, when large numbers of Saudis were being hurriedly trained for craft and operator jobs. The Direct Method called on instructors to communicate with students only in the target language. Word meanings were conveyed "directly" by linking words with actions, objects, gestures or situations. Instruction in speaking came first; reading and writing followed. There was little or no teaching of grammar.

A student at the Abqaiq Training Center recites an English lesson.

The 1950s and '60s brought a new awareness about language-learning problems, an interest in the scientific study of language problems, and the growth of linguistics as an independent discipline. Various new systems of teaching English as a Second Language (ESL) were developed during the period. One of them, the Structural Approach, came to the forefront in the 1960s. The Structural Approach stressed the study of grammatical items and structures, tenses, grammatical rules, and sentence patterns. Students repeated their lessons until they became second nature. The company's first in-house language series, the 12-volume *Aramco English Series*, produced by linguists Ron Goodison and Jim Sitar between 1960 and 1962, was a blend of the Direct Method and the new Structural Approach.

Jim Sitar, photographed in 1991.

Goodison, a linguist with a Ph.D. from Cornell University, and Sitar, an ITC language teacher, worked on the *Aramco English Series* for three years. The series contained a vocabulary of more than 3,500 words — from "*Abha*" to "zinc." There was one volume in the series for each trimester during the first four years of ITC English-language training. Book One began with the student being asked to repeat after the teacher, "a door, this is a door ... a book, this is a book." An appropriate picture accompanied each phrase. Being tailor-made for Aramco, the series differed from commercial English language textbooks previously used in the ITCs. The vocabulary included common words and expressions of the oil industry such as "pipeline," "refinery," "stabilizer" and "GOSP." The reading exercises contained information on topics ranging from Aramco's history and organizational structure to world history, geography and science. For example, in their reading exercises students learned where the Canary Islands were, who Christopher Columbus was, when the first Aramco refinery was built and what a district manager's responsibilities were. Each reading exercise was followed by a list of questions to which the student responded in written English. By the time a student got into the 12th and most advanced volume, he was

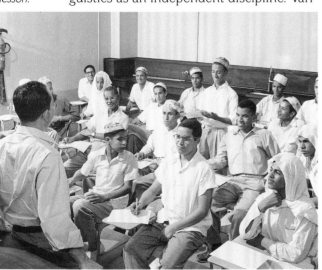

expected to be able to answer in writing questions such as "What are the five raw materials that are used in the manufacture of synthetic materials?"

The Structural Approach

Several textbooks using the new Structural Approach were introduced to Aramco ITCs in the 1960s as replacements for *Aramco English Series* texts. Two of the best-known replacement texts were *English Pattern Structure* and its companion, *English Pattern Practice*, by Lado and Fries. By 1966 the *Aramco English Series* textbooks were used only in the first two levels of the ESL program, but these texts continued to be used in levels one and two for more than a decade.

The English program developed a more academic focus during the 1960s, when the size of the work force was steadily declining. One of Training's major goals became the preparation of Saudis for supervisory or higher positions, jobs requiring enhanced reading and writing skills. To help satisfy this goal, in 1965 the standard ITC English class period in levels one through four was increased from one hour to two hours. Training's 1965 Accountability Report explained: "One-hour courses were found to be inadequate to prepare trainees for secondary-level courses."

The rapid expansion of the 1970s created an urgent need to once again train large numbers of Saudis as craftsmen and technicians. These jobs required more emphasis on speaking skills, particularly a vocabulary in the technical language of the trainee's specialty. Training was

Registration day at Dhahran ITC, 1956.

pressured to expand and speed up vocational training while continuing to get high-potential employees ready to pursue advanced degrees.

Enrollment in company training centers increased during the 1970s from less than 1,000 to nearly 10,000. During the decade, as many as 5,000 new Saudi employees were hired in one year, an average of about 100 per week, or 15 per day, nearly all requiring basic English-language training. Most of them were assigned full-time jobs while awaiting openings in the industrial training classrooms.

Muhanna Al-Muhanna

Despite the crowded conditions and arguments over teaching methods, knowledge was being transferred on a huge scale inside ITC classrooms. Trainees did learn, sometimes to their own surprise. Among the new recruits in April 1975 was Muhanna Al-Muhanna, age 16, who came to Aramco with a seventh-grade education and no understanding of English. He drew a pre-ITC work assignment in the Trainee Administration Unit in Dhahran, where the interviewing was done and the paperwork kept on new Saudi hires. One Saudi trainee in the unit was taking fourth-year English. "He was great in communicating by phone and in his writing ability," Muhanna said. "I was thinking, could I be like him when I finished English four?"

Muhanna's doubts eased one memorable evening a few months after he began studying English at the Dhahran ITC. On this particular evening, he was sitting on a bench with dozens of others at the outdoor movie in al-Munirah compound. They were watching a Hollywood film starring Egyptian-born actor Omar Sharif. Muhanna always enjoyed the scenery and action shots in English-language films and usually had a general idea of what the story was about, but he could not follow the dialogue. That changed in one surprising moment. "It was very interesting because all of a sudden I realized what we were trying to build on in the classroom. The dialogue began to make sense to me! It was real! I could tell what the teacher had told us was true. This was how they [Americans] talked and the way they pronounced words. It clarified things for me. The movie linked me to the classroom, and the classroom linked me to the movie."

Muhanna was a training success story. He learned English well enough to qualify for out-of-Kingdom training and to earn a college degree from Temple University in Philadelphia. He returned to Aramco to begin a management-level career in the Training Organization.

Training took some extraordinary steps to deal with the pressure of expansion in the 1970s. Rather than turn away trainees, they increased class size from an average of 16 students per class to more than 20 students. A chronic teacher shortage forced Training to relax its hiring standards somewhat. The preferred score for teacher candidates on the Michigan Proficiency Test was reduced from 90 to 85. The workload in the plants and refinery was such that many trainees could not attend ITC classes during the regular daytime hours. For their benefit, the ITCs began running English classes around the clock at the plant sites in Abqaiq and Ras Tanura. Classrooms were improvised in whatever space was available. One classroom was set up inside Shelter No. 7 at Abqaiq. Shelter No. 7 was a thick-walled, windowless building, stocked with food and water, where workmen were to retreat in case of a toxic gas leak or other calamity. The Ras Tanura ITC used a portable building next to a noisy operating unit as an around-the-clock classroom for refinery workers.

Standardized Testing

ITC instructors of that time had almost unlimited control over curriculum and testing in their classrooms. They developed their own curriculum from the materials at hand; did their own lesson plans; and wrote, administered and scored their own tests. There was little coordination between districts. What was taught at level-three English in one ITC might be taught at level four in another. In 1974, at the urging of Bill Griffin, then superintendent of Industrial Training, the Curriculum and

English language skills were required for the study of chemistry and other subjects.

Test Development Unit (CTDU) was formed in hopes it would bring some uniformity into the system.

With the backing of Bill O'Grady, director of Training, CTDU took responsibility for curriculum and testing in all ITCs. The unit began with a staff of six under the supervision of Charles D. Johnson, holder of a master's degree in linguistics and a doctorate in education from the University of Michigan. The unit worked out of leased offices on the top floor of the Saudi Cement Building in Dammam, adjacent to the company's Management Training facility. "I was told our job was to straighten things out, to get uniform curriculum and testing in the districts," Johnson recalled. The CTDU staff got together with ITC instructors and principals to work on a standard curriculum for each grade level. The negotiations proved to be delicate. The curriculum that was agreed upon made use of commercial, off-the-shelf material modified by some locally written, Aramco-specific instruction. It followed the Structural Approach to English-language training.

Developing standardized tests was another delicate matter. Test scores had become an unreliable indicator of trainee achievement. "If students got bad grades, it reflected on the teacher, or so many teachers feared," Johnson said. "So we had situations where no one flunked and nearly everyone got 100 [on tests]." The CTDU staff again negotiated with ITC teachers and principals on the design of tests, their frequency, and grading. They agreed on multiple-choice tests, mixed with some essay questions, which were to be administered in the districts under the

Instructor Donald Burr's English class in a U-shaped seating arrangement.

guidance of CTDU staff members. The test questions were written and stored at CTDU offices under strict security. Requests from ITC teachers and principals for an advance look at the test questions were firmly rejected. The essay questions were graded by a panel of ITC teachers. The multiple-choice portion of the tests was sent to the University of Petroleum and Minerals for machine scoring. Students were graded on the curve using norm-reference methods.

The Communicative Approach

In the summer of 1977, two Aramco trainers returned from refresher courses in Britain with word of two new ESL teaching methods. One of these was a newly developed teaching system called the Communicative Approach. The other was a concept known as Language for Special Purposes, or English for Specific Purposes. Wadie Abdelmalek, who succeeded Johnson as CTDU supervisor, first heard of these new developments during a course at Lancaster University led by Christopher N. Candlin. Khalil Nazzal, principal of the Dammam ITC, attended lectures on the same new ESL methods during a two-week course organized by David Wilkins at the University of Reading. Candlin and Wilkins were among a number of British linguists whose ideas contributed to development of the Communicative Approach.

The Communicative Approach moved away from grammar-based methods and toward instruction in the appropriate use of everyday speech. The main language skills of listening, speaking, reading and writing were practiced in conjunction with each other rather than being taught and

tested at separate times. The objective was "communicative competence" — making oneself understood, rather than grammatical correctness. For instance, it would be perfectly acceptable in the Communicative Approach to say "A hammer, please." Traditional grammarians would treat "A hammer, please" as a sentence fragment, and have the student use a more complex form, such as "May I have a hammer, please?"

The Communicative Approach called for extensive use of audio-visual teaching aids. Organized group activities and U-shaped classroom seating arrangements were suggested to improve interaction and communication between students. The closer the students' group activity came to mirroring needs on the job, advocates of the new method said, the greater would be the students' motivation to learn.

English for Specific Purposes (ESP)

The second new method, English for Specific Purposes (ESP), had been developed in Canada in the 1970s in response to the English-French language needs of that country. It was recognized in Britain and Europe as one way of catering to some of the special language needs occasioned by development of a European Common Market. English for Specific Purposes focused on equipping trainees with the communication skills they needed to do a specific job. In the case of industrial training, the program's vocabulary list would be restricted as much as possible to the language of the workplace. Documents such as operators' manuals, time sheets and material request forms would be considered teaching tools.

These down-to-earth teaching methods seemed like a breath of fresh air to Abdelmalek and Nazzal, two traditionally trained instructors. "I was impressed by this approach to language training," Abdelmalek said. "It was a big change from what we had been doing. As soon as I got back to Aramco, I made a recommendation to bring this man [Candlin] in to look at our training programs."

Nazzal thought the new methods were great. Returning to Aramco, he wrote a paper describing the Communicative Approach in laudatory terms and shared it with other ITC principals who "thought I was crazy at first." That attitude soon changed.

English Training Evaluated

On Abdelmalek's recommendation, Candlin was brought in as a consultant to evaluate Aramco's English program. Between November 30 and December 7, 1977, he toured ITCs in Abqaiq, Dammam, Dhahran, Mubarraz and Ras Tanura. He met in joint sessions with English teachers and individually with the principals at each center. He observed English classes at all levels, from 1-A through 7-B. He examined curriculum and testing material. On the final day of his tour, he discussed his findings with the director of Training, Bill O'Grady, and Ali Dialdin, then superintendent of Industrial Training.

"The overall impression," Candlin wrote in his evaluation report, "is of an extremely unwieldy program that is by far not as cost effective as it could be, given the amount of time that the students expend and the amount of hours invested by the teaching staff.

"There is a tendency," he wrote, "to reduce all students to a lowest common denominator by a process of identical teaching and identical materials throughout each level of the program."

Among his other criticisms: The "texts that the students read did not relate, in the main, to the oil industry;" training materials were out of date; "uninteresting and unmotivating" classrooms were mostly devoid of visual aids or mechanical aids such as tape recorders. He found teaching schedules dominated by "too frequent and detailed testing procedures," forcing teachers to hurriedly prepare students, often by rote, for the next test. Students on a regular track were tested six times, or once every six weeks, during the 36-week school year. Accelerated track students were tested every three weeks, a total of 12 tests in a single school year.

Candlin thought ITC instructors were generally rigid and authoritarian. Those teachers who were trained in the Middle East (about 70 percent of the English teachers at Aramco were Middle Easterners) often "carry over into their classroom

the traditional pattern of teaching in the Arab world, which is heavily centered upon the teacher and choral response in classes," he wrote. This tendency, he asserted, "promotes an atmosphere in which the learners are typically uncreative and, to a large measure, uninvolved in the teaching-learning process." He was glad to hear that a number of instructors were attempting to change the program and were interested in further teacher training.

Candlin recommended a number of changes in the English program. These changes included an immediate "communication needs analysis" covering both the ITCs and line organizations, a program of pre-service and in-service teacher training, more contact between Aramco ESL trainers and trainers elsewhere in the world, creation of "project-like" exercises for students that reflect the language skills needed on the job and language tests that measure the validity of exercises against known communication demands and updated training materials.

He left the general impression that what was really needed was a more thorough study of the ESL program by outside experts. The current program, he wrote, was "not likely to respond effectively to small and individual piecemeal amend-

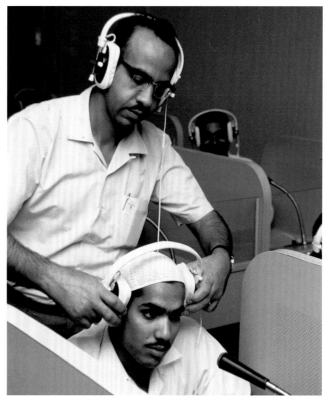

An instructor helps a new student in the English-Learning Lab, 1967.

Aramco teachers and staff met for a two-day seminar on new English-language teaching methods, 1978.

ments." The job is too big to be handled by present Training personnel, he said, and too broad to be changed immediately. "What is required is a systematic survey of the program done by a team of experts in the field, who would then be able to flesh out in more detail the recommendations which I have given in outline."

Another consultant, Larry Emigh, the long-time Abqaiq ITC principal, was brought out of retirement in September 1978 to survey the English program. Emigh was blunt in his written assessment of English-language training.

"The most serious problem in the English program is lack of well-qualified ESL teachers, along with a sizable increment of almost useless teachers who cannot be let go until replacements arrive," Emigh wrote.

"Total staff required to reduce classes to teachable size and replace unproductive teachers is 85 at present," an increase of about 15 percent in the teaching ranks, he declared. He saw little hope of recruiting so many more qualified teachers with the existing restrictions on family housing and with limited pay scales for teachers from Middle East countries.

Emigh described the English curriculum as a "hodgepodge" of "badly planned core courses" with "misalignments, gaps and overlaps" resulting from the addition and subtraction of textbooks over the years.

Candlin had also noticed an excessive number of English textbooks with little limit to the variety or number of textbooks that could be ordered. Training's book budget for 1977 alone was more than $300,000. Instructors took a little from this text and a little from that one, resulting in a buildup of books at all levels. Bill

Granstedt, then an English teacher at the Mubarraz ITC, believes about 50 different textbooks were being used in the English curriculum at the time.

Like Candlin, Emigh thought testing was too frequent and rigidly paced. He likened it to training "modeled on a factory assembly line." His prescription: "Drop all interim tests and let classroom teachers teach and give their normal quizzes internally."

The Mubarraz ITC, which Emigh found both extremely overcrowded and lacking in basic facilities, should be phased out and trainees moved to the 'Udhailiyah ITC, he wrote.

Emigh found at least one bright spot: The Curriculum and Test Development Unit had designed standard tests for all ITCs so that "a grade of 86 percent from Mubarraz means the same as an 86 from Dhahran. A major step forward," he said.

Gradual Changes Introduced

A thorough overhaul of the old monolithic, tightly paced English program had started even before Emigh completed his consultant's report. It began in September 1977 with the introduction of Ibrahim Shihabi's revised new English level 1-A course, the first new entry-level English course produced at Aramco in 15 years. Shihabi, then senior advisor in the Curriculum and Test Development Unit, labored five to six months, mostly on his own time at his home in Dhahran, to develop the new course. Beginner's English, as his course was called, followed the Structural Approach. It was truly a beginner's course. The primary objectives for the first semester were to teach trainees to distinguish the sounds of English, recognize the letters of the Roman alphabet and associate those letters with the sounds of the language.

Beginner's English was written in 1976 and revised in 1977. It took the place of the last *Aramco English Series* textbook still in use. The 12-volume *Aramco English Series*, the original company-produced English-language textbooks, had been completed in 1962. They originally covered the first four years of English training, but by 1976 they had been replaced by commercial texts at all levels except level 1-A. The last textbook in the venerable old series was replaced by *Beginner's English* starting with the September 1977 school term.

In March 1978, some two months after Candlin issued his evaluation report, about half of all Training Department employees, 240 teachers and staff members, representing all Aramco ITCs, were brought to Dhahran for a two-day training seminar. It was the largest gathering to date of Aramco trainers. The seminar focused on ways of adapting new teaching methods to fit Aramco's needs. Participants met at the Aramco Oil Exhibit building, then located on a site later used for the Dhahran hospital complex. The principal speaker was Raja Tawfig Nasr, Ph.D., professor in the College of Education at the American University of Beirut and director of the university's Centre for English Language Research and Teaching. Nasr had lectured at the University of Reading's 1977 summer program attended by Nazzal. Although a follower of the Communicative Approach, Nasr recommended it be used in combination with elements of the grammar-based Structural Approach.

Blending the Communicative with the Structural Approach

Only four months after the seminar, the Curriculum and Test Development Unit produced the first in-house textbooks blending the Communicative with the Structural Approach to English training. The combined methods were embodied in two textbooks: one called *Communicate*, for grade level 2-A, and a small companion text named *Measurement* — both issued in July 1978, and both written by Helen Sanchez under the direction of Abdelmalek. *Communicate's* 259 pages included novel features such as crossword puzzles and a vocabulary for automobile drivers — "fasten the seat belt" and "adjust the mirror" — plus illustrations of the driver's hand signals and common traffic signs. The course, however, contained little or no vocabulary specific to Aramco or to the oil industry. The companion text, *Measurement*, consisted of 10 lessons in 42 pages. It was not a how-to-measure book, but an explanation of English words commonly used in measuring length, width and height, and exercises on how to use such words in a sentence.

English for Specific Purposes (ESP) was also gaining momentum in the company's training circles. Advocates of ESP contended that training time could be shortened by concentrating on the specific skills, topics and vocabulary needed for certain jobs. They argued that English training in the upper levels was aimed at scholarship candidates, placing trainees who were not university-bound at a disadvantage. What the vast majority of Saudis needed to know, at least at the beginning of their careers, ESP advocates said, was how to communicate satisfactorily on the job. "Just give them enough English so we can communicate with them, and we'll do the rest" became a familiar refrain of some line organizations.

A text called *Technical English* for grade levels 6-A and 6-B, introduced in early 1978, was an outgrowth of the ESP movement. *Technical English* exposed students to a vocabulary of about 650 technical terms. Ibrahim Shihabi collected the technical vocabulary from manuals in the Industrial Training library, and a group in the English Curriculum and Test Development Unit wrote the textbook. Among the writers were Karen Evans, Pat Foreman, Shirley Hedjazi and Mary Jane Onnen. The vocabulary spanned the alphabet from "abrasion" to "zone" and included oil industry terms such as "bleeder," "catalyst," "fractionate," "pulsation," "viscous" and "volatile." Reading exercises ranged from an explanation of the Fahrenheit, Centigrade and Kelvin temperature scales, to an account of how a fractionat-

Employees gather in an ITC library, 1982.

Study period at the Dhahran ITC, 1983.

for college entrance exams. This type of material concentrated on the study of English literature and teaching for English literacy rather than on training for industrial purposes.

InterCom

About 50 different programs were evaluated. It came down to a choice between a series published by the America Book Co. called *English for International Communication* and a program put out by Longman's publishing house in England. The International Communication program, or InterCom as it was more frequently called, was selected. "InterCom had the advantage that it was comprehensive," Atiyeh said. "It was an integrated approach. It had all the language skills — speaking, reading, writing and listening — in one series."

InterCom consisted of a series of six books intended for secondary and adult ESL students. The series was written by a team of seven ESL trainers headed by Richard Yorkey, professor of applied linguistics and director of the Centre for TESL (Teaching English as a Second Language) at Concordia University in Montreal. Yorkey was formerly a professor of ESL teaching at the American University of Beirut. Of the six other authors, five were language professors — three at U.S. universities; one at National University in Colombia, South America; and another at the University of Puerto Rico. The other author on the team, William Woolf, was a curriculum development supervisor in a technical training program in Iran. Woolf was hired by Aramco as a consultant to help implement the InterCom program.

ing tower worked and a description of the fire hazards from various hydrocarbon vapors. The textbook was revised in 1979 to put more emphasis on ESL training and less on technical content. The Technical English 6-A and 6-B course was limited at first to trainees for the refinery and other industrial operations. It later became the sixth-level program for all trainees except those being prepared for college work.

In late 1977, the director of Training, Bill O'Grady, turned his attention to what Larry Emigh had described as the "hodge-podge" of training materials then in use. O'Grady created a four-member Curriculum Committee to search for a commercially available program that answered the needs of English levels 1-A through 7-B — no easy task, given the wide spectrum of education levels among ITC trainees.

The committee, composed of Elias Matouk, principal of the Ras Tanura ITC; Nimr Atiyeh, assistant principal, Dhahran ITC; William F. Granstedt, chairman, Mubarraz ITC; and Charles Lucas, chairman, Abqaiq ITC, met weekly over a period of several months. Nimr Atiyeh was the driving force, Granstedt said, the man with the most ideas on the committee. "We looked at nearly every English-teaching program published in the U.S. and U.K. at the time," Atiyeh said. "We were very thorough."

The majority of commercially available materials were quickly eliminated on grounds of difficulty, relevance, vocabulary or cultural appropriateness. Beginner and intermediate materials were usually produced with schoolchildren in mind, not the adult learner at Aramco. Most advanced-level material was intended for college students or students preparing

Students study in the ITC Library, Dhahran, 1968.

Many InterCom lessons involved short narratives about the comings and goings of a cast of more than 20 fictitious characters. The lessons were built around conversations among these characters and descriptions of their everyday activities — dining at restaurants, shopping, using the telephone, opening a bank account, etc. The vocabulary lesson in Book One began with the pronunciation of numbers and concluded in Book Six with words such as "acrylic," "kerosene," "refinery," "slalom" and "transparent."

InterCom was introduced in January 1978 at the Dhahran, Dammam and Ras Tanura ITCs as a pilot program for selected students with high grade averages. It met immediate resistance from some ITC trainers. They complained that InterCom was not job related and was too academic, too theoretical and too difficult. The grammar was sequential and not repeated at various grade levels, which, some trainers said, was a problem because students entered the ITCs at different grade levels. A revised version of the course was introduced in January 1980, but InterCom was never used as originally intended. Instead of integrating the various levels of Elementary English, it was used in levels 2-A through 4-B for classes of selected students with good grades, mostly high school graduates on the accelerated track. InterCom was also used to instruct advanced students being readied for out-of-Kingdom college assignments. In 1980 only 312 out of some 11,000 English-language students, were taking InterCom courses.

In 1979 the Training Department launched an experimental program to determine if trainees on the accelerated track could satisfactorily complete Elementary English training in three semesters instead of four. Ibrahim Isa, principal of the Dhahran ITC, and Nimr Atiyeh, his assistant principal, headed the program. More than 150 of the best students, all with a ninth-grade education or better, and some of the best ITC English teachers participated in the program. The feeling at the time, Atiyeh said later, was that this was a program that must not fail. "I met with the ITC teachers and told them this is not really a pilot program, this is an experiment that's got to succeed. We selected the best

Students at work on an English writing assignment, 1983.

teachers. We formed a teacher's committee. We met weekly and sent copies of our minutes to Bill O'Grady, the director of Training." The experiment did not fail. It succeeded by simply compacting the last three semesters of the English program into two semesters. "I don't think trainees really suffered," Atiyeh said. "It worked just as well in three years as four." The experiment was short lived since it was dropped after students in the original program completed English 4-B, but the concept of the program and its name endured. The Saudi Arab Management Committee (SAMCOM) borrowed the name "fast track" and applied it in 1979 to a new program for high school graduates who qualified for a company-sponsored college scholarship.

The line organizations' demands for faster and more thorough English-language training increased in proportion to the rate of the company's expansion and the numbers of new-hires added to the ranks. Bill O'Grady recalled that "the pressure kept building and building for us to come up with some better vocational training and do it faster. The line organizations kept asking, 'Why can't we have a core [language] program and then add on specialized training?' At one point we were talking about specialized English and math classes for the storehouses, for the shops, for the refinery, for the gas plants, and so on." The idea of trying to administer such an abundance of specialized programs gave him nightmares, he said.

The company's decision in 1980 to begin a large-scale craft training program for Saudis spotlighted the issue. The goal of the program was to develop as quickly as possible a corps of several thousand highly skilled Saudis in the metals/machinists and electrical crafts fields. For that purpose, the company decided to build large Area Maintenance Training Centers at Abqaiq, Mubarraz and Ras Tanura. The Training Department would be responsible for phase one of craft training, during which time the future craftsmen would receive instruction in English, math and science in addition to basic craft training. Afterward, line organizations would provide recruits with specialized, in-depth technical training. Line organization trainers wanted the trainees to come to them equipped with a basic craftsman's vocabulary.

Upon Further Review

A needs analysis study conducted in early 1980 by a group of ESL experts from San Diego State University derailed resistance to this type of specialized program. The ESL experts studied the needs of both the ITCs and line organizations and concluded that the English program should be loaded with additional vocational content. In March 1980, Training requested and executive management approved a budget of $2 million to develop a special English-language curriculum primarily for the craft training program. Later that same month, Training advertised in U.S. publications for bids on development of a new Aramco-specific English-language program. The new program was to include specialized technical-language courses for craft and operator trainees — just what line organizations had been asking for. The contract for creation of a new vocationally-oriented English program was awarded in June 1980 to Pacific

English was the language of instruction in industrial shops.

American Institute (PAI) of Corte Madera, California, a suburb of San Francisco.

The PAI team was headed by Louie Trimble, a retired University of Washington linguist who specialized in curriculum building, and his wife, Mary, an applied linguist and coauthor, with her husband, of many professional articles. The Trimbles and a team of 10 to 15 other PAI people began working on this project out of an office in Dhahran in the summer of 1980. They focused on seven job categories: metals, machine tools, heavy-equipment, electrical maintenance, instrument repair, oil operations and gas operations.

Team members visited both onshore and offshore Aramco work sites from 'Udhailiyah in the south to Safaniya in the north and interviewed more than 320 employees to compile data on the specific listening, speaking, reading and writing skills needed in the seven job categories. They followed workmen for hours, taking notes and making tape recordings of actual on-the-job conversations. In a matter of weeks, the team compiled and sent to PAI's California office enough boxes of material "to make a wall for a good-sized house," according to Julia Schinnerer Erben, the first on-site supervisor representing Aramco at PAI headquarters. Schinnerer Erben was a linguist herself and a former student of Candlin's at Lancaster.

Vocational English Language Training (VELT)

Over the next several years, and at a cost estimated to be in excess of $2 million, the PAI team produced the English-language program known as VELT, an acronym for Vocational English Language Training. It was an all-out attempt to respond to the line organizations' requests for a job-related English-language training program. As a method of teaching English, it combined the two newest and most popular ESL trends of the time, the Communicative Approach and English for Specific Purposes.

The VELT program consisted of four levels (A, B, C, D), with each level corresponding to a full school year. Level A was a core course for all craft and operator trainees. Trainees were then routed onto separate tracks of increasingly specialized language training. Level B separated students into two tracks: those training for jobs in the mechanical fields, and those training in electronics, instrument repair or plant operations. At Level C, trainees separated into three tracks — one for metals and mechanical trainees, another for electrical and instrument repairmen, and the third for gas-oil plant operators. Level D divided students into five specialty tracks: metals and mechanical workmen, electrical maintenance workers, instrument repairmen, gas plant operators and oil plant operators.

The VELT core course was introduced as a pilot program for 165 students at four ITCs in February 1981, only seven months after the contract with PAI had been signed and before the second semester of the Level A core course was written. The rush to introduce VELT in a piecemeal fashion was "one of the things that eventually got us into trouble," Erben said. The entry-level VELT core course was introduced into classrooms on a large scale in September 1981, but the entire four-year program was not fully implemented until 1984.

With the introduction of VELT, the ITCs, which for many years operated only one English program for all trainees, now had three basic English programs running simultaneously. Besides VELT, there was the Regular English program, with a new name for what was essentially an extension of the old Elementary English course. It catered to Saudis destined for nontechnical jobs in areas such as accounting, administration, finance, materials supply, medical and purchasing. Finally, there was InterCom for high school graduates with exceptionally good potential for out-of-Kingdom training leading to managerial or professional jobs.

In the first-year core course, VELT students learned to understand simple instructions, ask and answer questions such as "Where is the wrench?" and receive and give uncomplicated messages. They learned to read warning signs and notices such as "danger" and "no smoking" and to complete simple forms and checklists. By their fourth and final year in VELT, trainees were supposed to be able to understand lectures, films and slide shows; receive and give instructions and requests; read

manuals; and read and write messages. For instance, fourth-year students in the gas-oil operator course were expected to be able to read a manual in English about the treatment of water by reverse osmosis and then draw a diagram illustrating the process.

A VELT classroom could be distinguished from other classrooms by the abundance of training aids. Standard equipment for the Level A core course included a slide projector, a projector table, a projection screen, a chalkboard with hooks for flash card materials, 10 wall charts, a cursive-writing chart, a counting chart, a large wall clock and a large thermometer hanging on the wall. The teachers' materials list for the same course included 123 transparencies, 359 flash cards, 74 picture cards, a cassette recorder and three audio cassettes, 10 small clocks and 10 small thermometers. All this was in contrast to the nearly complete lack of audio-visual aids noted by Candlin in his 1977 evaluation report of other English programs.

The first American oil men in Saudi Arabia found that many Saudis were unfamiliar with common Western hand tools. That was still true 40 years later, at least in the opinion of VELT's creators, despite the mingling of cultures and the massive introduction of modern technology. VELT instructors were issued a brightly colored toolbox holding about a dozen common hand tools, including a hammer, saw, file, pliers, socket wrench, tape measure and scissors. Early in the core course, the instructor carried the toolbox into class, held up each item, and identified it for the students: "This is a hammer." He described how the tool worked, and had the students

"This is a hammer," the instructor says, holding up a hammer in a VELT class.

repeat its name. Instructors were also issued a hard hat, a workman's jacket, safety gloves and safety glasses. Standing before the class, the instructor put on and took off each item while saying "put on … (safety gloves)" and "take off … (safety gloves)." Then he distributed the items to students and told each student to "put on those … (safety gloves)" and "take off those … (safety gloves)."

The arrival of VELT was greeted with great enthusiasm at Aramco and with interest in ESL circles worldwide. It was hailed as a breakthrough in English training; major international textbook publishers inquired about publishing rights. But the warm feelings quickly began to cool. VELT, once put into practice, was revealed to have serious shortcomings, even for the line organizations it was designed to please.

Testing procedures were among the first things changed. Originally, instead of multiple-choice tests as in the other English programs, VELT students underwent a "criterion-referenced evaluation." A student might be handed a map of al-Khobar and asked to explain to someone, in English, how to get from King Faisal Street to the bank on Second Street. This required several skills, including map reading, and enough English-language speaking ability to communicate directions to someone else. Company trainers, however, considered that scoring such a test was far too subjective, compared to the purely objective multiple-choice tests given elsewhere in Aramco. New VELT tests were written in-house by the Curriculum and Test Unit. These tests were 80 percent objective, consisting primarily of multiple-choice questions and questions requiring short written answers.

The use of computers enabled students to study independently, 1985.

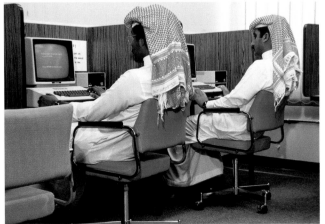

Enthusiasm for VELT Cools

The VELT program generated a list of other problems and complaints. The many specialized tracks gave administrators scheduling nightmares, just as O'Grady had feared. Some classes had only one or two students at a time. Supervisors reported that VELT had a negative effect on morale because trainees considered it a career-limiting program. A survey of VELT students in Ras Tanura confirmed these reports. The majority of students surveyed felt that by being enrolled in VELT, with its narrow focus on a technical vocabulary and limited attention to grammar, they would be less likely to qualify for college scholarships or be promoted into supervisory positions where good reading and writing skills were necessary. This perception seemed contrary to the company's announced policy of training all Saudis to their maximum potential.

Although VELT students showed superior speaking and listening skills compared to other English students in the early levels, that advantage disappeared in higher levels. Tests showed they were less developed in reading, writing and syntactic skills than students in Regular English. A study of trainees in Ras Tanura revealed that, contrary to all expectations, VELT trainees were not stronger in technical on-the-job training assignments than students from the Regular English program.

VELT teachers interviewed in 1983, two years after the program was introduced, estimated that anywhere from seven to 100 percent of the program needed to be revised. The average estimate was 45 percent. Line organization supervisors, responding to a letter from the Training Department, agreed.

In addition to its other drawbacks, VELT was a victim of unfortunate timing. Circumstances changed by 180 degrees between the time VELT was commissioned and the time it was fully implemented. VELT was commissioned during a period of unbridled expansion. The pool of available Saudi talent was small. The typical recruit had only an elementary school education and no previous English-language training. VELT was a training program for technicians. It aimed to impart as quickly as possible the basic English skills the trainee needed to do useful work in the field. But before VELT became fully implemented, the expansion was over and the company was cutting back its work force. A surplus of talent was available along with more time to train them. More high school graduates were being recruited than ever before. Almost all of them had basic English-language training in government schools. The top-level Saudi Arab Manpower Committee was pushing Saudization, particularly at the professional and management level. Training once again focused on developing the Saudi work force to its maximum potential.

Replacing VELT

A five-member VELT Revision Task Force was created and began work on September 3, 1983, under the direction of Ahmad A. Ajarimah, supervisor of Training's Academic Curriculum Unit. Ken Bus, a test writer and former ESL teacher in Iran and at the Dhahran ITC, was named Task Force coordinator. Charter members of the Task Force, in addition to Bus, were Abdul Gasim, George Scholz, Michel Bekhazi and Hamdi Al-Awadi. The team was soon enlarged and eventually reached a peak of 16 people. The multinational staff included curriculum writers, test writers, and illustrators, plus some of the better ITC English teachers. Teachers were not usually asked to join such Task Force groups, but, in this case, the teachers "proved to be one of our strengths," Bus said.

The original idea was to simply revise VELT, as the Task Force name suggested, but the more the Task Force looked into the problem, the less workable that idea seemed. No matter how much the course was revised, VELT graduates would not be able to change jobs or transfer across organization lines without taking more English courses, nor could they be promoted to supervisory positions without further language training, particularly lessons in reading and writing. In addition, VELT training was not suitable for personnel in support organizations such as Industrial Security and Community Services. Line organization managers, recognizing some of these drawbacks, had enrolled more than twice

Many levels of job skills required technical English.

as many of their Saudi employees in the Regular English course than in VELT.

Ajarimah and Bus decided an entirely new English-language core course was needed. "It looked like an opportunity to do something, not only about VELT, but about the outdated Regular English program and InterCom, which had only a very small number of students," Bus said. "Our conviction was that we could replace all the first four levels of English with one program."

As a model for this comprehensive new program, the Task Force turned to work done in the mid-1970s by the Council for Cultural Cooperation of the Council of Europe. In the interests of greater unity and understanding among the various countries of Europe, the council had commissioned a small multinational group of experts to carry out a number of fundamental studies of foreign-language learning. "Theirs was a very well regarded and respected work," Bus said. "It was a study of how people learn languages and the stages of learning they go through. It was a very concentrated work. A lot of time and manpower went into it." These studies provided a framework for Aramco's new English program.

The Task Force took two years, under a lot of deadline pressure, to complete an entire new core English program. Their first year was especially difficult. The first 50 lessons produced by the Task Force were rejected when presented to management for review. "The reviewers were unhappy with the quality. We had to start all over again," Bus said. George Scholz, in charge of curriculum writers, established a brisk pace for the project. He required each writer to complete at least one lesson a day.

Each level of the new program was introduced into ITC classrooms almost as soon as it was completed. The Level 1-A book was finished just in time for the September 1984 semester. The completed book was brought to Ajarimah for final review late in the afternoon of the day before it was due to go to the printer. "I spent all night on it," Ajarimah said, "and at 7:00 the next morning the very capable editor, Virginia Freeman, and I went over it. We had some serious problems, too. For instance, they got all confused between imperial and metric measurements."

The group decided to call their new program the Basic English Program. "We kicked around a lot of names; that just seemed the simplest and most direct," Bus said. He would perhaps have preferred a more rousing acronym for the program than BEP.

The Basic English Program

The VELT Revision Task Force changed its name to the Basic English Program Project. Its leadership also changed. Ajarimah left the Academic Curriculum Unit in 1984 to become assistant superintendent of the Program Development and Evaluation Division. Nimr Atiyeh took over the Curriculum Unit. Mark Mendizza, principal of the Ras Tanura ITC, became the project leader. Bus stayed with the project as a test writer. Mendizza was replaced as project leader in 1985 by Virginia Charlton. Charlton was succeeded, in turn, by Joe Rodriguez from Abqaiq in late 1985, just as the final two books were being completed.

Virginia Charlton

The complete new four-year program was contained in 16 trainee textbooks and teachers' guides. It encompassed more than 300 lessons, 1,029 instructional hours and 1,235 class periods. It included a target vocabulary of 2,350 words, and 747 other words students should recognize. In all, a graduate of the program was expected to recognize more than 3,000 common

English words, names and abbreviations and to use 2,350 of them.

BEP was a mixture of the structural and communicative systems. It was primarily structural in the first level, gradually becoming more communicative until the communicative became dominant in level four. It was Aramco-specific, but deliberately avoided the narrow English for Specific Purposes approach. The program borrowed concepts from the old Regular English program, and retained a few topics from VELT, such as safety instruction and lessons on how to fill out forms.

The new program devoted about 30 percent of the time in level one to grammatical structure. Speaking and listening dominated levels two and three; reading and writing were emphasized in level four. In total, during all four school years, BEP gave nearly equal time to listening, speaking, reading and writing.

"I think the general consensus of the group was that students learn better if they have the structures identified and they can hang other things onto those structures," Ajarimah said. "That was the rationale, and it was actually supported by some research done by a British linguist [H.G. Widdowson] in Kuwait. He found that if students in Kuwait did not have the grammatical structures in focus, they did not respond well. They were able to learn more effectively with the grammatical structures in focus up front."

An important change was made in testing procedures. Norm-referenced testing, the so-called "grading on the curve" method, was dropped and replaced with the W.H. Angoff rating and standards-setting methods. Some ITC principals had manipulated the curve to make their pass-fail ratios look good. The new testing procedures cured this problem. In the Angoff method, a committee of trainers agreed on a standard pass-fail level: a certain number of correct answers equaled a specific grade, leaving no more room for manipulation. Unlike the "curve" grading method, where students at the bottom of the curve always failed, under the new method "all could pass, and all could fail. We would not compromise on that," Ajarimah said.

Business economics class, 1986.

Lesson one in Book One of Beginning English opened with students chorusing after the teacher, "Good morning (or good afternoon). How are you? I'm fine, thank you." Later in the first lesson, trainees were introduced to the English alphabet, reciting with the instructor all the letters of the alphabet. By the end of their first semester, trainees were supposed to have an English vocabulary of more than 130 words, and knowledge of numerals from zero to 100. They were expected to know words and phrases such as: "Aramco employee," "ID card," "teacher," "trainee," "telephone number," "worker" and "workshop," plus polite expressions such as "thank you." By the middle of the four-year program, trainees had been introduced to work-related terms such as "stabilizer," "GOSP," "tower," "storage tank," "kerosene," "butane," "propane," "methane," "evaporate" and "tower." Near the end of the BEP program they were reading brief articles on subjects such as how to use a tape recorder, and they were writing answers to questions about the articles.

According to program standards, the BEP graduate would be proficient enough in spoken and written English to make himself understood by his supervisor, fellow workers, visitors and friends. He would be able to follow conversations between native English-language speakers. He could read company publications such as the *Highlights and Notices*, training manuals, maintenance and operating manuals, memos, maps, directories and inventories. He would be capable of writing short, informal memos of 25 to 35 words; completing forms; making simple log entries; and making lists in alphabetical or numerical order.

The switch of students into BEP from VELT and Regular English began in 1984 and was completed in 1986. It meant an extra semester of classroom work for some trainees. A conversion plan drafted by the Training Department required trainees who had passed only Level 1-A in VELT or Regular English to remain Level 1-A students in BEP. They did not advance to the next higher level, although a trainee who had completed a level-B semester in the old programs could advance into the next higher level of BEP. No new InterCom classes were started at the elementary English level. The handful of InterCom students were allowed to complete the four-year basic English course, and then InterCom was adapted for use by students being prepared for out-of-Kingdom training.

The conversion into BEP was especially difficult for trainees from the VELT program. One can only wonder what effect it had on their careers. Years later, some former ITC students talked of this difficult transition.

"I got a good grade, a 96, in VELT," recalled R.N. Al-Rumaizan, an inspector III, but "when I got into academic English (BEP), I got a 72 because I did not know grammar. I went from being a top student to being a poor student."

Students responded to multiple-choice questions, 1985.

Two other former VELT students, M.A. Al-Doukhy, an engineering aide, and A.M. Al-Shammary, an inspector III, had similar stories. After switching from VELT to regular English, their grades plummeted and they had to work extra hard to avoid flunking out.

"Technical training was definitely easier," Al-Doukhy said. "We found it difficult when we transferred from VELT. So many problems. We were lost in the beginning. One teacher forced me to read all the time [to catch up on grammar with others in his class]."

Al-Shammary also played catch-up. He found the Jobs Skills Training he took in the late 1980s more relevant to his work than VELT. When a VELT teacher showed trainees tools from his toolbox, Al-Shammary said, he was showing them tools most of them already worked with in their jobs.

BEP remained basically unchanged for more than a decade. In terms of longevity, it was the most successful core English program ever created for Aramco. Thousands of Saudis learned to read, write and speak English using this program.

"I think," Ajarimah said, "the design of the program was better than the design of other programs, in the sense that it had materials designed to meet the specific needs of the organization. It also had a full battery of tests that went with it, so the validation of the testing instruments was very clear. The instructional objectives of the program were the guide for the teachers. In previous programs, with the exception of VELT, there was nothing called an instructional objective. The only thing you had was a book. If the book went out of print, the instructional objectives went with the book, the tests went with the book, so we lost everything. This was not the case with a program owned by the company. It had instructional objectives, instructional materials, testing instruments. It was really comprehensive."

Appendix C
The Role of Testing
in Saudi Aramco Training

"I think what I most appreciated about the people in Saudi Aramco is that they were willing to validate [tests]. They did not just take things on the surface, by intuition. They wanted to do research. You don't find many organizations willing to spend the effort to do that."
– Shoukry D. Saleh, Ph.D.

The first Aramco hiring halls were crowded, clamorous places. By early morning on a typical day, jobseekers filled the hiring offices in Dammam and Abqaiq and overflowed onto the streets outside. "There'd be a lot of commotion and noise," a former Aramcon said. "If there was anyone who knew half a dozen words of English, why we'd grab onto him."

Back then, recruiters filled their quotas by choosing almost at random from the milling crowd. As the recruiter prepared to make his picks, the crowd surged forward, jostling and shouting. Jobseekers tried in various ways to catch the recruiter's attention. Legend has it that one young Saudi climbed on a chair and shouted the only English words he knew, "I am 17." Actually he was only about 11 years old, but 17 was the minimum age for full-time employment. He got a job.

Hiring in the early days was done on an as-needed basis. No one considered what the new man was capable of doing, much less asked what he wanted to do. A new-hire might work a week or two, decide he didn't like the job, and leave, only to return a few days later hoping to be picked for a job he liked better. At times during the 1930s and '40s the turnover rate in the Saudi work force reached almost 70 percent.

Muhammed Al-Mughamis, who became an Aramco executive, experienced the capricious nature of the hiring process.

He was one of perhaps 30 young job hopefuls standing outside the employment office at Abqaiq one day in the early 1950s. A recruiter came to test the candidates on their multiplication tables. The result would determine who got a job. Al-Mughamis was doing fine until:

"I remember his asking me what 13 x 13 was. I couldn't answer. Then he said 18 x 18, which was worse! Then he said, 'You've flunked.'

"A boy that age has pride in what he does, and it was an insult to fail. I think I memorized the advanced portion of the multiplication tables from 11 x 11 through 20 x 20. Two weeks later I returned to take the test again. The first thing he hit me with was 13 x 13, and the answer was there. Apparently this man was programmed.

Aramco job applicants wait their turn in Dammam, early 1950s.

He asked me the same questions he had asked two weeks ago! I felt fine. I picked up the small acceptance paper and went to the [employment] office. Everybody was surprised that I made it." But no one will ever know how many other bright young men did not make it.

By the 1950s Aramco's work structure included 1,400 specific jobs covering about every profession and vocation found in the oil industry. Although the company had no shortage of job applicants, it had a chronic problem matching the applicant to the job. Choosing a new-hire for one of those 1,400 jobs was a matter of trial and error.

The first person to do something about qualifying Saudi recruits and finding an appropriate training track for them was a young industrial psychologist, Jack Rushmer, son of Ras Tanura district manager Larry Rushmer. In 1950 he set out to develop a learning abilities and aptitude test for Saudis similar to those used by U.S. industries to screen job applicants. Rushmer worked on the project for two years, guided in technical matters by a New York consulting firm. He was assisted in Dhahran by another young Aramco industrial psychologist, John Tarvin, a recent University of Tennessee graduate.

Developing a test for Saudi job applicants presented some unusual challenges. For one thing, almost none of the Saudi job applicants could read and write, so the questions had to be read to each applicant by a test examiner. This created the problem of how to score an applicant's response. "Developing scoring standards was particularly tricky," Tarvin said. "The applicants really came from all over the country. We had to have scoring standards

A new recruit takes a spatial-abilities test in Dhahran, 1950.

that would recognize a correct oral response from people who spoke different dialects and came from different parts of the Kingdom."

The Saudi Test for Job Assignment (STJA)

Rushmer and Tarvin created what was known as the Saudi General Classification Test, later renamed the Saudi Test for Job Assignment (STJA). It proved to be a fairly reliable index to the job-learning abilities of Saudi recruits.

STJA had five test sections: antonyms, arithmetic, similarities, story comprehension and a paper-folding or spatial relationship test. Test scores were arranged in five equal percentage groups, with 100 percent as a perfect score. A score of 60 percent or above was considered good. Only the lowest-scoring recruits, those in the bottom 20th percentile, were eliminated from job consideration. Those in groups three and four who scored between the 20th and 60th percentile were considered for low-level jobs. The overall score was linked to learning ability. The results in each separate category served as an indicator of aptitudes.

Nearly all Saudi recruits hired during the mid- to late 1950s had to take the STJA, as did some other employees who, it was felt, might be in the wrong job. At first, testing was done only in Dhahran. Specially trained Arab examiners administered the tests in a portable building, an old field kitchen with a picture of a camel on the front. Applicants were told to report to the "camel house" located in a lot behind the Dhahran dining hall. Up to a dozen applicants were tested at a time and their responses to questions written down by the examiners. Saudis named it the *"mukh"* or "brain" test. Test centers were set up later in Abqaiq and Ras Tanura.

Supervisors used the test results to supplement their own judgment, based on job performance, in selecting employees for training, promotion and other placement decisions. A 1958 follow-up study by the Industrial Relations Department's Planning Division seemed to show a marked relationship between an employee's STJA score and his training achievements.

According to Tarvin, Rushmer's accomplishments were never fully appreciated. "I'd like to see Rushmer get some recognition for what he did," Tarvin said. "I can't think of anything that did any more for Aramco. It saved the company millions of dollars in wasted time and misdirected training effort."

By 1960, improvements and expansion of the Saudi public school system had made STJA virtually obsolete. Whereas in the 1950s a score of 20 correct answers out of a 100 was considered acceptable for low-level jobs, in 1960 the average job candidate got 80 percent of the answers correct. Hiring was usually limited to those who had scored in the top eight percent. Clearly, the old test had lost its ability to discriminate among applicants. The company needed a more finely tuned measurement of mental ability.

The Job Aptitude Test (JAT)

A new, paper-and-pencil version of the STJA was devised. The test was updated using U.S. military placement tests as a model, and renamed the Job Aptitude Test (JAT). It was first administered in 1961. Unlike the old oral test, JAT was a multiple-choice test written in Arabic. It was designed to measure overall mental ability plus mechanical and clerical aptitude. The JAT battery consisted of nine tests that provided scores in eight categories. The tests were:
- Speed and Accuracy of Marking, checking aptitude for clerical tasks.
- Number Checking, comparing two columns of figures.
- Coding of Symbols, comparing sets of symbols.
- Mental Arithmetic, computation without the aid of pencil and paper.
- Minnesota Paper Form Board, a spatial-abilities test using two-dimensional forms.
- Block Counting, a test of ability to see how three-dimensional structures were assembled by counting the number of blocks in them.
- Non-Verbal Reasoning, measuring abstract reasoning not dependent on verbal learning — considered the most difficult of the nine tests.
- General Training Aptitude (GTA), the test used to screen job applicants. (The scores of two tests, Verbal Problem Solving and Arithmetical Problem Solving, were combined to get the GTA score.)

The GTA, believed to be the best measure of general mental ability and training success, consisted of 120 multiple-choice questions, each with five possible answers. It was divided into two sections — one section of number problems and a second of verbal problem solving. Applicants were given 65 minutes to complete the two sections. Questions ranged from easy to difficult. Here are two sample questions from the numbers section:
1. "A car agent had 60 cars. He sold 15 cars. How many cars remain?"
2. "A clerk can complete 65 forms in 40 minutes. Another clerk can complete the same information on 44 forms in 32 minutes. If both work together, how long would it take them to complete 273 forms?"

The following two questions are taken from the verbal problem solving section:
1. "Which of the following five words is the opposite of the word 'bad'?" Fair, Good, Okay, Done, Passing.
2. "A lawyer is to a court as a teacher is to a _____."

An Arabic version of the IBM answer sheets was used at first for JAT testing, but it was soon dropped in favor of questions and multiple-choice answers written in the test booklet. Candidates had found directions for using the IBM answering sheets complicated and confusing. As much time had been spent in preparing the candidates for testing as in the testing itself.

A Saudi youth compares sets of symbols.

A follow-up study completed more than a decade later showed the GTA to be a good predictor of performance in an academic training program, but not a very good predictor of how well a man would do on the job. The study, completed in 1976, followed the careers of 167 men hired in 1964 and 1965. It showed that only 13 percent of those who scored in the lowest one-fifth on the GTA test had qualified for any type of out-of-Kingdom training, while more than 50 percent of those who scored in the top 20 percent on the GTA had been out-of-Kingdom trainees. Nevertheless, the study found very little difference in progress up the job ladder between the groups. The low GTA scorers advanced an average of 3.5 job codes in 10 years, while the top scorers moved up 4.2 job codes. All others fell somewhere in between. This slight trend was not judged to be a significant predictor of job advancement.

Shoukry D. Saleh

Almost all improvements in the company's testing program during the next three decades, the 1970s, '80s and '90s, involved a consultant, Shoukry D. Saleh, Ph.D., professor of management sciences at the University of Waterloo in Ontario, Canada. His background and experience in the testing field were uniquely suited to Aramco's needs. Saleh was born in Egypt, earned a bachelor's degree at Cairo University, and did his graduate work at Case

Driver trainees learn to read numbers.

Western Reserve University in Cleveland, Ohio, where he came to the attention of some Aramco executives attending refresher courses. He not only had the expertise to analyze the company's screening and placement tests, he could also read, write and speak Arabic and had a thorough understanding of the Arab culture.

Those who worked with him described Saleh as very quiet, very courteous and very productive. He put out voluminous, complicated statistical analyses and reports within the short span of his consultant's contract, usually about six weeks. He did many test reviews and evaluations for Aramco over a period of more than 20 years and has been called the "father and grandfather of modern aptitude testing at Saudi Aramco."

Saleh first came to Aramco as a consultant in 1973, during a year-long sabbatical from his teaching post at Waterloo University. He had been retained by the Industrial Relations Department to review the JAT battery and devise new test forms. The tests had not been updated since the JAT battery was introduced 12 years earlier, and the cutoff scores had not changed since 1963 when Stanley Bolin developed new test norms. All applicants, with the exception of college graduates and direct-hires with proven skills, had to score above a certain level on the GTA test in order to qualify for employment. The other tests in the JAT battery were designed to reveal the applicants' work-related aptitudes.

Saleh spent nearly a year reviewing and revising the JAT battery. He compared the test scores of more than 1,300 job applicants in 1972 with the scores of applicants from 1963. He found that the 1972 candidates scored higher in all test categories, a tribute to the continued expansion and improvement of the Saudi government school system. Saleh also detected a strong correlation between performance on the JAT and performance in Aramco's ITC classrooms. In a report to the Executive Management Committee in October 1973, Saleh praised the JAT battery, especially the GTA, as a predictor of success in training.

"All correlations are significant, indicating the validity of all tests with respect to the training criteria. The GTA score is the most valid one, regardless of the area of training. The magnitude of correlations for this score is impressive. The GTA correlations range from .65 to .74. This evidence should prove clearly that the battery is very valuable for screening out the unfit for the kind of training provided at the ITCs."

Saleh had been unable to determine the validity of JAT as a predictor of performance on the job. He was stymied by a lack of data, primarily due to poor record keeping and the problem of different supervisors rating job performance in different ways. What was only a fair performance to one supervisor might be rated as good by another supervisor. Even though no significant predictor of job performance could be identified, Saleh did a factor analysis and a cluster analysis of the test battery. On the basis of these analyses, he recommended changes to update and streamline the JAT battery. He suggested that three of the nine tests in the battery be dropped and that scores on the remaining six tests be combined in order to produce scores in three test categories. Scores would be recorded for the GTA test, the Clerical Aptitude test and the Mechanical Aptitude tests. He also recommended that the GTA test be reduced from 120 problems to 90 problems and the time allowed for completion be cut from 65 minutes to 40 minutes.

Saleh also proposed new cutoff scores in all three test categories. He proposed the following minimum scores for applicants with nine years or more years of schooling: 46 on the GTA, 105 on the Clerical and 48 on the Mechanical Aptitude test. For those with six to eight years of school, he recommended a minimum score of 35 on the GTA, 78 on the Clerical and 39 for the Mechanical Aptitude test. He also suggested that all applicants be required to exceed the cutoff score on one of the two aptitude tests in addition to the GTA in order to be accepted for employment.

These rigorous new standards, Saleh predicted, would upgrade the quality of new employees by screening out about two-thirds of the sixth graders, one-half of the seventh graders and one-fourth of eighth graders who applied for employment. The company accepted most of his recommendations, including cutoff scores

A Saudi recruit competes against the clock on a manual-dexterity test in the early 1950s.

for the Clerical and Mechanical Aptitude tests, but it did not accept his recommended GTA cutoff score, nor his recommendation that all candidates be required to exceed the cutoff score on at least one aptitude test. The minimum GTA score required for employment was set at 31 for all applicants, well below the 35 suggested by Saleh for applicants with a sixth-grade education.

Saleh also suggested that the JAT battery should be administered by the Training Department's newly created Curriculum and Test Development Unit (CTDU), rather than the Employment Division of the Saudi Arab Employment Department, which had been giving the JAT since the test battery's inception. His suggestion was accepted, and in January 1975 the CTDU began to administer the tests to job candidates referred to them by the Employment Department.

Minimum Test Scores Abandoned in 1975

Just three months later, in March 1975, the company abandoned the requirement that applicants achieve a minimum test score to qualify for employment. The pressure of expansion caused the company to offer employment to virtually anyone with at least a sixth-grade education. The JAT battery continued to be administered to new employees, but only as a guide to where they might be placed in training or to what kind of job they should be assigned, and not as a qualification for employment.

The lifting of the GTA test requirements had an immediate impact in recruiting and the need for training. More than 2,000 new Saudi employees were hired in 1975. About 50 percent of them scored below the GTA 31 level, the previous minimum score required for employment. The Training Department reported academic performance of new-hires in the "B" program for trainees with a sixth- to eighth-grade education was "significantly poorer" than it had been only a year earlier. Eight out of 10 "B" program trainees with GTA scores of 25 or below either flunked or got a D, the lowest passing grade, in ITC English classes. The "B" program attrition rate soared. More than one-third of the trainees who started "B" program English classes in March 1976 dropped out within six months. On the other hand, elimination of GTA test requirements seemed to have no effect on the "A" program trainees, those who had nine to 11 years of schooling. Seventy-four percent of the "A" trainees who scored below 30 on the GTA received a passing grade of C or better in English in their first term.

In a report to executive management, the Training Department suggested three reasons for this apparent anomaly. The report pointed out that the typical "A" trainee was more mature than the "B"

Saudi recruit completes Mechanical-Aptitude test.

trainee, had up to three years more schooling and, unlike the average "B" trainee, had probably studied English at a Saudi government school before coming to Aramco.

Test Score Requirements Reinstated in 1976

The total abandonment of screening tests lasted 18 months. In October 1976, the company reinstated GTA test-score minimums for some employment candidates, but at a lower level than in previous years. A new GTA minimum of 26 was instituted for entry into training programs, five points below the previous minimum. Recruits with a lower GTA could still qualify for employment by scoring above 78 on the Clerical test or above 38 on the Mechanical Aptitude test. Someone with a low GTA score but with good scores on one of the other tests would be placed in an on-the-job training program where his work and his aptitudes were clearly related and where academic training could be taken at a slower pace. Even the requirement for a sixth-grade education was waived in some cases. Applicants with proven job skills, such as welders or pipe fitters, did not take the JAT battery and did not need a sixth-grade certificate, and neither did those who fit into a category called "Specific Direct-Hire with No Minimum Qualifications." Specific Direct-Hires were assigned for long terms to low-level work such as janitors, food servicemen, desert convoy drivers, and roustabouts. They were usually placed where they could learn a job simply by doing it, although some received formal on-the-job training or part-time ITC/ITS instruction.

Saleh returned to Aramco in 1979 at the request of the company's Saudi Arab Manpower Committee (SAMCOM). The top-level committee asked Saleh to validate the performance of the GTA. Saleh's review raised doubts about the continued effectiveness of the GTA. He found the GTA still adequate for screening purposes, but his report emphasized that "there is room for improvement." Saleh made many improvements in the GTA himself. He changed the test forms. He reduced the number of questions in the math section from 40 to 30. He wrote new items to replace 15 questions in the verbal section of the GTA and modified eight other questions. He rearranged all 50 questions in the verbal section and put them in order of difficulty. He reorganized the test booklet and the booklet's directions for test administrators, and recommended that a new testing manual be written for test administrators.

Saleh appeared before SAMCOM in August 1979 to read a report on his latest performance review. In it he included a warning to Aramco about lowering hiring standards too far. "It is of great importance to realize the limits to which the standards may be lowered without seriously affecting the functioning of the organization," he wrote. "Accepting a large number of below-average individuals for skilled jobs would affect not only the quality of performance, but also morale in general. Fast development and promotion for such individuals would not be realistic in view of their ability, even though, at the present time, there is a high level of expectation in the company for fast development of Saudi employees. A conflict could result in this situation, either by promoting such individuals regardless of their ability and readiness or by holding them back regardless of expectations."

He recommended new GTA cutoff scores, and this time his recommendations were adopted. The GTA cutoff for "B" group trainees was set at 31, up from a score of 26. Saleh predicted that raising the qualifying score would reduce the ITC failure rate by half. A cutoff of 37 was adopted for "A" trainees, and a minimum GTA score of 43 was established for

Trainee works on an English examination, 1951.

employees to qualify for Training's "fast track," a college scholarship program for high school graduates who had a grade average of 85 or higher.

The higher GTA cutoff scores reflected an improvement in the educational level of applicants. At the same time, it exposed the weakness of the JAT battery. "The JAT was really very elementary," Saleh recalled. "There was not much differentiation at higher levels. It was not effective screening for better-educated Saudi recruits."

The General Aptitude Test Battery (GATB)

SAMCOM's aggressive Saudi development policy created a real need for an instrument with the ability to predict performance on the job as well as a test battery that could be used to screen Saudi job applicants at many different education levels, from sixth-grade dropouts to high school graduates. As Bill Valbracht, supervisor of the Test and Evaluation Unit, put it, "Good science told us there was a better way" than the JAT battery. The most obvious choice was the General Aptitude Test Battery (GATB). "Everyone knew of the success of the U.S. military and U.S. Employment Service with the GATB," Valbracht said.

The GATB was generally recognized as the best-validated multiple-aptitude test battery then in existence. It had been developed by the U.S. Employment Service in the late 1940s, and had been used for more than 30 years by state government employment offices in the U.S. It had been the subject of extensive validation studies to determine its success as a placement tool for about 450 different occupations. It had been translated into at least 12 languages and was being used for job placement and vocational counseling in 35 countries. Studies by Michigan State University researcher J.E. Hunter showed a significant linear relationship between GATB scores and future job performances in 95 percent of 23,400 cases involving 3,300 GATB and job performance relationships.

In 1980, executive management approved a pilot program to evaluate GATB at Aramco. That same year the GATB was translated into Arabic for the first time by Ali M. Hijjeh and Yusif Ali Safi of the

Testing and Evaluation Unit. During the next year, they developed test administration and scoring procedures, trained test administrators and scheduled and supervised the administration of the GATB. By November 1981 the GATB had been administered on a pilot basis to about 3,000 Aramco employees.

Paul J. Kay, supervisor of the Testing Unit, wrote a letter of appreciation in 1981 thanking Hijjeh for his work on the GATB. (Safi had left the company.) "You sold that program to all of the area managers and substantially effected its general adoption," Kay wrote. He praised Hijjeh for his "leadership in upgrading vocational testing and evaluation."

Between April 1981 and May 1983, Saleh visited Aramco three times. In 1981, he revised the Arabic translation of the GATB. In 1982 he wrote a report that defined the form and structure the GATB was to take and which served as a blueprint for revision of the test and for the definition of descriptive data. In 1983 he conducted a study among highly skilled craftsmen in the mechanical maintenance, electronics, instrumentation and metals fields that indexed the GATB's reliability and validity and confirmed a relationship between GATB results and job performance.

Parallel testing of the GATB and the GTA was conducted between May and October 1983, to establish equivalency and to check the GATB. Finally, on October 15, 1983, management approved the GATB for company-wide application as a replacement for the JAT battery.

The GATB comprised 12 separate tests that gave a measure of eight different aptitudes. (The tests were titled: Name Comparison, Computation, Three-Dimensional Space, Vocabulary, Tool Matching, Arithmetic Reasoning, Form Matching, Mark Making, and two each for Manual and Finger Dexterity.) The tests produced a "profile" measuring general learning ability, verbal aptitude, spatial aptitude, form perception, speed of perception, motor coordination, manual and finger dexterity. The dexterity tests were performance tests, while the others were paper-and-pencil tests. Speed counted. The shortest of the 12 subtests took one minute; the longest, seven minutes. While candidates sat for 2½ hours, only 47 minutes were for actual performance. The remainder of the time was devoted to instruction and practice.

A new screening test score called the GATB Selection Score (GSS) was developed by the Testing & Evaluation Unit in 1983 as a replacement for the venerable GTA test score. "We knew the GTA was a good predictor of success, particularly in the ITC, and to some degree in the ITS," Valbracht said. "So we looked for portions of the GATB that could be combined together statistically to mirror the GTA." They found the solution in a combination of three general-knowledge tests in the GATB — the Computation, Vocabulary, and Arithmetic Reasoning tests. Scores on these three GATB tests could be combined so they closely mirrored GTA scores on roughly the same scale and curve spread.

The minimum acceptable GSS fluctuated over the years, depending on Aramco's needs and the availability of recruits in the labor pool. In May 1983 a minimum score of 40 was established for non-high-school graduates to qualify for employment. High school graduates also took the GSS, but only for record purposes after they were on the payroll. In May 1984 the minimum was raised to 45 for those without a high school diploma. At that same time a minimum GSS score of 50 was established for job applicants who had finished high school.

In April 1985 the minimum was raised from 50 to 60 for high school graduates

Wadie Abdelmalek, left, and M.U. Takiyyuddeen review machine-scored test results, Dhahran, 1980.

and from 45 to 55 for non-high-school graduates. In October 1987 it was reduced from 60 to 55 for high school graduates as part of the company's new Apprenticeship Program. In July 1988, it was lowered to 50 for selected high school graduates who had GATB verbal scores of 100 or more.

Ali Dialdin, general manager of Training & Career Development, credits GATB with reducing the time it took to complete the Apprenticeship Program, introduced in 1988, by a full year. It eliminated the need to give trainees introductory courses in various job skills specialties. When the Job Skills Training programs began in the early 1980s, Dialdin recalled, "we gave the trainee about six weeks of exposure to instrumentation training, six weeks of electrical, and so on, for a total of about a year, to see what his inclination was. It was a trial-and-error method. But now we had refined the GATB to the point where it was predictive of the individual. That enabled us to adopt a two-year instead of a three-year Apprenticeship Program. Imagine — one year of a man's training and a lot of savings to the company!"

Aramco was the first company in the Middle East to adopt the GATB. "I think it really improved the human intake at the worker's level," Saleh said. "It was of great value to the company. It saved on both time and money."

As successful as the GATB was in pinpointing individual aptitudes, it did not measure an individual's likes and dislikes, a void that Dialdin wanted to fill. "I wanted something to go along with the GATB to show me the person's desire in addition to his aptitude. A guy could have the aptitude to be a medical doctor, for example, but not the interest. In such a case it would be a waste of time to try to train him to be a doctor."

The Self-Directed Search (SDS)

Accordingly, in late 1989 Valbracht wrote to Saleh in Canada asking him to recommend an instrument to measure individual interests and serve as a complement to the GATB. One approach, using an Arabic version of a popular U.S. personality test, the California Psychological Inventory, had already been tried on Saudi gas and oil plant operators, with unsatisfactory results.

Shoukry Saleh, 1996.

Saleh recommended an interest inventory known as the Self-Directed Search (SDS), developed by a U.S. industrial psychologist, John L. Holland. He classified people into six basic types — Realistic, Investigative, Artistic, Social, Enterprising and Conventional — and drew up a list of questions to reveal which categories an individual favored. For example, a person who said he liked to fix cars, use tools and build things would be a Realistic type. He leaned toward the technical side of the work force. A person who said he liked to operate a computer, take inventory and fill out forms would be a Conventional type, and a candidate for clerical work. Those were much simplified examples. The actual inventory measured preferences in several categories, subtly balancing the test-takers' likes and dislikes against the interests of successful people in more than 1,100 job categories.

Holland's work drew high praise from others in the field. One writer, F.H. Borgen, called SDS "the single most influential approach in vocational behavior in the past 15 years."

Saleh found SDS "easy to administer, easy to understand and easy to adapt to a new culture." It had already been translated from English for use in at least 10 countries, including Japan, Nigeria and Guyana, in addition to several European countries.

Saleh himself translated SDS into Arabic. In doing so, he adapted several questions to fit the Saudi culture and environment. For example, the question as to whether a person liked to "Fill out income tax forms" was changed to "Fill out job application forms," because there was no income tax in Saudi Arabia. "Take charge of a political campaign" was changed to "Take charge of a social project," a more appropriate activity in Saudi Arabia. "Raise dairy cows or beef cattle" was changed to "Fix house furniture."

SDS Adapted to Become Vocational Interest Test (VIT)

The Arabic version of SDS was first administered in March 1990. SDS became known at Saudi Aramco as the Vocational Interest Test (VIT), although it was not really a test. It was simply an attempt to measure a person's likes and dislikes. As Saleh put it, "You are collecting information about people and helping them to get information about themselves. It is self-revealing, in a sense."

In the next five years, the VIT was administered by Saudi Aramco, along with the GATB, to more than 10,000 Saudis. The GATB alone had been taken by more than 25,000 Saudis since it was introduced at Aramco.

After nearly 25 years as a consultant on testing matters, Saleh found Saudi Aramco to be one of the top companies in the world in testing and placing of personnel.

"I think what I most appreciated about the people in Saudi Aramco is that they were willing to validate tests," he said. "They did not just take things on the surface, by intuition. They wanted to do research. You don't find many organizations willing to spend the effort to do that. It's expensive, but, in my view, it definitely paid off. The cost is peanuts compared to the value of improving human resources and manpower in the company."

Index

Saudi Employee Firsts in Saudi Aramco

The first Saudi...

Crew to move a derrick – 9
Teachers in company schools – 12-13
Jabal School graduates – 25-26
15-year service awards – 31
Government scholarship students – 38
Scholarship recipients at Aramco – 46
Work force supervisors – 46
GOSP operations crew – 51
Practical nurses – 51
Firemen and watertenders – 53
Crew to drill an oil well – 63
University graduate on company scholarship – 63
Supervisor at a GOSP – 63
Assistant foreman – 63
Operator at the Ras Tanura Refinery – 63
Refinery supervisor – 63
Nurses supervisor – 63
Crafts supervisor – 63
Senior-staff personnel – 64
Board of Directors members – 81
Foremen in the crafts field – 86
Management candidates – 91
Employees enrolled at UPM – 95
Degree in pharmacy – 101
Crew at the fluid hydroformer – 101
Crew at the Abqaiq LPG plant – 101
Department manager – 103
ITC principal – 103
Computer operator – A-2
Apprentices – 116
Company vice president – 132
Women's training program – 143
Senior vice president – 162
Director of Career Development – 177
President of Aramco – 192
Ph.D. in geology from a Saudi Arabian university – 193
Board of Directors chairman – 199
President and chief executive officer of Aramco – 199

Note: Page numbers in italics refer to photographs and captions

A

Abadan, 22, 23, 32
'Abd al-'Aziz, Ibrahim ibn, 51
'Abd al-'Aziz Al Sa'ud, King of Saudi Arabia, 1, *4*, 10, 29, *29*, 31, 43, 58, 64, 200
'Abd Allah, 'Abd al-Monim, 86
'Abd Allah, Mubarak ibn, 51
'Abd Allah, Sulayman, 26
'Abd Allah, Tahir ibn, 51
'Abd al-Karim, Sa'id ibn, 51, 63
Abdalla, Khamis, 185
Abdelmalek, Wadie, 118, 151, 202, B-6, B-7, B-9, *C-8*
Abdi, Omar, *190*
Al-Aboudi, Fahad S., 119, 235
Abqaiq. *See also* Southern Area
   administration building in, 136
   Area Maintenance Training Center in, 185, 193, 195, 201, B-12
   clerical training in, 216
   Computer Applications Group in, A-3
   as district training center, 39, 55, 70, 162
   early industrial training in, 33, 39, *50*, 53
   English-language instruction in, 21, *B-3*
   General Industrial Training (GIT) centers in, 46, 54, 67
   GOSP operations in, *50*, 51
   housing in, 28, *29*
   Industrial Training Center (ITC) in, 68, *68*, 70, 87, 104, 106, 124, 145, 150, 163, 170, 193, 195, 201, 213, A-7, B-5
   Industrial Training Shop (ITS) in, 106, 139, 150, 163
   Job Skills Training at, 79, 201
   as largest crude oil processing center, 228
   Liquefied Petroleum Gas (LPG) Plant in, 101
   maintenance crew in, *139*
   maintenance training in, 182-183, 186, A-3
   Management Training/Continuing Education Center in, 139, 150
   in 1950s, 61
   in 1970s, 123, 138
   On-the-Job Training unit in, 157
   pipeline to, 30
   railroad to, 31
   and satellite ITCs, 101
   Saudi drillers near, *35*
   Senior-staff school in, 37, 57, 143
   supervisory training in, 46, 88
   test center in, C-2
   training facilities closed, 106, 201
   training rig in, *185*
   Voluntary School in, 32, 39, 46
   wildcat wells in 1930s, 9
Abqaiq Field, 58
Abqaiq No. 1 (oil well), 12
Abu Ali, 196
Abu Hadriya, 9, 29
Abu Hadriya No. 1 (oil well), 12
'Abu Zaid, Issam, 206
Abubshait, Khalid, 109, 118, 119, *190*
   profile of, 121, *121*
Aburayyan, Ali I., 116
Academic Curriculum Unit, 209, 214, B-15, B-16
Academic programs. *See also* Arabic-language instruction;
      English-language instruction; Mathematics instruction
   General Industrial Training (GIT), 45-46, 50-51
   in Industrial Training Centers (ITCs), 79, 88, 195
Accidents. *See* Industrial accidents; Vehicle accidents
Accounting Department, 20, 21, 47
Accrediting Council for Continuing Education and Training (ACCET), 240, 242
Adam, Tom, A-8
Adamopoulos, Ted, A-8
Adenese, 27

Adrian College, A-9
Advanced Industrial Training Center, 38-39, *54*, *59*, 70, *79*, 100
Advanced Industrial Training Program, 45, *54*, 56-57, *57*, *58*, 70
Advanced Trade Training, 43, 45
Advanced Training Division, 130
'Adwan, Shaykh 'Abd Allah ibn, 93
Afaleq, Ibrahim 'Abd Allah, profile of, 21, *21*
Al-Agnam, Salman, *214*
Agricultural project. *See* al-Kharj agricultural project
Ahmad, Muhammad, 26
Ahmed, Mahdi ibn, *16*
'Ain Dar, 29
Air base in Dhahran, 31, 43, 210
Airline crash, 104
Ajarimah, Ahmad, 209, 210-212, 223, 235, 236, 238, *238*, A-9, B-15, B-16, B-18, 240
Al-Ajmi, Ali A., 119
Al-Ajmi, Nassir M., 6, 46, 163, 197, 199, 228-229, *229*
   profile of, 33, *33*
Al-Ajmi, Sami, *221*
AJTSs. *See* Aramco Job Training Standard
'Akif, Ibrahim, 142
Aleppo College, 46, *46*, 63
Alexander, Robin "Pinkie," 36
Alexander, Dr. T. C. "Alex," 16, *16*, 36
Ali, 'Abd Allah ibn, 46
Al-Ali, Hadi, A-13
Ali, Hilal ibn, 86
Ali, Ibrahim ibn, 51
Ali, Khalil Younis, 86
Ali, Muhammad ibn, 46
Allegheny College, 80
Allen, Jim, 84, 97, 110
Alturki, Khalid, 143
Alturki, Sally, 143
Ambah, Salih, 94
American Book Co., B-10
American children, education of. *See* Senior-staff schools
American families, 10, 35
American Graduate School of International Management, 75, 170
American Language Academy (Tampa), 170
American Language Institute (San Diego State University), 170
American Manufacturers' Association, 96
American University of Beirut (AUB), 42, 43, 46, *46*, *57*, 63, 101, 149, 237, 241, B-9, B-10
American University in Cairo (AUC), 125, 146, 237
Al-Amiri, Ahmad ibn 'Abd Allah, 53
AMTCs. *See* Area Maintenance Training Centers
Anderson, Adrien Louis "Andy," 14
Anglo-Iranian Oil Company, 22, 23, 32, 58
Angoff, W. H., B-16
Anti-Communism, 58
Antioch College, 80, 103
Antonelli, Vincent, A-6
APEP. *See* Associate Professional English Program
APP. *See* Associate Professional Program
AppleWorks course, A-10
Apprenticeship programs, 23, 32, 111, 116-22, *116-119*, 143, 179, 204-206, *204*, *205*, *208*, 209-214, 219, 224, 225, 230, 236, 238, 242, A-9 to A-10, A-13, C-9
Arab Contracting Department, 15
Arab-Israeli wars, 54, 73, 100, 105, 110, 125
Arab League, 91
Arab League Petroleum Congress, 81, 91
Arab Preparatory School, 24. *See also* Jabal School
Arab Refugee Scholarship, 102
Arab Trade Preparatory School, 25. *See also* Jabal School
Arab Trade School, 38-39, 43-44
Arabian American Oil Company. *See* Aramco
Arabian Drilling Co., 38
Arabian Gulf, 3, 58, 104, 124, 133
Arabian Peninsula, 1, 6, 73, 228